Pathogenic microorganisms exploit a number of different routes for transmission and this book demonstrates how the spread of disease can be prevented through the practices of disinfection and control of microbial growth.

The book is organized into four parts. The first part addresses the processes of infectious disease transmission and considers how best to minimize the spread of disease. The second part deals with the prevention of infectious diseases that are transmitted by water or food. Transmission by aerosols, environmental surfaces and medical devices is considered next. The fourth and final part discusses some general mechanisms of disinfection.

This book includes contributions from leading scientists, who provide a wide-ranging synthesis of the problems and prospects for containing the spread of human infectious diseases.

Edited by CHRISTON J. HURST

Modeling disease transmission and its prevention by disinfection

CAMBRIDGE
UNIVERSITY PRESS

CAMBRIDGE UNIVERSITY PRESS
Cambridge, New York, Melbourne, Madrid, Cape Town, Singapore,
São Paulo, Delhi, Dubai, Tokyo

Cambridge University Press
The Edinburgh Building, Cambridge CB2 8RU, UK

Published in the United States of America by Cambridge University Press, New York

www.cambridge.org
Information on this title: www.cambridge.org/9780521121163

First published 1996
This digitally printed version 2009

A catalogue record for this publication is available from the British Library

Library of Congress Cataloguing in Publication data

Modeling disease transmission and its prevention by disinfection /
    edited by Christon J. Hurst.
        p. cm.
    Includes index.
    ISBN 0 521 48131 7 (handbook)
    1. Disinfection and disinfectants. 2. Communicable diseases –
    Transmission. 3. Communicable diseases – Prevention. I. Hurst,
    Christon J.
        [DNLM: 1. Disease Transmission – prevention & control.
    2. Disinfection.   WA 110 M6888 1996]
    RA761.M59   1996
    614.4'8–dc20   96-13529 CIP
    DNLM/DLC
    for Library of Congress

ISBN 978-0-521-48131-1 Hardback
ISBN 978-0-521-12116-3 Paperback

For Pei-Fung, and our children Rachel and Allen

Contents

List of contributors xi

Preface xv

Part 1 Health and disease 1

1 *The transmission and prevention of infectious disease*
Christon J. Hurst and Patricia A. Murphy 3

2 *Strategies for modeling microbial colonization of the human body in health and disease*
Robin A. Ross and Mei-Ling T. Lee 55

Part 2 Preventing disease transmission by water and food 73

3 *The role of pathogen monitoring in microbial risk assessment*
Joan B. Rose, John T. Lisle and Charles N. Haas 75

4 *Estimating the risk of acquiring infectious disease from ingestion of water*
Christon J. Hurst, Robert M. Clark and Stig E. Regli 99

5 *Bacterial resistance to potable water disinfectants*
Mic H. Stewart and Betty H. Olson 140

6 *Preventing foodborne infectious disease*
Christon J. Hurst 193

Part 3 Preventing disease transmission by aerosols, surfaces and medical devices   213

7 *Disinfection of microbial aerosols*
Scott Clark and Pasquale Scarpino   215

8 *Transmission of viral infections through animate and inanimate surfaces and infection control through chemical disinfection*
Syed A. Sattar and V. Susan Springthorpe   224

9 *The role of chemical disinfectants in controlling bacterial contaminants on environmental surfaces*
Donna J. Gaber, Timothy M. Cusack and Elizabeth Scott   258

10 *Sterilization and disinfection of medical devices*
Aaron B. Margolin and Virginia C. Chamberlain   285

Part 4 General mechanisms of disinfection   311

11 *Ultraviolet light disinfection of water and wastewater*
Peter F. Roessler and Blaine F. Severin   313

12 *Thermal inactivation of microorganisms*
Guy Le Jean and Gérard Abraham   369

Index   397

Contributors

Dr Gérard Abraham
Laboratoire de Maîtrise des Technologies Agro-Industrielles (LMTAI), Pôle Sciences et Technologie de l'Université de la Rochelle, Avenue Marillac, 17042 La Rochelle Cedex 1, France

Dr Virginia C. Chamberlain
Center for Devices and Radiological Health, US Food and Drug Administration, 20850, Washington DC, USA

Dr Robert M. Clark
Risk Reduction Engineering Laboratory, United States of America Environmental Protection Agency, Cincinnati OH 45268, USA

Dr Scott Clark
Departments of Environmental Health, and Civil & Environmental Engineering, University of Cincinnati, Cincinnati OH 45267-0056, USA

Dr Timothy M. Cusack
Reckitt & Colman Inc., L & F Products Division, One Philips Parkway, Montvale NJ 07645-1810, USA

Dr Donna J. Gaber
Reckitt & Colman Inc., One Philips Parkway, Montvale NJ 07645-1810, USA

Dr Charles N. Haas
Environmental Studies Institute, Drexel University, Building 29-W, Philadelphia PA 19104, USA

Dr Christon J. Hurst
Risk Reduction Engineering Laboratory, United States of America

Environmental Protection Agency, 26 Martin Luther King Drive West, Cincinnati OH 45268, USA

Dr Mei-Ling T. Lee
Channing Laboratory, Brigham & Women's Hospital and Harvard Medical School, 180 Longwood Avenue, Boston MA 02115, USA

Dr Guy Le Jean
Laboratoire d'Energetique et de Thermique Industrielle de l'est Francilien (LETIEF), URA CNRS 1508, IUT de Créteil, Université Paris-XII – Val de Marne, 94010 Créteil Cedex, France

Dr John T. Lisle
Department of Marine Sciences, University of South Florida, 140 South 7th Avenue, St Petersburg FL 33701, USA

Dr Aaron B. Margolin
Department of Microbiology, University of New Hampshire, Spaulding Life Sciences Building, Durham NH 03824, USA

Dr Patricia A. Murphy
Environmental Criteria and Assessment Office, United States of America Environmental Protection Agency, 26 Martin Luther King Drive West, Cincinnati OH 45268, USA

Dr Betty H. Olson
Program in Social Ecology, University of California, Irvine CA 92717, USA

Dr Stig E. Regli
Office of Water, United States of America Environmental Protection Agency, 401 M Street, SW, Washington DC 20460, USA

Dr Peter F. Roessler
Analytical Microbiology Research and Development, Amway Corporation, 7575 Fulton Street East, Ada MI 49355–001, USA

Dr Joan B. Rose
Department of Marine Sciences, University of South Florida, 140 South 7th Avenue, St Petersburg FL 33701, USA

Dr Robin A. Ross
Channing Laboratory, Brigham & Women's Hospital and Harvard Medical School, 180 Longwood Avenue, Boston MA 02115, USA

Dr Elizabeth Scott
Consultant in Food and Environmental Hygiene, Newton MA 02159–2535, USA

Dr Syed A. Sattar
Department of Microbiology & Immunology, Faculty of Medicine, University of Ottawa, 451 Smyth Road, Ottawa Ontario K1H 8M5, Canada

Dr Pasquale Scarpino
Departments of Environmental Health, and Civil & Environmental Engineering, University of Cincinnati, Cincinnati OH 45267–0056, USA

Dr Blaine F. Severin
Director of Environmental Technology, Michigan Biotechnology Institute, 3900 Collins Road, Lansing MI 37610, USA

V. Susan Springthorpe
Department of Microbiology & Immunology, Faculty of Medicine, University of Ottawa, 451 Smyth Road, Ottawa Ontario K1H 8M5, Canada

Dr Mic H. Stewart
Water Quality Division, Metropolitan Water District of Southern California, 700 Moreno Avenue, La Verne CA 91750, USA

Preface

We have always had theories about how diseases are transmitted, just as we have always ascribed to methods for preventing disease transmission. Fortunately, our science has progressed to the point that we no longer believe influenza to be caused by the influence of the stars, and we no longer carry nosegays of flowers as protection against the evil vapors once believed to transmit the plague. Instead, we have come to learn that both of these diseases, as well as many others, result from our becoming infected by pathogenic microorganisms. We have also come to understand that not all microorganisms cause disease, and in fact our bodies are naturally colonized by nonpathogenic microorganisms whose presence serves to help protect us against becoming colonized by pathogens. Similarly, we sometimes add nonpathogenic microorganisms as a means of preserving foods against the activity of other organisms that might cause spoilage or disease.

Most of the routes by which pathogens are transmitted involve a period of time when those organisms are exposed to the environment, affording us the opportunity to prevent their transmission through use of disinfection practices. Disinfection can occur naturally since, with the passage of time, any population of microorganisms will die away under conditions that do not favor their replication. The ancient discovery that immersing objects in fire had purifying properties has led to our use of heat treatments to destroy microbial contaminants on objects and in foods. From the knowledge that sunlight had the capability to destroy the causes of infectious diseases, we have progressed to the development of artificial sources of ultraviolet, microwave and gamma irradiation for use in destroying pathogens. Old habits of attempting to purify objects by either burying them in soil, or casting them into water, have led to the development of chemical disinfectants. Our ability to model the processes of disease transmission and disinfection helps us to understand these processes, and affords us knowledge that aids us in achieving our goal of reducing disease-related suffering.

I wish to thank Gerard N. Stelma and Elizabeth C. Martinson for editing my chapter on diseases associated with foods. The United States of America's Environmental Protection Agency was not involved with the editing of this book.

Christon J. Hurst
Cincinnati, Ohio, USA

PART 1 HEALTH AND DISEASE

The transmission and prevention of infectious disease

CHRISTON J. HURST and
PATRICIA A. MURPHY

Introduction

The purposes of this chapter are threefold. First, it introduces the subject of modeling the epidemic versus endemic propagation of infectious disease through populations of individuals, for which a compartmental modeling approach is used, and the mathematical products are formally known as 'disease transmission models'. Second, it explains the routes by which the causative microbial agents are transferred between individuals who comprise the population. Third, it describes the use of disinfection to prevent or control the occurrence of those human diseases that are caused by pathogenic microorganisms. Other disease prevention measures are also mentioned. These other measures include physical barrier concepts, immunization and the use of antibiotics. Finally, the effectiveness of utilizing disease prevention measures is presented both by examining historical data for human death rates due to endemic typhoid fever in the United States during the early 20th century, and by using a mathematical model developed for epidemic disease.

Terminology

Pathogenic organisms are termed 'infectious agents' for epidemiological purposes, and this sets them apart from other causes of disease such as exposure to chemical agents. There are three basic considerations involved in the cycle of infectious disease transmission. First, there must be a source or reservoir of the infectious agent and the agent must escape from that reservoir. The source may be a human, some other animal, or the environment. Second, the agent must be conveyed to a susceptible host by some path or route. Third, there must be a susceptible host, and unless the causative microorganism produces only skin surface infections, the agent must successfully gain access to the interior of the host's body through some portal of entry. Portals of entry are the openings by which the organism gets past the barrier posed by intact skin. Common portals of entry include the respiratory tract, the gastrointestinal tract, the genitourinary tract, the eye, the ear and wounds to the skin such as tears, abrasions, burns and punctures (Isenberg & D'Amato, 1991). Absence of any one of these three considerations – existence of a reservoir, means of conveyance and availability of a susceptible new host – precludes disease transmission. Thus, all three considerations present opportunities to prevent the occurrence of infection, and this knowledge allows us to identify places in the cycle of disease transmission where specific measures (e.g. disinfection) can be used to interrupt the cycle.

It may be helpful to discuss and classify infectious agents by their mode of transmission, because the mode of transmission influences our choice of methods for disease control. Epidemiologists have tended to discuss the diseases transmitted between humans as being either 'direct' or 'indirect'. This classification possibly represents a misnomer, but the following is an attempt to explain the dichotomy. The term 'direct' transmission is used to indicate that the pathogens are passed either by direct physical contact between the human reservoir and the new host, or through an environmental route that is very short in terms of travel time and distance, such as coughing or sneezing directly onto someone. In these instances, our preventative efforts are directed toward the source of the disease. 'Indirect' transmission refers to those diseases for which the causative agent is carried to a susceptible host by an insect vector (vectorborne), an aerosol (airborne), solid environmental surfaces termed fomites, food, or water (all the last three described as vehicleborne). If transmission is 'indirect', then our preventive efforts are aimed toward various points along the environmental route.

Alternatively, the routes by which infectious microorganisms are transferred from one individual to another can be divided into those whose transmission relies upon direct physical contact versus those for which the transmission has some environmental component. Only a relative minority of diseases result from infectious agents that are transferred by direct physical contact between individuals. The majority of infectious agents are instead transmitted by indirect contact, meaning that the infectious agent moves through the environment for at least a short time or distance before it encounters a new host. This latter category therefore includes all of the 'indirectly' transmitted diseases plus the majority of 'directly' transmitted diseases. The exposure of the infectious agents to the environment can be viewed as beneficial, as it often allows us to use disinfection as a means of preventing transferral of the agents. This chapter employs the alternative approach to classifying disease transmission routes, i.e. direct physical contact versus environmental.

Two additional concepts need to be mentioned before delving further into this chapter. The first concept is that of primary versus secondary transmission. Primary transmission is the initial introduction of an infectious agent into a group of susceptible individuals. Secondary transmission is the subsequent transfer of the infectious agent from the initially infected individuals to other members of the group. Primary and secondary transmission of a disease can occur by different routes. An example used later in this chapter is the disease typhoid, which often has water as the vehicle of primary transmission into a population, and for which food contaminated by handlers may then serve as a vehicle of secondary transmission within the population. The second concept is prevention of disease, which is the hallmark of public health. Disease prevention can be divided into three separate phases, termed primary, secondary and tertiary. Primary prevention is preclusion of disease either by reducing the exposure of new susceptible 'host' individuals to the disease agent, or by altering the susceptibility of the host. Secondary prevention consists of early detection of the disease which hopefully leads to successful early treatment. Tertiary prevention consists of treatments intended to alleviate disability resulting from the disease and attempts to restore effective functioning of the individual.

Modeling the transmission of disease through populations: epidemic versus endemic

Modeling the transmission of infectious disease through populations of animals and humans has been explored by a number of researchers. Figure 1.1 presents a very basic example of the models developed for these applications. Such models are commonly refered to as 'Disease transmission models', and represent a type of compartmental model. In Figure 1.1 the individual compartments are shown as boxes representing segments of the population being studied. The different compartments are connected by solid arrows that represent the directions in which individual members of the population move from one compartment to another. This movement is expressed in the form of daily rates of movement, which are sometimes described as the 'force' of flow through the model. The model presented in Figure 1.1 has four compartments: susceptible (those individuals who are susceptible to infection by the pathogenic microorganism whose affect is being studied); infectious (those individuals who have become infected and are in a state where they can transmit the infectious agent to other individuals); immune (those individuals who have successfully completed convalescence from the infection and who are, for at least a time, resistant to reinfection); and removed (individuals who at least temporarily are excluded from the population under study). In this example, the removed individuals are those who have died from the infection. Figure 1.1 also shows the point in the model at which transferral of an infectious agent

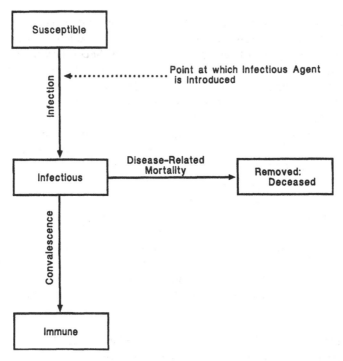

Figure 1.1   Basic compartment model appropriate for describing the epidemic propagation of disease through a population and the response of individuals to that disease. Models of this type are commonly termed 'disease transmission models'. This model also shows the point at which an infectious agent (causative microorganism) is transmitted to susceptible individuals.

occurs, represented as an arrow with a dotted line. For ease of comparison, all of the disease transmission models presented in this chapter will be drawn with a common format. Those boxes representing 'included' individuals will be listed vertically, with the box representing the immune individuals at the bottom. Boxes representing 'removed' individuals will be separated horizontally either to the right or to the left of those boxes which represent 'included' individuals.

The model presented in Figure 1.1 was developed for epidemic disease transmission, and accordingly contains relatively few compartments. Figure 1.2 presents a model developed for endemic disease transmission. This takes us a large step forward to a model that is more complex and more complete. In Figure 1.2, individuals can move in and out of three removed categories: susceptible, infectious and immune. The epidemic disease model presented in Figure 1.1 allows for disease-related mortality. The endemic disease model shown in Figure 1.2 likewise includes disease-related mortality, and additionally allows for both natural mortality and vaccine-related mortality. The model shown in Figure 1.2 also enables the addition of susceptible individuals to the population through new births, and returns immune individuals to the susceptible compartment through the eventual waning of immunity.

It is clear that models for disease transmission can grow very complex. Black & Singer (1987) have discussed the relative merits of elaboration versus simplification in mathematical disease transmission models, and concluded that extensive elaboration is not always necessary. Disease transmission models found in the literature can vary with respect to both the compartments they contain and the indicated movement of individuals. Such variation reflects the intended application of a given model, i.e. whether it is for endemic versus epidemic disease, and whether it represents a disease exposure which is pertinent to only a subset of the population, such as infants, or to the entire population.

Figure 1.3 represents an epidemic disease transmission model drawn for this chapter to illustrate a series of equations published by Anderson & May (1985), who developed the equations through observation of human populations. This model differs in three ways from what is presented in Figure 1.1. First, the model shown in Figure 1.3 includes a compartment for infants who are temporarily protected from infection because of maternal antibodies. These maternal antibodies can be acquired either through the

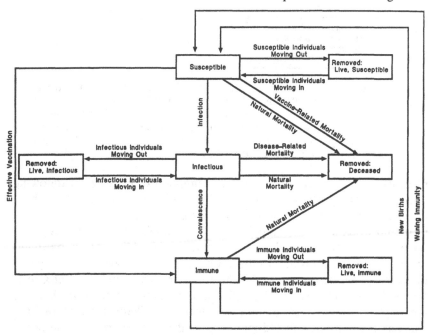

Figure 1.2 Compartment model appropriate for describing endemic disease propagation in a dynamic population.

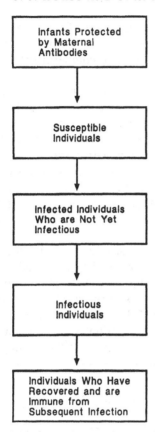

Figure 1.3   Compartment model for epidemic transmission of disease in a human population, including compartments for both infants not yet susceptible to infection, due to protection by maternal antibodies, and individuals who are in an incubation period following infection but prior to development of a state of infectiousness. This figure has been drawn to illustrate equations presented by Anderson and May (1985).

placenta prior to birth, or through consumption of breast milk following birth. Mathematically, the result of adding this compartment would be a time delay between when new individuals are added to the population by birth and when those individuals become susceptible to infection. Second, the model by Anderson & May (1985) contains an additional compartment for those individuals who have been infected, but who are not yet capable of transmitting the disease organisms to others. This compartment affects the model in two ways: it prolongs the duration of the epidemic by delaying the onset of infectiousness; and it reduces the likelihood that random contacts between an infected individual and other members of the population could result in transmission of the infection. Third, the Anderson & May model does not include disease-related mortality. Some excellent additional references on modeling human disease transmission are those by Anderson (1982, 1994), Black & Singer (1987), and Doege & Gelfand (1978).

Figure 1.4 shows a model developed by Miller (1979) which describes the epidemic transmission of infection through livestock animals. Miller's model includes vaccination as a means of preventing an initial infection in otherwise susceptible animals. The immunity conferred is not permanent, which allows previously ill or vaccinated animals that remain unslaughtered to return to the pool of susceptibles. Removal of animals in Miller's model occurs through selective slaughter of diseased animals and their contacts, a consciously introduced preventive measure which cannot be used with humans. Miller's model also allows for restocking of the study population, likewise something which cannot be done with human populations. Miller's model includes only one compartment for removed individuals, with movement both into that compartment by slaughter, and back out to the pool of susceptible individuals by restocking. Presumably, a single compartment was used for the sake of simplicity. It is important to understand that, in

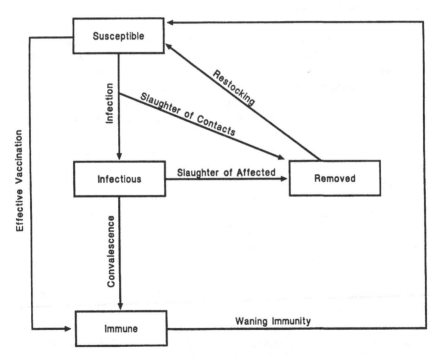

Figure 1.4   Compartment model applicable for long-term epidemic exposure in a livestock population. This model includes removal of both infected and potentially infected individuals, restocking, vaccination as a means of preventing infection, and waning immunity due to passage of time. Redrawn from Miller, 1979.

actuality, the newly restocked individuals cannot come from the same population that was slaughtered, and so we must assume that they come from some outside reserve. A sampling of additional recent examples of models for the transmission of disease through animal populations includes rinderpest in cattle and wild-life, jointly described by James and Rossiter (James & Rossiter, 1989; Rossiter & James, 1989), pseudorabies in swine (Smith & Grenfell, 1990), and rabies in raccoons (Coyne, Smith & McAllister, 1989). Modeling the transmission of disease through plant populations can be almost as simple, but will not be addressed in this book. Cvjetanović (1982) has written an excellent reference on the dynamics of using compartment models to study the transmission of disease through populations. A good general discussion of population modeling appears in the book by Lotka (1956). The remainder of this chapter will only address disease transmission as it relates to humans.

How do infectious agents get transferred between individuals?

The disease transmission models presented in Figures 1.1–1.4 show how diseases are propagated in a population of humans or animals. In Figure 1.1, a side arrow shows the point of disease transmission, where infectious agents are transferred to a susceptible individual. This brings us back to important concepts in disease transmission: there must be a source of the disease agent, there must be a susceptible host and there must be an effective transfer of the agent to the host.

Disease transmission routes can be divided into two broad categories, vertical and horizontal. Vertical transmission occurs between mother and child, either prior to or during the birth process (Mims, 1981; Watson *et al.*, 1993). Horizontal transmission encompasses all other routes, including both environmental and nonenvironmental. Usually, the mother becomes infected with a disease agent that she acquires horizontally, and she then transmits that agent to her fetus vertically.

Diseases can also be classified according to whether their transmission is considered to be 'direct' or 'indirect'. If the infectious agent is incapable of maintaining viability on its own outside the host, or if it can maintain viability in the open environment for only an extremely short period of time, then the disease associated with that infectious agent is assumed to be directly transmitted. Examples of directly transmitted diseases are the bacterial disease gonorrhea, and the viral disease acquired immune deficiency syndrome. Both gonorrhea and acquired immune deficiency syndrome happen to be sexually transmitted, but the concept of directly transmitted disease is not limited to sexual transmission, as it also includes some diseases transmitted through nasal secretions and saliva. Organisms that are transmitted indirectly are those that can maintain viability in the environment long enough to have sat around, or to have been blown or carried around, before a new host is contacted. Examples of indirectly transmitted diseases are vaccinia, caused by viruses that are released from lesions on the skin and transferred by fomites; and influenza, transmitted by aerosols generated during coughing and sneezing.

All of the infectious agents that are transmitted 'indirectly', plus the majority of 'directly' transmitted infectious agents, are exposed to the environment for some time during their transfer to a new susceptible host. Environmental exposure can occur when the organisms are in aerosols, when they are in contaminated food or water, or when they are on the surface of fomites, the last category consisting of solid environmental objects including toys, clothing and blankets. The major factor that must be considered on the potential for any given infectious agent to be successfully transmitted by an environmental route is the ability of that microorganism to remain viable outside the body of its last host animal, surviving for example in air or water, long enough to encounter a new susceptible host organism.

A disease-causing microbe is less likely to be transmitted by an environmental route that exposes the organism to conditions under which it cannot easily survive. Thus, those infectious agents which are likely to have the greatest success at being transmitted in aerosols or on fomites will logically be the organisms that have evolved resistance to desiccation. Accordingly, it is noted without surprise that both the smallpox and influenza viruses have evolved an ability to survive for long periods of time in air at low relative humidity levels (Harper, 1961). Two other viral diseases that are also suspected of being transmitted via fomites are hepatitis caused by the hepatitis A virus, which in one disease outbreak was possibly acquired from the handling and smoking of cigarettes after infected patients had been cared for in a hospital (Doebbeling, Li & Wenzel, 1993), and gastro-enteritis caused by *Rotavirus* (Keswick *et al.*, 1983), which is common to children at day-care centers. Appropriately, both hepatitis A virus and *Rotavirus* are capable of prolonged survival when exposed to drying on fomites (Abad, Pintó & Bosch, 1994).

Members of the viral genus *Enterovirus* present a good contrast to the smallpox, influenza, hepatitis A and rotaviruses. Generally, the entero-viruses tend to lose viability very rapidly in drying soils and wastewater sludges (Hurst, 1991), and at least one of them, human poliovirus 1, has been shown to possess only minimal ability to survive in air at low relative humidity levels (Harper, 1961) or during drying on fomites (Abad *et al.*, 1994). Their susceptibility to desiccation suggests that, on the whole, enteroviruses are bad candidates for being transmitted by aerosols or on fomites. However, enteroviruses can be detected in wastewater (Melnick, 1947; Melnick *et al.*, 1954a,b), where they occur as contaminants during epidemic as well as non-epidemic times (Melnick, 1947), and they are capable of prolonged survival both in environmental waters (Hurst, 1991; Hurst, Wild & Clark, 1992; Moore, 1993) and in saturated soil (Hurst, 1991). These facts about enteroviruses, when taken in combination, make them good candidates for transmission by water. Indeed, polioviruses have been transmitted by water (Mosley, 1966).

This should not lead us to expect that all of the viruses which can be found in water will necessarily be transmitted by the water route. It is true that both enteroviruses and the human immunodeficiency viruses, which cause the acquired immune deficiency syndrome, may be found in wastewater (Preston et al., 1991). The knowledge that the enteroviruses can be transmitted by water represents a logical extension from the fact that enteroviruses have evolved stability in water. In contrast, the human immunodeficiency viruses do not survive well in water (Moore, 1993) and therefore are not good candidates for being transmitted by water. Evolutionary pressures can explain these findings. The human immunodeficiency viruses are effective at being transmitted by sexual contact, and thus have not needed to evolve stability in environmental waters.

The hepatitis A virus is of particular interest, because it is environmentally stable both in water and when exposed to low relative humidity levels on fomites. This knowledge suggests that the hepatitis A virus has the capability of being transmitted by many different routes.

Most environmentally transmitted human illnesses result from encountering human or animal pathogens. There are additional environmentally associated illnesses which result from the acquisition of microorganisms that are natural environmental inhabitants (Grimes, 1991). Included in the latter category are a number of bacterial organisms such as *Clostridium perfringens*, which causes gas gangrene in wounds when contact is made, or gastroenteritis when it is consumed (Allen & Baron, 1991); *Clostridium botulinum*, which causes botulism (Allen & Baron, 1991); *Clostridium tetani*, which causes tetanus (Allen & Baron, 1991), *Legionella pneumophila*, which causes legionellosis (Breiman, 1993; Fields, 1993); and *Vibrio vulnificus*, which causes septicemia (O'Neill, Jones & Grimes, 1990). Several genera of endemic fungi that reside in the soil can cause diseases if inhaled (Sternberg, 1994). Among these fungi are *Aspergillus*, which causes aspergillosis; *Coccidioides*, which causes coccidioidosis, and *Histoplasma*, which causes histoplasmosis. An interesting issue regarding environmentally transmitted diseases is the role that they have played in military engagements (Doyle and Lee, 1986; Poupard, Miller & Granshaw, 1989).

Figures 1.5 through 1.8 illustrate the various routes of disease transmission such as insect vectors, water and aerosols. These routes of disease transmission can also be viewed as potential compartment models and subjected to mathematical probability analysis. Subsequent sections of this chapter will elaborate on each figure. Efforts at modeling portions of these transmission routes will be presented in Chapters 4 and 6.

Disease transmission by physical contact

Many diseases are aquired by direct physical contact with a human or animal harboring an infectious agent. Figure 1.5 presents the basic paths by which this type of transmission occurs.

Direct contact between humans

Vertical disease transmission is considered to represent a form of transmission by direct contact. Horizontal transmission of disease can also represent transmission via direct contact, encompassing those organisms transferred in semen (Hamed *et al.*, 1993), saliva (Yeung *et al.*, 1993), transplanted body tissues (Gode & Bhide, 1988), blood and blood products. The touching of contaminated skin (Hlady *et al.*, 1993; Jernigan *et al.*, 1993) and wounds can also result in transmission by direct contact. The category of transmission by direct contact includes infections acquired from sexual partners. These include bacterial diseases such as syphilis (Sims *et al.*, 1992), viral diseases such as hepatitis B (Hou *et al.*, 1993) and herpes simplex (Bryson *et al.*, 1993), and the chlamydial disease caused by *Chlamydia*

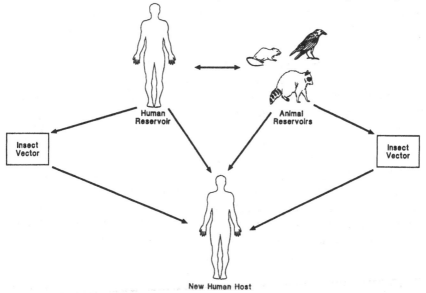

Figure 1.5   Routes by which infectious agents are transmitted to susceptible individuals, requiring direct physical contact between the human or animal reservoir, possibly including an insect vector, and the human recipients, who then serve as new hosts for the microorganisms. The animals symbolically represented are raccoons as a source of rabies, rodents as a source of plague and birds as a source of encephalitis.

trachomatis (Viscidi *et al.*, 1993). For the purpose of this chapter, diseases transmitted by bites from humans, vertebrate animals and arthropods are also considered to represent transmission by direct physical contact.

Direct contact between animals and humans

Examples of illnesses transmitted from vertebrate animals to humans by direct physical contact with contamination on the animals' skin are the bacterial disease salmonellosis (Svitlik *et al.*, 1992) and the viral disease caused by simian immunodeficiency virus (Centers for Disease Control, 1992*b*). Rabies is an example of a disease whose transmission results from animals biting humans (Cartter *et al.*, 1992), as represented by raccoons in Figure 1.5. Diseases that can be transmitted from animals to humans are called zoonoses. Animals are the natural hosts for the causative microorganisms. Humans usually serve only as incidental hosts, and are nonessential for the normal cycle of transmission.

Arthropod vectored diseases

A large number of diseases are transmitted via arthropod vectors (Rehle, 1989; Pratt & Smith, 1991; Cotton, 1993). These diseases have been included under this section of the chapter because their transmission involves a chain of physical contact, with humans or other vertebrates as the reservoir, invertebrates serving as the vector, and humans being the susceptible host. A variety of insects are known to serve as arthropod vectors, and some of the causative microorganisms mentioned below are capable of replicating in their vectors. Chiggers are capable of transmitting the rickettsial disease scrub typhus (Twartz *et al.*, 1982). Fleas are perhaps most notorious, because they transmit the bacterial disease plague (Doll *et al.*, 1992). Biting flies can transmit bacterial diseases such as tularemia (Stewart, 1991). Mosquitoes are responsible for the transmission of viral diseases such as dengue (Centers for Disease Control, 1991*b*) and encephalitis (Hlady, 1993; Anders *et al.*, 1994). Ticks are associated with the transmision of bacterial illnesses such as Lyme disease (Wallach *et al.*, 1993) and tularemia (Stewart, 1991), and the viral disease caused by tickborne encephalitis virus (Ramelow *et al.*, 1993). Flies that do not bite may be associated with the transmission of gastrointestinal diseases, as will be discussed later in this chapter. In these last instances the flies serve only as passive carriers or vehicles when they transfer microorganisms from feces to food.

The above diseases, plus a number of others that are transmitted by physical contact, are summarized in Table 1.1 along with the names of the causative microorganisms.

As alluded to above, many viral diseases that some people consider

Table 1.1. *Examples of diseases transmitted by direct physical contact*

Transmission chain	Disease	Causative microorganism(s)
Human to human	Bacterial	
	Syphilis	*Treponema pallidum*
	Gonorrhea	*Neisseria gonorrhoeae*
	Fungal	
	Candidiasis	*Candida*
	Viral	
	Acquired immune deficiency syndrome	*Lentivirus*
	Congenital rubella	*Rubivirus*
	Hepatitis	*Deltavirus, Flavivirus, Orthohepadnavirus*
	Mononucleosis	*Cytomegalovirus, Lymphocryptovirus*
Animal to human	Bacterial	
	Tularemia	*Francisella tularensis*
	Viral	
	Rabies	*Lyssavirus*
Arthropod vectored[a]	Bacterial	
	Lyme disease	*Borrelia burgdorferi*
	Plague	*Yersinia pestis*
	Tularemia	*Francisella tularensis*
	Rickettsial	
	Rocky mountain spotted fever	*Rickettsia*
	Typhus	*Rickettsia*
	Viral	
	Colorado tick fever	*Coltivirus*
	Dengue	*Flavivirus*
	Encephalitis	*Alphavirus*
	Yellow fever	*Flavivirus*

Note: For this and other tables in the book, if a microorganism is identified by both genus and species names, then the listed finding applies to that particular species. If a genus name is listed without an accompanying species name, then the finding applies to more than a single species within that genus.

[a] The causative microorganisms listed for the arthropod vectored diseases may replicate in those vectors. However, for these microorganisms, replication in that vector is not a required part of their life cycle. This contrasts with the causative microorganisms listed in Table 1.4., which do require replication in an intermediate host.

to be directly transmitted are in fact normally transmitted through the environment. These include rotaviral gastroenteritis (Sattar *et al.*, 1986; Sawyer *et al.*, 1988), smallpox (Poupard *et al.*, 1989) and the so-called 'childhood diseases': chickenpox, measles, mumps and pertussis. These diseases will be described in subsequent sections of the chapter, because they are normally environmentally transmitted.

Disease transmission by environmental water routes

Water serves as the vehicle for a large proportion of environmentally transmitted diseases. Bacterial diseases spread by this route include the notorious typhoid (Johnson, 1916; Kehr & Butterfield, 1943) and cholera (Glass, Libel & Brandling-Bennett, 1992). In particular, outbreaks of cholera often take the form of massive, wide-ranging epidemics, termed pandemics, which have caused fear in human populations since at least 1817 (Glass & Black, 1992). The most recent cholera pandemic occurred in the Americas, with the causative bacteria perhaps arriving there from another part of the world in contaminated ship ballast water (McCarthy & Khambaty, 1994). This latest pandemic resulted in approximately 350 000 cases of illness and nearly 4000 deaths (Swerdlow & Ries, 1992). These two bacterial illnesses were the driving force behind the development of formal drinking water treatment utilities during the middle and end of the 19th century. Yet, as we approach the 21st century, we are faced with the fact that disease microorganisms consumed in conventionally treated drinking water may account for 35% of the gastrointestinal illnesses among consumers even in developed countries (Payment *et al.*, 1991). It is for this reason that many governments have developed surveillance programs for waterborne disease (St Louis, 1988; Tulchinsky *et al.*, 1988; Moore *et al.*, 1993).

Public health emphasis is often placed on diseases associated with the ingestion of water, since this route of transmission has resulted in massive outbreaks. However, it is important to remember that ingesting water is not the only way in which we acquire diseases that are transmitted via contaminated water. We also acquire diseases from occupational or recreational activities performed in water, as well as from the consumption of water-contaminated shellfish and food crops. The various paths of disease transmission associated with waterborne microbial contaminants are illustrated in Figure 1.6.

Animal reservoirs of microbial contaminants

Vertebrate animals can be an important source of microbial contaminants in surface water via two different paths. The first of these is the deposit by animals of fecal material and urine directly into the water (Pinfold *et al.*,

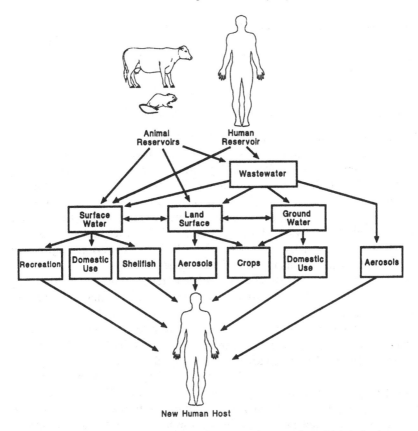

Figure 1.6   Water-related environmental routes by which infectious agents are transmitted to susceptible individuals. The animals symbolically represented are cows as a source of cryptosporidiosis and rodents as a source of giardiasis and campylosis.

1993). The second is the deposit of fecal material and urine on the land, and subsequent washing of the organisms into surface waters by overland runoff (Sonzogni *et al.*, 1980; Baxter-Potter and Gilliland, 1988). Bacterial diseases of humans that may originate from animal reservoirs include campylosis contributed by beavers, migratory birds, muskrats and other rodents (Pacha, Clark & Williams, 1985; Pacha *et al.*, 1987, 1988), and leptospirosis contributed by livestock (Alexander, 1991). In humans, protozoans are responsible for causing giardiasis (Erlandsen *et al.*, 1988; Healy, 1990) and cryptosporidiosis (Hayes *et al.*, 1989). These protozoans are capable of cross infecting a variety of animals including beavers and livestock, with a result that these animals may serve as reservoirs of the diseases, and transmit them to humans via surface waters (Pacha *et al.*, 1987; Hayes *et al.*, 1989; Erlandsen *et al.*, 1990; Healy, 1990; Isaac-Renton *et al.*, 1993;

MacKenzie *et al.*, 1994). Rotaviruses, which cause waterborne gastro-enteritis, can also cross infect humans and a variety of animals, including cattle (Clark *et al.*, 1986). This suggests that infected livestock could be a reservoir of waterborne viral gastroenteritis.

Human reservoirs of microbial contaminants

Humans also contribute microbial contaminants directly to water, during the course of recreational activities. This can include contamination of both swimming pools (Grabow, 1991) and beaches (Fattal, Peleg-Olevsky & Cabelli, 1991). Protozoan contaminants contributed in this way include the causative agents of cryptosporidiosis (Joce *et al.*, 1991; Sorvillo *et al.*, 1992; Bongard *et al.*, 1994) and giardiasis (Porter *et al.*, 1988). Viral contaminants contributed to water during recreational activities include causative agents of gastroenteritis (Holmes *et al.*, 1989) and pharyngoconjunctival fever (Ormsby & Aitchison, 1955; Martone *et al.*, 1980).

Microorganisms are shed in human fecal material (Abraham *et al.*, 1993), making wastewater a major source of microbial contaminants. Additionally, animals such as rats, which dwell in sewerage collection systems, directly contribute contaminants to the wastewater. Wastewater and the pathogens that it contains are disposed of by direct discharge into surface waters (McFeters, Barry & Howington, 1993; Fiksdal *et al.*, 1994), injection into ground water (Dryden & Chen, 1978) and application onto land surfaces (Burge & Marsh, 1978; Kladivko & Nelson, 1979b; Abernathy, Zirschky & Borup, 1984). Microbial contaminants placed onto the land surface are not assured of remaining on the land, because surface-applied wastewater can percolate into the ground water (Edmonds, 1976; McPherson, 1979) and run off as surface water (McPherson, 1979). Humans may subsequently encounter these pathogens either by contact with, or by consumption of, contaminanted surface waters and ground waters.

Inhalation of aerosols represents another route of exposure to microbial contaminants contained in wastewater. Liquid (droplet) aerosols are generated during the processes of wastewater aeration (Sawyer *et al.*, 1993) and the spray application of wastewater sludge suspensions onto land (Sorber *et al.*, 1984). Aerosols generated during wastewater treatment may serve as a source of disease in wastewater workers (Clark *et al.*, 1985). However, aerosols generated during community wastewater treatment processes do not seem to serve as a source of disease for members of the general population (Katzenelson, Buium & Shuval, 1976; Ward *et al.*, 1989).

The interconnected flow of water and its microbial contaminants

Microbial contaminants are not necessarily stationary within the environment. As water flows, it can transport microbial contaminants. For this reason, horizontal lines connect surface water, the land surface and ground water in Figure 1.6. Wastewater contaminants applied to the land surface may be carried into surface waters by runoff (Kladivko & Nelson, 1979a). Microorganisms contained in wastewater that is applied to the land surface for irrigation of crops (Walker & Demirjian, 1978) or for ground-water recharge (Bouwer & Rice, 1984) may be carried into ground water by both vertical and horizontal infiltration (Allen & Morrison, 1973; Powelson, Gerba & Yahya, 1993). Once in the ground water (Abbaszadegan *et al.*, 1993), natural flow can carry these microbial contaminants into surface waters (Collins, 1994) and they can emerge onto the land surface via springs (Bergeisen, Hinds & Skaggs, 1985). Surface waters can dramatically return to the land surface during floods. Flooding also brings attendant concern about mosquito-transmitted illnesses (Atchison *et al.*, 1993; Cotton, 1993; Anders *et al.*, 1994), because flooding increases water surface area, which in turn supports the life cycle of mosquitoes.

Diseases acquired from microbially contaminated surface water

Participation in recreational surface-water activities is frequently associated with the acquisition of infectious disease (Cabelli, 1989; Cheung *et al.*, 1991: Corbett *et al.*, 1993). Seyfried *et al.* (1985a,b) polled visitors at beaches in Ontario, Canada and found crude morbidity rates of 69.6 per 1000 for swimmers versus only 29.5 per 1000 for nonswimmers. The symptoms experienced by ill swimmers included respiratory, gastrointestinal, eye, skin and allergic ailments. Specific examples of diseases acquired by human activities in water include the bacterial disease leptospirosis (Alexander, 1991), and the viral diseases gastroenteritis (Koopman *et al.*, 1982) and pharyngoconjunctival fever (McMillan *et al.*, 1992).

The domestic use of contaminated water clearly results in disease transmission, recently reviewed by Cliver (1990). It is possible to find microbial contaminants associated with water taps even when the water lines feeding the taps apparently are free of the suspect organism. This latter type of contamination has been described by Grundmann *et al.* (1993). The most commonly suspected mode of transmission associated with domestic water use is through ingestion of the contaminated water. Bacterial diseases acquired in this way include campylosis (Tauxe *et al.*, 1988), cholera (Glass & Black, 1992), typhoid (Johnson, 1916) and tularemia (Stewart, 1991). Viral diseases acquired from drinking water include diarrhea (Tao *et al.*, 1984), hepatitis (Viswanathan, 1957; Mosley, 1966; Bosch *et al.*, 1991) and poliomyelitis (Mosley, 1966). Recently, the most notable protozoan disease acquired by the ingestion of water is cryptosporidiosis (Rose, 1988; MacKenzie *et al.*, 1994). Another interesting route by which diseases can be acquired from tap water is through the use of contaminated water when wounds are washed (Lowry *et al.*, 1991). Prevention of these diseases is the reason that we practice drinking-water treatment at community levels (Stetler, Waltrip & Hurst, 1992). The subject of treating drinking water prior to its ingestion as a means of preventing disease transmission is further addressed in Chapter 4.

Ingestion of shellfish harvested from waters that have been contaminated by discharged wastewater represents another means through which people acquire microbial illnesses from polluted water (Higashihara *et al.*, 1991). Crustaceans can accumulate pathogenic microorganisms from the environment and then potentially pass these organisms on to humans who consume the crustaceans (Davis & Sizemore, 1982; Roman *et al.*, 1991). Molluscs are divided into two categories. The first is the bivalves, such as oysters and hardshell clams. The second group is the gastropods, such as the conch. Human pathogenic viruses, such as the hepatitis A virus, can be detected in bivalve molluscs (Goswami, Koch & Cebula, 1993), and the ingestion of contaminated bivalves has caused outbreaks not only of viral disease (De Leon & Gerba, 1990; Desenclos *et al.*, 1991), but also of bacterial diseases such as cholera (Eichold *et al.*, 1993). It is presumed that the bivalve molluscs accumulate these microbial contaminants during their natural process of filter feeding. Gastropod molluscs do not filter feed, and thus would not be expected to accumulate pathogens by this mechanism.

Diseases acquired from microbially contaminated land surfaces

Wastewater is sometimes used for the irrigation of vegetable crops. As mentioned previously, this practice presents a health hazard, because it may contaminate those crops with potentially pathogenic microorganisms that have been shed in feces (Hopkins *et al.*, 1993; Rosas, Báez & Coutiño, 1984). Human consumption of crops that were irrigated with wastewater has been implicated in the spread of cholera (Glass & Black, 1992; Glass *et al.*, 1992). At least one research group has tried to ameliorate this problem by disinfecting sewage effluents that are intended for agricultural usage (Acher, Fischer & Manor, 1994). Particulate aerosols may be generated from wastewater sludges that have been applied onto land surfaces either for the purpose of composting that sludge, or for final sludge disposal. These aerosols present a pathogenic hazard, because they can contain both bacterial (Edmonds & Littke, 1978) and fungal contaminants (Millner, Bassett & Marsh, 1980), and they may be transported by the wind.

Diseases acquired from microbially contaminated ground water

It is often thought that ground water is intrinsically pristine and pure, but ground water can be contaminated with microorganisms that are pathogenic for humans (Schwartzbrod *et al.*, 1985). Domestic use of contaminated ground water has resulted in at least one outbreak of the bacterial illness cholera (Blake *et al.*, 1977). Ingestion of contaminated ground water has also resulted in outbreaks of the protozoan illnesses cryptosporidiosis (D'Antonio *et al.*, 1985) and giardiasis (Dennis *et al.*, 1993), and the viral illnesses gastroenteritis (Hejkal *et al.*, 1982; Lawson *et al.*, 1991; Murphy, Grohmann & Sexton, 1983) and hepatitis (Bloch *et al.*, 1990; Hejkal *et al.*, 1982). A heavy reliance is often placed on the use of ground water for crop irrigation (Task Committee on the Status of Irrigation and Drainage Research, 1984), and this reliance might result in the microbial contamination of crops.

The above diseases, along with a number of others transmitted through environmental routes associated with contaminated water, are summarized in Table 1.2 along with the names of the causative microorganisms.

Disease transmission by environmental routes other than water

Diseases are often transmitted by environmental routes other than water. These routes are illustrated in Figure 1.7. Some of these diseases, such as leptospirosis (Alexander, 1991), tularemia (Stewart, 1991) and histoplasmosis, can originate from animal reservoirs. Other illnesses transmitted to humans by these routes are doubtlessly of human origin (Vazquez *et al.*, 1993). Non-biting flies represent a particularly interesting means of facilitating disease transmission between humans. These invertebrates acquire pathogenic microorganisms from feeding on fecal material, and then passively carry those microorganisms to human foods. Microorganisms carried in this way include the bacteria that cause cholera, shigellosis (Echeverria *et al.*, 1983) and typhoid (Johnson, 1916). Because enteric viruses also are shed in fecal material (Melnick *et al.*, 1954a; Abraham *et al.*, 1993), they likewise can be passively acquired by flies (Melnick *et al.*, 1954a). It is possible that the natural curiosity of pet animals, such as dogs, may result in these vertebrates similarly acting as passive carriers of fecally shed human pathogens (Engleberg *et al.*, 1982).

Diseases acquired from microbially contaminated medical devices

The use of contaminated syringes and syringe needles, as well as the related issue of improperly used multidose medication vials, may inadvertently result in the transmission of disease between humans (Hlady *et al.*, 1993; Levy, Moll & Jones, 1993). Contaminated needles clearly have been implicated in the transmission between humans of the viral diseases acquired immune deficiency syndrome (Farzadegan *et al.*, 1993) and hepatitis

Table 1.2. *Examples of diseases transmitted by environmental water routes*

Exposure route	Disease	Causative microorganism(s)
Recreation	Bacterial	
	Cholera	*Vibrio cholerae*
	Leptospirosis	*Leptospira interrogans*
	Protozoan	
	Entamoebiasis (Amoebic dysentery)	*Entamoeba histolytica*
	Amoebic encephalitis	*Naegleria*
	Viral	
	Meningitis	*Enterovirus*
	Pharyngoconjunctival fever	*Mastaadenovirus*
	Viral encephalitis	*Enterovirus*
	Viral gastroenteritis	*Astrovirus, Calicivirus, Coronoavirus, Rotavirus*
Domestic use	Bacterial	
	Campylosis	*Campylobacter*
	Cholera	*Vibrio cholerae*
	Paratyphoid	*Salmonella paratyphi*
	Shigellosis (bacterial dysentery)	*Shigella*
	Tularemia	*Francisella tularensis*
	Typhoid	*Salmonella typhi*
	Protozoan	
	Cryptosporidiosis	*Cryptosporidium parvum*
	Entamoebiasis	*Entamoeba histolytica*
	Giardiasis	*Giardia lamblia*
	Viral	
	Encephalitis	*Enterovirus*
	Gastroenteritis	*Astrovirus, Calicivirus, Coronavirus, Rotavirus*
	Hepatitis	*Calicivirus, Hepatovirus*
	Meningitis	*Enterovirus*
Ice	Viral	
	Gastroenteritis	*Calicivirus*
Shellfish (crustacean)	Bacterial	
	Enteritis	*Vibrio*
Shellfish (molluscan)	Bacterial	
	Enteritis	*Vibrio*
	Viral	
	Gastroenteritis	*Calicivirus*
	Hepatitis	*Hepatovirus*
Crops (vegetables)	Bacterial	
	Enteritis	*Salmonella, Vibrio*
	Viral	
	Hepatitis	*Hepatovirus*
Aerosols	Bacterial	
	Legionellosis	*Legionella pneumophila*

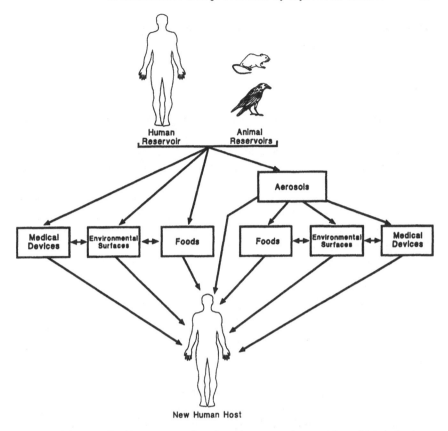

Figure 1.7 Non-water-related environmental routes by which infectious agents are transmitted to susceptible individuals. The animals symbolically represented are rodents as a source of hemorrhagic fever and birds as a source of histoplasmosis.

(Centers for Disease Control, 1992*a*; Clayton *et al.*, 1993). Accidents involving contaminated needles have also been implicated in the transmission of viral disease from animals to humans (Centers for Disease Control, 1992*b*). Contaminated optical instruments have been implicated in the transmission of viral keratoconjunctivitis (Jernigan *et al.*, 1993). Contaminated endoscopes and bronchoscopes have, respectively, been implicated in the transmission of *Pseudomonas* bacterial infections and mycobacterial pseudoinfections (Alvarado *et al.*, 1991). (A pseudoinfection is one in which the organism is present but is not producing detectable deleterious effects upon the host nor eliciting a detectable response from the host.) Vascular catheters can be colonized by bacteria (Raad *et al.*, 1993) and, in particular, catheter-related sepsis can be a serious problem for neonates (Salzman *et al.*, 1993).

Diseases acquired from microbially contaminated environmental surfaces

Environmental surfaces, termed fomites, can become contaminated with pathogenic microorganisms. At least some viral (Keswick *et al.*, 1983) and chlamydial (Falsey & Walsh, 1993) pathogens may survive on environmental surfaces, and susceptible individuals may then acquire these disease agents as a result of touching the contaminated surfaces (Sattar *et al.*, 1986, 1993). Salmonellosis is one example of the bacterial diseases that presumably may be acquired via fomites (Barrett *et al.*, 1992). Published examples of viral diseases transmitted via fomites include smallpox acquired from blankets (Poupard *et al.*, 1989), aseptic meningitis from shared drinking cups (Alexander *et al.*, 1993) and hepatitis that possibly was acquired from the handling and smoking of cigarettes in hospital areas (Doebbeling *et al.*, 1993).

Diseases acquired from microbially contaminated food

Microbially contaminated foods represent a major source of human illness, including a large number of viral diseases (Cliver, 1990). Microbial contamination of foods can occur within the environment before the food is harvested. Foods contaminated by water in this way include bivalve molluscs, crustaceans and vegetables. Foods can also be contaminated in the environment by processes that are not directly associated with contaminated water, examples of which include the bacterial agents of salmonellosis which may be acquired from ingesting eggs (Ching-Lee *et al.*, 1991; Ewert *et al.*, 1993; Hedberg *et al.*, 1993), and campylosis which may be acquired from both poultry (Wegmüller, Lüthy & Candrian, 1993) and dairy products (Pearson *et al.*, 1993).

Outbreaks of disease have also been associated with the ingestion of foods that were contaminated during the course of handling. These outbreaks often result from ingesting food that was either uncooked or inadequately cooked. Most of these illnesses seem to originate from handlers who had intestinal infections, and, as a result, the infections acquired from these contaminated foods are usually intestinal in nature. Recent examples of bacterial illness transmitted in this way include gastroenteritis resulting from consumption of contaminated coconut milk (Taylor *et al.*, 1993), orange juice (Birkhead *et al.*, 1993) and undercooked ground beef (Centers for Disease Control, 1993*a*; Le Saux *et al.*, 1993). An example of intestinal protozoan illness associated with uncooked food contaminated by an infected handler is giardiasis acquired from raw sliced vegetables (Mintz *et al.*, 1993). Viral intestinal illness has been associated with the consumption of uncooked contaminated foods, recent examples of which are fresh cut fruits (Herwaldt *et al.*, 1994), and ice consumed in beverages (Khan *et al.*, 1994). Not all handler-associated foodborne illnesses originate from intestinal infections. An interesting example of this is streptococcal pharyngitis acquired from consumption of macaroni and cheese that had been contaminated by a food handler's skin lesion (Farley *et al.*, 1993).

Thermal abuse, a term describing food that has been held at temperatures high enough to enable the growth of contaminating pathogenic bacteria or fungi, has been associated with enteritis (Luby *et al.*, 1993). Thermal abuse of foods is not usually implicated in protozoan or viral illnesses, since human enteric protozoa will not usually grow in stored foods, and human enteric viruses cannot grow in foods, since these viruses require viable animal host cells in order to replicate.

The environmental transfer of microbial contaminants

Horizontal connecting lines have been drawn in Figure 1.7 to show that microbial contaminants can be transferred from one type of contaminated material to another, creating a cycle of contamination. An interesting example of this is provided by an outbreak of giardiasis, for which it appears that raw food initially became contaminated by being washed in tap water, after which the contaminated food was chopped on a cutting board. Other foods appear to have become contaminated when they were later prepared on that same cutting board, which was never adequately washed (Grabowski *et al.*, 1989). Adequate washing of the cutting board, to physically remove contaminating microorganisms, or disinfection of the cutting board would have broken the cycle of contamination.

Diseases acquired from aerosolized microorganisms

Aerosols serve as the route of transmission for many infectious diseases, and microbial agents can be transferred between animals, from animals to humans, and between humans. Indeed, high levels of microorganisms can be found in aerosols (Westwood & Sattar, 1976).

Examples of diseases known to be transmitted between animals via aerosols are foot-and-mouth disease and Newcastle disease (Gloster, 1983), both of which are caused by viruses. Examples of diseases transmitted by aerosols from animals to humans are plague (Opulski *et al.*, 1992), tuberculosis (Centers for Disease Control and Prevention, 1993c) and tularemia (Stewart, 1991), each of which is bacterial in nature. Examples of viral diseases that can be transmitted from animals to humans via aerosols are hemorrhagic fever (Barry *et al.*, 1994) and respiratory illness (Centers for Disease Control and Prevention, 1993b).

Perhaps the most notorious bacterial disease transmitted between humans via aerosols is tuberculosis (Hewlett *et al.*, 1993; Jereb *et al.*, 1993;

Table 1.3. *Examples of diseases transmitted by environmental routes other than water*

Exposure route	Disease	Causative microorganism(s)
Aerosols		
	Bacterial	
	Anthrax	*Bacillus anthracis*
	Diphtheria	*Corynebacterium*
	Legionellosis	*Legionella*
	Pertussis	*Bordetella pertussis*
	Plague	*Yersinia pestis*
	Tuberculosis	*Mycobacterium tuberculosis*
	Tularemia	*Francisella tularensis*
	Chlamydial	
	Psittacosis	*Chlamydia psittaci*
	Fungal	
	Aspergillosis	*Aspergillus*
	Coccidiosis	*Coccidioides*
	Histoplasmosis	*Histoplasma capsulatum*
	Viral	
	Bronchiolitis	*Paramyxovirus*
	Chickenpox	*Varicellovirus*
	Common cold	*Coronavirus, Rhinovirus*
	Hemorrhagic fever	*Arenavirus, Filovirus*
	Influenza	*Influenzavirus A, B, and C*
	Measles	*Morbillivirus*
	Parainfluenza	*Paramyxovirus*
	Rubella	*Rubivirus*
Medical devices	Protozoan	
	Cryptosporidiosis	*Cryptosporidium parvum*
	Giardiasis	*Giardia lamblia*
	Pneumocystis pneumonia	*Pneumocystis carinii*
Environmental surfaces	Bacterial	
	Leptospirosis	*Leptospira interrogans*
	Viral	
	Common cold	*Coronavirus, Rhinovirus*
	Conjunctivitis	*Mastaadenovirus*
	Diarrhea	*Rotavirus*
	Meningitis	*Enterovirus*
	Smallpox (eradicated)	*Orthopoxvirus*
Food	Bacterial	
	Brucellosis	*Brucella*
	Campylosis	*Campylobacter*
	Cholera	*Vibrio cholerae*
	Enteritis	*Escherichia*
	Listeriosis	*Listeria*
	Salmonellosis	*Salmonella*
	Shigellosis	*Shigella*
	Tularemia	*Francisella tularensis*

Table 1.3. *Cont.*

Exposure route	Disease	Causative microorganism(s)
	Protozoan	
	Cryptosporidiosis	*Cryptosporidium parvum*
	Giardiasis	*Giardia lamblia*
	Sparganosis	*Spirometra*
	Taeniasis	*Taenia*
	Trichinosis	*Trichinella*
	Viral	
	Gastroenteritis	*Calicivirus*
	Hepatitis	*Calicivirus, Hepatovirus*
	Poliomyelitis (carried passively by flies)	*Enterovirus*

Tabet *et al.*, 1994). *Chlamydia*, which can cause pneumonia and bronchitis, likewise may be transmitted between humans by aerosols (Grayston *et al.*, 1993). Examples of viral diseases transmitted between humans via aerosols are chickenpox (Sawyer *et al.*, 1994), gastroenteritis (Monto & Koopman, 1980; Sawyer *et al.*, 1988), pneumonia and bronchiolitis (Emory University & Centers for Disease Control, 1993). The common cold represents an example of the relative importance of aerosols versus fomites in the transmission of at least some diseases, in that Jennings *et al.* (1988) have demonstrated rhinoviruses, one of the viral groups that can cause common colds, to be transmitted more efficiently via aerosols than by fomites. Natural environmental microorganisms represent an additional source of human illnesses that are acquired from aerosols. This category is represented by the bacteria responsible for legionellosis (Joly, 1993; Srikanth & Berk, 1993).

Perhaps the most important determinant of the probability that a disease will be transmitted by aerosols is the capability of the causative microorganisms to survive in aerosolized droplets or on aerosolized particulates. Variables that are important with regard to the survival of viruses in aerosols include air temperature and relative humidity (Theunissen *et al.*, 1993).

Many different microorganisms are capable of surviving in aerosols, and the gravitational settling of aerosols may result in microbial contamination of food, environmental surfaces and medical devices. Additionally, microorganisms contained in aerosolized saliva and nasal secretions may be directed onto foods, environmental surfaces and medical devices during coughing and sneezing.

A summary of some diseases that are transmitted through environmental routes other than contaminated water can be found in Table 1.3 along with the names of the causative microorganisms.

Disease transmission by routes that require an intermediate animal host

Some of the helminthic illnesses affecting humans are normally transmitted only through an intermediate animal host. These helminths complete their life cycle in that intermediate animal host. This route of disease transmission is illustrated in Figure 1.8. Diseases of this nature include schistosomiasis, for which snails serve as the intermediate animal host (Wolfe *et al.*, 1993) and which humans acquire from physical contact with contaminated water. Biting flies serve as the intermediate host for the helminthic disease onchocerciasis (Dietz, 1982). Taeniasis is caused by tapeworms, whose intermediate hosts may be domesticated animals such as cattle and pigs, and which humans acquire from ingesting the infected animals (Barbier *et al.*, 1990; Ash & Orihel, 1991). Examples of helminthic diseases whose transmission between humans normally requires an intermediate animal host are listed in Table 1.4 along with the names of the causative microorganisms.

Means of preventing or ameliorating the outcome of disease transmission

Microbial illnesses are an important cause of morbidity and mortality in human populations (Feachem, Hogan & Merson, 1983; Bukh *et al.*, 1993; Watson *et al.*, 1993). There are several means by which we try either to prevent disease transmission, or to lessen the severity of its impact upon the affected individuals. Figure 1.9 illustrates the points at which immunization, disinfection, physical barriers and antibiotics are employed for these purposes.

Immunization

Immunization is a very potent technique by which we try to prevent the transmission of disease (Anderson & May, 1982; Behbehani, 1991; Service, 1994). There are two basic approaches employed for immunizing individuals. The first of these is passive immunization, which results from the administration of immune globulin preparations by injection. Preformed antibodies present in the injected material provide immediate but short-lived protection. Passive immunization is sometimes performed before an individual is exposed to the disease-causing microorganism, in which case it represents primary prevention. Passive immunization may also be performed following exposure, in which case it represents secondary prevention. Pre-exposure treatment is most likely to be used in those instances where the anticipated disease has a short incubation period. Post-exposure

Table 1.4. *Examples of diseases requiring
transmission through an intermediate animal host
for completion of helminthic life cycles*

Disease	Causative microorganism(s)
Onchocerciasis	*Onchocerca volvulus*
Schistosomiasis	*Schistosoma*
Sparganosis	*Spirometra*
Taeniasis	*Taenia*

Human Reservoir

New Human Host

Figure 1.8   Route of transmission for infectious agents that require an
intermediate animal host for completing successive steps in helminthic
life cycles. Some of these organisms are transmitted to humans through
consumption of the infected animals; others represent transmission
through either insect bites or exposure to contaminated water. The
animals symbolically represented are cows and pigs as a source of taeniasis
and snails as a source of schistosomiasis.

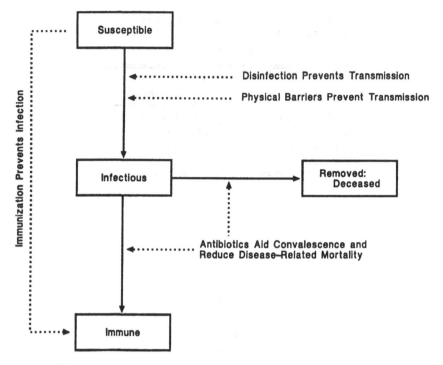

Figure 1.9 Basic compartment model from Figure 1.1, modified to demonstrate the points at which different approaches may be used singly or in combination, either to prevent disease transmission or to lessen the consequences of disease.

treatment is used in those instances where the disease has a long incubation period, or the exposed individual is considered to be at risk of a severe disease outcome.

The second approach is active immunization, often termed 'vaccination'. This involves administering antigenic materials that cause the recipient to develop a natural immune response. Active immunization is normally used before exposure to the causative microorganism as a means of primary prevention. The protection resulting from active immunization usually lasts for a much longer period than does the protection afforded by passive immunization. The United States Government has recently published an informative advisory document on immunization practices for humans (Centers for Disease Control and Prevention, 1994). Perhaps the most impressive success achieved by active immunization has been the eradication of smallpox disease (Al-Awadi & Mahler, 1980). Active immunization is now being used in an effort to eradicate poliomyelitis (Kim-Farley et al., 1984).

There are, however, drawbacks to reliance upon vaccination for disease control. The process of developing, commercially producing, distributing and successfully administering vaccines is both complex and costly. The complexity of this process limits the number of diseases against which we can attempt to protect people. The cost factor sadly limits the extent to which vaccination programs are utilized. Problems can result from vaccination, with instances of vaccine-related illness and even deaths known to occur among vaccine recipients (Centers for Disease Control, 1991c). Also, live attenuated organisms that may be administered for vaccination purposes can occasionally revert to virulent forms (Abraham *et al.*, 1993), and cause illness instead of preventing illness. Immunity acquired through vaccination is not necessarily permanent, and can wane with time so that immunized individuals again become susceptible to illness caused by the targeted pathogen.

Infection control practices

There are other approaches, aside from vaccination, that can be used to prevent infection. Among these are two categories of infection control practices (Centers for Disease Control, 1991a). The first represents the establishment of physical barriers for preventing the transfer of infectious agents. The second consists of disinfection. Both categories of infection control practices represent primary prevention.

Use of physical barriers to prevent disease transmission

The concept of physical barriers includes quarantine practices, whereby infected individuals are isolated from susceptible individuals. Quarantine isolation has been used for humans, most notably for diseases such as tuberculosis and leprosy. Quarantine can be effective, but may be expensive and is considered stigmatizing.

Practices as simple as wearing surgical gloves (Denman *et al.*, 1993) represent another aspect of the physical barrier concept.

The United States Department of Health and Human Services divides physical barriers for laboratory situations into two groups, primary barriers and secondary barriers (Centers for Disease Control and Prevention & National Institutes of Health, 1993). According to this classification scheme, primary barriers protect individuals and their immediate environment. Primary barriers consist of protective clothing such as gloves and masks, as well as personal safety equipment such as respirators equipped with filters (Adal *et al.*, 1994). Secondary barriers include physical containment practices and design factors intended to limit dispersion of the pathogens, among which are the use of separate rooms for handling contaminated materials, washing and other decontamination techniques that are used for work surfaces and implements, and disinfection practices including the sterilization of contaminated materials (Ozanne, Huot & Montpetit, 1993).

These and other physical barrier concepts can also be applied for public health purposes, to prevent disease transmission in our daily living situations. Those practices performed on an individual basis include the washing of hands (Barrett *et al.*, 1992), dishes and eating implements. We often use soap as an aid when washing. However, washing with just water alone can have some beneficial effect (Mbithi, Springthorpe & Sattar, 1993). Interestingly, Schürmann & Eggers (1985) have shown that, at least in the case of decontaminating skin, the use of moistened sand may be even more effective than washing with ordinary soap. Other physical barrier practices employed in our daily lives include the use of condoms to prevent the transmission of sexually transmitted diseases, and the use of protective clothing, window screens and mosquito netting to reduce our exposure to insects and any diseases they may be carrying. Physical barrier techniques can be used at the household level for treating drinking water. This includes the use of commercial drinking water filtration devices (Payment *et al.*, 1991) and natural coagulants (Al Azharia Jahn & Dirar, 1979; Lund & Nissen, 1986).

Use of disinfection to prevent disease transmission

Disinfection, the destruction of pathogenic microorganisms, is something which, when given sufficient time, can occur under natural environmental conditions that do not allow growth of the pathogenic organisms (Kehr & Butterfield, 1943; Hurst, 1991; Hurst, *et al.*, 1992; Kaspar & Tamplin, 1993). This can be termed natural disinfection. Sunlight can be an effective disinfectant (Fujioka & Narikawa, 1982; Acher *et al.*, 1994) as are manufactured sources that generate microwave (Welt *et al.*, 1994), ultraviolet (Dizer *et al.*, 1993; Lindenauer & Darby, 1994), or gamma radiations (Farooq *et al.*, 1993; Harewood, Rippey & Montesalvo, 1994). Disinfection usually represents primary prevention. Under special medical situations, where an individual is already infected, disinfection could be considered a means of secondary prevention by either reducing the extent of infection or precluding dispersal of the pathogens. Washing with an artificial detergent, such as ordinary soap, can provide some chemical disinfection activity (Baker, Harrison & Miller, 1941). Other hand washing agents and detergents can be more effective in aiding the physical removal or destruction of microorganisms (Baker *et al.*, 1941; Gordon *et al.*, 1993; Mbithi *et al.*, 1993; Sattar *et al.*, 1993). The simple practice of using chlorine bleach when washing clothing also has a disinfection effect.

Disinfectants may be used to destroy pathogens in drinking water, a practice that can be used either at the municipal level with formal treatment facilities (Stetler *et al.*, 1992), or at individual and household levels. Disinfection of drinking water on an individual or household basis can be accomplished by boiling the water (Rice & Johnson, 1991; Akhter *et al.*, 1994; Fayer, 1994), distilling water (Van Steenderen, 1977), or adding chemicals such as chlorine or iodine (Ellis, Cotton & Khowaja, 1993; Powers *et al.*, 1994). Disinfection is also performed on wastewaters to protect the quality of receiving waters (Haas, 1986). The subject of using disinfection practices to reduce disease transmission will be addressed extensively in other chapters of this book.

Use of antibiotics to preclude or lessen the severity of disease

The successful development of antibiotic drugs represents one of the miracles of modern science and medicine (Taylor, 1942). These drugs are normally used for secondary prevention, to reduce the severity of illness in diseased patients. However, in some instances antibiotic drugs are used prophylactically to preclude the initiation of an infection. Prophylactic use of antibiotics represents primary prevention. Despite the miracle that these antibiotic compounds represent, microorganisms are capable of evolving drug resistance (Coronado *et al.*, 1993). Therefore, time has proven Taylor (1942) to have been overly optimistic when he stated that, with discovery of the organic arsenic compound salvarsan and the sulfonamides, human-kind had successfully conquered bacteria.

Demonstrating the effectiveness of techniques for reducing the transmission of infectious agents

Drinking water treatment will be used as an example of our effectiveness at reducing the incidence of infectious disease in human populations. The bacterial disease typhoid is well known to be transmitted by the ingestion of contaminated water, which often represents its primary route of transmission into a population of susceptible individuals. This organism may be harbored by human carriers of the disease, individuals who experience prolonged infection without necessarily displaying the symptoms normally associated with this illness. Human carriers may represent a secondary route of transmission for typhoid, the means by which the infectious agent is transferred within a human population following primary waterborne transmission. These carriers can pass the pathogen on to other members of a population through vehicles such as contaminated food. Typhoid caused extremely high incidences of waterborne disease and death worldwide in the 19th century and early part of the 20th century. Although this disease still occurs both endemically and epidemically throughout the world, in developed countries its occurrence has almost completely been staunched by the introduction of adequately operated drinking water treatment facilities. Table 1.5 presents the change in urban typhoid death rates in the United States during the early 20th century, serving as an example of the effectiveness of community drinking water treatment for reducing waterborne disease. In this case, the treatment used was water filtration, which is a form of physical barrier. In some of the cities, the water may have been chlorinated following filtration; a combined-treatment practice whose usage began during the later part of the period represented in Table 1.5 (Holmquist, 1924). The data listed in Table 1.5 are from a publication by Johnson (1916) for cities with accurate reporting as to cause of human death. These data show that in the year 1900, for the represented cities, the percentage of urban population served by water filtration utilities was 8.7%, and the concurrent annual typhoid death rate was 36 per 100 000 population. During the 13 years that followed, the percentage of population served by water filtration in those cities gradually increased to 48% and the annual typhoid death rate gradually decreased to 16 per 100 000 population. These data dramatically demonstrate how drinking water treatment can

Table 1.5. *Relationship between increase in percentage of United States city population supplied with filtered drinking water and decrease in overall city population typhoid death rates*

Year	Percentage of city population served by water filtration utilities	Typhoid fever death rates[a] for same cities
1900	8.7	36
1901	10.8	34
1902	11.9	37
1903	13.3	38
1904	16.0	35
1905	17.4	30
1906	20.5	33
1907	23.2	32
1908	23.3	25
1909	30.1	21
1910	34.6	24
1911	37.2	20
1912	42.4	16
1913	48.0	16

Note: Data are presented for those United States cities having accurate registration of cause of human death.
[a] Death rates given per 100 000 population per year.
Source: Johnson (1916).

be effective for reducing waterborne disease even when secondary transmission of the disease can occur within the community.

The second approach for demonstrating the effectiveness of drinking water treatment is the disease transmission modeling represented in Tables 1.6 and 1.7. The basic disease transmission model presented in Figure 1.1 was used for this purpose, with exclusion of mortality, to estimate the social impact of a waterborne outbreak in a community of 10 000 people. It was assumed for the purpose of this modeling that: (1) all of the individuals in the community were susceptible to infection at the time of the community's initial exposure to the causative microorganism; (2) an individual became infectious one day after becoming infected; (3) the probability of an infectious individual recovering from the infection and subsequently becoming immune was 20% on any given day during their period of infectiousness; and (4) on each day that an individual was infectious, they had a 2% chance of transmitting the infection to another individual. The likelihood of the disease being transmitted to another individual was proportionately reduced during the course of the epidemic as the percentage of susceptible individuals in the population decreased. Running of the model was discontinued when the movement of individuals from the infectious category to the immune category had become less than 0.5 per day, at which point movement would have rounded down to zero individuals per day.

For the models presented in Tables 1.6 and 1.7 it was assumed that exposure to the primary route of transmission, presumably microbial contaminants in the drinking water, occurred for a single day and that the initial infection rate was 10%. The model shown in Table 1.6 assumes that physical barrier techniques and disinfection practices were not employed to reduce either primary (waterborne) or secondary transmission of the disease. In this case, 1000 individuals initially become infected. The epidemic ends with a total of 1097 individuals having been infected, two of whom remain infected at day 31. Ninety-seven of these infections result from secondary transmission of the infectious agent within the community. Table 1.7 shows what would happen to this same population assuming that drinking water treatment was 80% effective at reducing the incidence of initial infection within the population, and that the use of physical barriers and disinfection practices was similarly 80% effective at reducing secondary transmission. Using these assumptions for the efficiency of physical barriers and disinfection, Table 1.7 indicates that a total of only 203 individuals would become infected during this outbreak, with two of those individuals remaining infected at day 22, and only three having been infected as a result of secondary transmission.

Table 1.6. *Model of community-wide waterborne disease epidemic caused by single-day ingestion exposure to contaminated water with subsequent person-to-person secondary transmission*

| | Number of individuals[a] | | |
Day number	Susceptible	Infectious	Immune (recovered)
1	10 000	0	0
2	9000	1000	0
3	8982	818	200
4	8967	669	364
5	8955	547	498
6	8945	448	607
7	8937	366	697
8	8930	300	770
9	8925	245	830
10	8921	200	879
11	8917	164	919
12	8914	134	952
13	8912	109	979
14	8910	89	1001
15	8908	73	1019
16	8907	59	1034
17	8906	48	1046
18	8905	39	1056
19	8904	32	1064
20	8903	27	1070
21	8903	22	1075
22	8903	18	1079
23	8903	14	1083
24	8903	11	1086
25	8903	9	1088
26	8903	7	1090
27	8903	6	1091
28	8903	5	1092
29	8903	4	1093
30	8903	3	1094
31	8903	2	1095

Note: Total cases of infection = 1097

[a]Assumptions:

(1) No initial immunity within the community.

(2) Infected persons become infectious on the next day.

(3) Each day 20% of the previous day's infectious individuals have recovered ([number recovered on day n] = [number recovered on day $n-1$] + [0.2 × number infections on day $n-1$]).

(4) Each day 2% of the infectious individuals transmit the infection to a single contact, with the assumption that that contact is susceptible ([number infectious on day n] = [[0.8 × number infectious on day $n-1$]] + [[0.02 × number infectious on day $n-1$] × [portion of initial population still susceptible on day $n-1$ expressed as a decimal value]]). This equation is valid only if there are individuals in the infectious category for day $n-1$.

(5) On any given day, everyone who is neither infectious nor immune is considered to be susceptible ([number susceptible on day n] = [10 000 − {number infectious on day n + number immune on day n}]).

(6) Reinfection does not occur during the course of the epidemic.

Table 1.7. *Model of the same waterborne disease outbreak as in Table 1.6 with the assumption that both the initial transmission and secondary transmission were reduced by 80% achieved through the use of physical barrier and disinfection practices*

	Number of individuals[a]		
Day number	Susceptible	Infectious	Immune (recovered)
1	10 000	0	0
2	9800	200	0
3	9799	161	40
4	9798	130	72
5	9797	105	98
6	9797	84	119
7	9797	67	136
8	9797	54	149
9	9797	43	160
10	9797	34	169
11	9797	27	176
12	9797	22	181
13	9797	18	185
14	9797	14	189
15	9797	11	192
16	9797	9	194
17	9797	7	196
18	9797	6	197
19	9797	5	198
20	9797	4	199
21	9797	3	200
22	9797	2	201

Note: Total cases of infection = 203
[a]Assumptions:
1–3, 5 and 6 same as stated for Table 1.6.
(4) Each day 0.4% of the infectious individuals transmit the infection to a single contact, assuming that contact is susceptible ([number infectious on day n] = [{0.8 × number infectious on day $n-1$}] + [{0.02 × 0.2 × number infectious on day $n-1$} × {portion of initial population still susceptible on day $n-1$ expressed as a decimal value}]). This equation is valid only if there are individuals in the infectious category for day $n-1$.

Summary

The main purpose of this chapter was to demonstrate how mathematical models are used for describing the propagation of diseases through populations, and to explain the various routes by which the associated infectious agents are transferred between individuals. Most disease transmission routes require that the causative microorganisms be exposed to the environment. The major factor in determining whether an organism can be transmitted via a particular environmental route such as water, aerosols, or contact with the surface of objects, is whether or not the organism has evolved the ability to survive environmental exposure under the conditions it would encounter along that route. As an example, microorganisms that lack stability in environmental waters are unlikely to be transmitted by water. Some diseases can be transmitted by more than a single route. Tularemia, for example, can be acquired by ingestion of contaminated food or water, inhalation of microbial aerosols, contact of bare skin with contaminated materials, and even through insect bites.

In this chapter the means have been described by which humans attempt to prevent the transmission of infectious agents of disease, and an approach has been demonstrated that can be used to model the effectiveness of these prevention techniques. For this purpose, historical data for typhoid was used to show how intercepting a primary route of disease transmission can dramatically reduce the incidence of that disease within a community, even when there exists another, secondary route of transmission for the disease. We presented the use of a disease transmission model to illustrate how this type of disease reduction can be accomplished.

References

Abad, F. X., Pintó, R. M. & Bosch, A. (1994). Survival of enteric viruses on environmental fomites. *Applied and Environmental Microbiology*, **60**, 3704–10.

Abbaszadegan, M., Huber, M. S., Gerba, C. P. & Pepper, I. L. (1993). Detection of enteroviruses in groundwater with the polymerase chain reaction. *Applied and Environmental Microbiology*, **59**, 1318–24.

Abernathy, A. R., Zirschky, J. & Borup, M. B. (1984). Land application of wastewater. *Journal, Water Pollution Control Federation*, **56**, 620–1.

Abraham, R., Minor, P., Dunn, G., Modlin, J. F. & Ogra, P. L. (1993). Shedding of virulent poliovirus revertants during immunization with oral poliovirus vaccine after prior immunization with inactivated polio vaccine. *Journal of Infectious Diseases*, **168**, 1105–9.

Acher, A. J., Fischer, E. & Manor, Y. (1994). Sunlight disinfection of domestic effluents for agricultural use. *Water Research*, **28**, 1153–60.

Adal, K. A., Anglim, A. M., Palumbo, C. L., Titus, M. G., Coyner, B. J. & Farr, B. M. (1994). The use of high efficiency particulate air-filter respirators to protect hospital workers from tuberculosis: a cost-effectiveness analysis. *New England Journal of Medicine*, **331**, 169–73.

Akhter, M. N., Levy, M. E., Mitchell, C., Boddie, R., Donegan, N., Griffith, B., Jones, M. & Stair, T. O. (1994). Assessment of inadequately filtered public drinking water – Washington, D.C., December, 1993. *Morbidity and Mortality Weekly Report*, **43**, 661–9.

Al-Awadi, A. R. & Mahler, H. (1980). Declaration of global eradication of smallpox. *Weekly Epidemiology Record*, **20**, 148.

Al Azharia Jahn, S. & Dirar, H. (1979). Studies on natural water coagulants in the Sudan, with special reference to *Moringa oleifera* seeds. *Water SA* (South Africa), **5**, 90–7.

Alexander, A. D. (1991). *Leptospira*. In *Manual of Clinical Microbiology*, 5th edn, ed. A. Balows, W. J. Hausler Jr, K. L. Herrmann, H. D. Isenberg & H. J. Shadomy, pp. 554–9. Washington: The American Society for Microbiology.

Alexander, J. P. Jr, Chapman, L. E., Pallansch, M. A., Stephenson, W. T., Török, T. J. & Anderson, L. J. (1993). Coxsackievirus B2 infection and aseptic meningitis: a focal outbreak among members of a high school football team. *Journal of Infectious Diseases*, **167**, 1201–5.

Allen, M. J. & Morrison, S. M. (1973). Bacterial movement through fractured bedrock. *Ground Water*, **11** (2), 6–10.

Allen, S. D. & Baron, E. J. (1991). *Clostridium*. In *Manual of Clinical Microbiology*, 5th edn, ed. A. Balows, W. J. Hausler Jr, K. L. Herrmann, H. D. Isenberg & H. J. Shadomy, pp. 505–21. Washington: The American Society for Microbiology.

Alvarado, C. J., Stolz, S. M., Maki, D. G., Fraser, V., Jones, M., O'Rourke, S. & Wallace, R. J. Jr (1991). Nosocomial infection and pseudoinfection from contaminated endoscopes and bronchoscopes – Wisconsin and Missouri. *Morbidity and Mortality Weekly Report*, **40**, 675–8.

Anders, J., Shireley, L. A., Volmer, L., Forsch, K., Osterholm, M. T., Schell, W. L., Davis, J. P., Rowley, W. A., Currier, R., Wintermeyer, L. A., Kramer, W. L., Safranek, T. J., Alfano, D., Haramis, L. D., Kottkamp, W., Frazier, C. L. & Satalowich, F. T. (1994). Rapid assessment of vectorborne diseases during the midwest flood – United States, 1993. *Morbidity and Mortality Weekly Report*, **43**, 481–3.

Anderson, R. M. (1982). Directly transmitted viral and bacterial infections of man. In *The Population Dynamics of Infectious Diseases: Theory and Applications*, ed. R. M. Anderson, pp. 1–37. London: Chapman & Hall.

Anderson, R. M. (1994). Mathematical studies of parasitic infection and immunity. *Science*, **264**, 1884–6.

Anderson, R. M. & May, R. M. (1982). Directly transmitted infectious diseases: control by vaccination. *Science*, **215**, 1053–60.

Anderson, R. M. & May, R. M. (1985). Vaccination and herd immunity to infectious diseases. *Nature*, **318**, 323–9.

Ash, L. R. & Orihel, T. C. (1991). Intestinal helminths. In *Manual of Clinical Microbiology*, 5th edn, ed. A. Balows, W. J. Hausler Jr, K. L. Herrmann, H. D. Isenberg & H. J. Shadomy, pp. 782–95. Washington: The American Society for Microbiology.

Atchison, C. G., Wintermeyer, L. A., Kelly, J. R., Currier, R., Vogel, C., Goddard, J. H., Weaver, M., Culp, J., Zwick, J., Eckoff, R. D., Hausler, W. J., Moyer, N. P., Rowley, W. A., Schwartz, G. & Knutsen, J. (1993). Public health consequences of a flood disaster – Iowa, 1993. *Morbidity and Mortality Weekly Report*, **42**, 653–6.

Baker, Z., Harrison, R. W. & Miller, B. F. (1941). Action of synthetic detergents on the metabolism of bacteria. *Journal of Experimental Medicine*, **73**, 249–71.

Barbier, D., Perrine, D., Duhamel, C., Doublet, R. & Georges, P. (1990). Parasitic hazard with sewage sludge applied to land. *Applied and Environmental Microbiology*, **56**, 1420–2.

Barrett, B., Khurana, K., Afanador, J. E. & Fleissner, M. L. (1992). Iguana-associated salmonellosis – Indiana, 1990. *Morbidity and Mortality Weekly Report*, **41**, 38–9.

Barry, M., Bia, F., Cullen, M., Dembry, L., Fischer, S., Geller, D., Hierholzer, W., McPhedran, P., Rainey, P., Russi, M., Snyder, E., Wrone, E., Gonzalez, J. P., Rico-Hesse, R., Tesh, R., Ryder, R., Shope, R., Quinn, W. P., Galbraith, P. D.,

Cartter, M. L., Hadler, J. L. & DeMaria, A. Jr (1994). Arenavirus infection – Connecticut, 1994. *Morbidity and Mortality Weekly Report*, **43**, 635–6.

Baxter-Potter, W. R. & Gilliland, M. W. (1988). Bacterial pollution in runoff from agricultural lands. *Journal of Environmental Quality*, **17**, 27–34.

Behbehani, A. M. (1991). The smallpox story: historical perspective. *American Society for Microbiology News*, **57**, 571–6.

Bergeisen, G. H., Hinds, M. W. & Skaggs, J. W. (1985). A waterborne outbreak of hepatitis A in Meade county, Kentucky. *American Journal of Public Health*, **75**, 161–4.

Birkhead, G. S., Morse, D. L., Levine, W. C., Fudala, J. K., Kondracki, S. F., Chang, H.-G., Shayegani, M., Novick, L. & Blake, P. A. (1993). Typhoid fever at a resort hotel in New York: a large outbreak with an unusual vehicle. *Journal of Infectious Diseases*, **167**, 1228–32.

Black, F. L. & Singer, B. (1987). Elaboration versus simplification in refining mathematical models of infectious disease. *Annual Reviews of Microbiology*, **41**, 677–701.

Blake, P. A., Rosenberg, M. L., Costa, J. B., Ferreira, P. S., Guimaraes, C. L. & Gangarosa, E. J. (1977). Cholera in Portugal, 1974 I. Modes of transmission. *American Journal of Epidemiology*, **105**, 337–43.

Bloch, A. B., Stramer, S. L., Smith, J. D., Margolis, H. S., Fields, H. A., McKinley, T. W., Gerba, C. P., Maynard, J. E. & Sikes, R. K. (1990). Recovery of hepatitis A virus from a water supply responsible for a common source outbreak of hepatitis A. *American Journal of Public Health*, **80**, 428–30.

Bongard, J., Savage, R., Dern, R., Bostrum, H., Kazmierczak, J., Keifer, S., Anderson, H. & Davis, J. P. (1994). *Cryptosporidium* infections associated with swimming pools – Dane County, Wisconsin, 1993. *Morbidity and Mortality Weekly Report*, **43**, 561–72.

Bosch, A., Lucena, F., Diez, J. M., Gajardo, R., Blasi, M. & Jofre, J. (1991). Waterborne viruses associated with hepatitis outbreak. *Journal, American Water Works Association*, **83** (3), 80–3.

Bouwer, H. & Rice, R. C. (1984). Renovation of wastewater at the 23rd avenue rapid infiltration project. *Journal, Water Pollution Control Federation*, **56** (1), 76–83.

Breiman, R. F. (1993). Modes of transmission in epidemic and nonepidemic *Legionella* infection: directions for further study. In *Legionella: Current Status and Emerging Perspectives*, ed. J. M. Barbaree, R. F. Breiman & A. P. Dufour, pp. 30–5. Washington: The American Society for Microbiology.

Bryson, Y., Dillon, M., Bernstein, D. I., Radolf, J., Zakowski, P. & Garratty, E. (1993). Risk of acquisition of genital herpes simplex virus type 2 in sex partners of persons with genital herpes: a prospective couple study. *Journal of Infectious Diseases*, **167**, 942–6.

Bukh, J., Wantzin, P., Krogsgaard, K., Knudsen, F., Purcell, R. H., Miller, R. H. & Copenhagen Dialysis HCV Study Group (1993). High prevalence of hepatitis C virus (HCV) RNA in dialysis patients: failure of commercially available antibody tests to identify a significant number of patients with HCV infection. *Journal of Infectious Diseases*, **168**, 1343–8.

Burge, W. D. & Marsh, P. B. (1978). Infectious disease hazards of landspreading sewage wastes. *Journal of Environmental Quality*, **7**, 1–9.

Cabelli, V. J. (1989). Swimming-associated illness and recreational water quality criteria. *Water Science and Technology*, **21** (2), 13–21.

Cartter, M. L., Hadler, J. L., Smith, M. G., Sorhage, F. E., Spitalny, K. C., Debbie, J. G., Morse, D. L., Hunter, J. L., MacCormack, J. N., Smith, K. A., Halpin, T. J., Jenkins, S. R. & Haddy, L. E. (1992). Extension of the raccoon rabies epizootic – United States, 1992. *Morbidity and Mortality Weekly Report*, **41**, 661–4.

Centers for Disease Control (1991a). Recommendations for preventing transmission of human immunodeficiency virus and hepatitis B virus to patients during exposure-prone invasive procedures. *Morbidity and Mortality Weekly Report*, **40** (RR-8), 1–9.

Centers for Disease Control (1991b). Imported dengue – United States, 1990. *Morbidity and Mortality Weekly Report*, **40** (RR-14), 519–20.

Centers for Disease Control (1991c). Vaccinia (smallpox) vaccine: recommendations of the immunization practices advisory committee (ACIP). *Morbidity and Mortality Weekly Report*, **40** (RR-14), 1–10.

Centers for Disease Control (1992a). Hepatitis B and injecting-drug use among American indians – Montana, 1989–1990. *Morbidity and Mortality Weekly Report*, **41**, 13–14.

Centers for Disease Control (1992b). Seroconversion to simian immunodeficiency virus in two laboratory workers. *Morbidity and Mortality Weekly Report*, **41**, 678–81.

Centers for Disease Control (1993a). Preliminary report: foodborne outbreak of *Escherichia coli* 0157:H7 infections from hamburgers – Western United States, 1993. *Morbidity and Mortality Weekly Report*, **42**, 85–6.

Centers for Disease Control and Prevention (1993b). Hantavirus infection – Southwestern United States: interim recommendations for risk reduction. *Morbidity and Mortality Weekly Report*, **42** (RR-11), 1–13.

Centers for Disease Control and Prevention (1993c). Tuberculosis in imported nonhuman primates – United States, June 1990–May 1993. *Morbidity and Mortality Weekly Report*, **42**, 572–6.

Centers for Disease Control and Prevention (1994). General recommendations on immunization: recommendations of the advisory committee on immunization practices (ACIP). *Morbidity and Mortality Weekly Report*, **43** (RR-1), 1–38.

Centers for Disease Control and Prevention & National Institutes of Health (1993). *Biosafety in Microbiological and Biomedical Laboratories*, 3rd edn. Washington: US Department of Health and Human Services Public Health Service, Centers for Disease Control and Prevention & National Institutes of Health.

Cheung, W. H. S., Hung, R. P. S., Chang, K. C. K. & Kleevens, J. W. L. (1991). Epidemiological study of beach water pollution and health-related bathing water standards in Hong Kong. *Water Science and Technology*, **23**, 243–52.

Ching-Lee, M. R., Katz, A. R., Sasaki, D. M. & Minette, H. P. (1991). Salmonella egg survey in Hawaii: evidence for routine bacterial surveillance. *American Journal of Public Health*, **81**, 764–6.

Clark, C. S., Linnemann, C. C. Jr, Gartside, P. S., Phair, J. P., Blacklow, N. & Zeiss, C. R. (1985). Serologic survey of rotavirus, Norwalk agent and *Prototheca wickerhamii* in wastewater workers. *American Journal of Public Health*, **75**, 83–5.

Clark, H. F., Offit, P. A., Dolan, K. T., Tezza, A., Gogalin, K., Twist, E. M. &

Plotkin, S. A. (1986). Response of adult human volunteers to oral administration of bovine and bovine/human reassortant rotaviruses. *Vaccine*, **4** (March), 25–31.

Clayton, S., Yang, H., Guan, J., Lin, Z. & Wang, R. (1993). Hepatitis B control in China: knowledge and practices among village doctors. *American Journal of Public Health*, **83**, 1685–8.

Cliver, D. O. (1990). *Report of WHO Collaborating Center on Food Virology Literature Review – 1990*. Madison, Wisconsin: Food Research Institute, University of Wisconsin.

Collins, J. M. (1994). Ocurrence of stormwater bacteria transported to coastal sediments via ground water. *Abstracts of the 94th General Meeting*, p. 428. Washington: American Society for Microbiology.

Corbett, S. J., Rubin, G. L., Curry, G. K., Kleinbaum, D. G. & Sydney Beach Users Study Advisory Group (1993). The health effects of swimming at Sydney beaches. *American Journal of Public Health*, **83**, 1701–6.

Coronado, V. G., Beck-Sague, C. M., Hutton, M. D., Davis, B. J., Nicholas, P., Villareal, C., Woodley, C. L., Kilburn, J. O., Crawford, J. T., Frieden, T. R., Sinkowitz, R. L. & Jarvis, W. R. (1993). Transmission of multidrug-resistant *Mycobacterium tuberculosis* among persons with human immunodeficiency virus infection in an urban hospital: epidemiologic and restriction fragment length polymorphism analysis. *Journal of Infectious Diseases*, **168**, 1052–5.

Cotton, P. (1993). Health threat from mosquitos rises as flood of the century finally recedes. *Journal of the American Medical Association*, **270**, 685–6.

Coyne, M. J., Smith, G. & McAllister, F. E. (1989). Mathematical model for the population biology of rabies in raccoons in the mid-Atlantic states. *American Journal of Veterinary Research*, **50**, 2148–54.

Cvjetanović, B. (1982). The dynamics of bacterial infections. In *The Population Dynamics of Infectious Diseases: Theory and Applications*, ed. R. M. Anderson, pp. 38–66. London: Chapman & Hall.

D'Antonio, R. G., Winn, R. E., Taylor, J. P., Gustafson, T. L., Current, W. L., Rhodes, M. M., Gary, G. W. Jr & Zajac, R. A. (1985). A waterborne outbreak of cryptosporidiosis in normal hosts. *Annals of Internal Medicine*, **103** (6 pt 1), 886–8.

Davis, J. W. & Sizemore, R. K. (1982). Incidence of *Vibrio* species associated with blue crabs (*Callinectes sapidus*) collected from Galveston Bay, Texas. *Applied and Environmental Microbiology*, **43**, 1092–7.

De Leon, R. & Gerba, C. P. (1990). Viral disease transmission by seafood. In *Food Contamination from Environmental Sources*, ed. J. O. Nriagu & M. S. Simmons, pp. 639–62. New York: John Wiley & Sons.

Denman, S., Dwyer, D. M., Israel, E. & Vacek, P. (1993). Handwashing and glove use in a long-term-care facility – Maryland, 1992. *Morbidity and Mortality Weekly Report*, **42**, 672–5.

Dennis, D. T., Smith, R. P., Welch, J. J., Chute, C. G., Anderson, B., Herndon, J. L. & von Reyn, C. F. (1993). Endemic giardiasis in New Hampshire: a case–control study of environmental risks. *Journal of Infectious Diseases*, **167**, 1391–5.

Desenclos, J.-C. A., Klontz, K. C., Wilder, M. H., Nainan, O. V., Margolis, H. S. & Gunn, R. A. (1991). A multistate outbreak of hepatitis A caused by the consumption of raw oysters. *American Journal of Public Health*, **81**, 1268–72.

Dietz, K. (1982). The population dynamics of onchocerciasis. In *The Population*

Dynamics of Infectious Diseases: Theory and Applications, ed. R. M. Anderson, pp. 209–41. London: Chapman & Hall.

Dizer H., Bartocha, W., Bartel, H., Seidel, K., Lopez-Pila, J. M. & Grohmann, A. (1993). Use of ultraviolet radiation for inactivation of bacteria and coliphages in pretreated wastewater. *Water Research*, **27**, 397–403.

Doebbeling, B. N., Li, N. & Wenzel, R. P. (1993). An outbreak of hepatitis A among health care workers: risk factors for transmission. *American Journal of Public Health*, **83**, 1679–84.

Doege, T. C. & Gelfand, H. M. (1978). A model for epidemiologic research. *Journal of the American Medical Association*, **239**, 328–30.

Doll, J., Fink, T. M., Levy, C., Sands, L., Roberto, R., Smith, C., Dixon, F. R., Pierce, J., Brus, D., Montman, C., Brown, T., Reynolds, P., Eidson, M., Sewell, C. M., Lanser, S., Tanner, R., Nichols, C. R., Akin, D. R., Bohan, P., Emory, W. & Fulgham, R. (1992). Plague – United States, 1992. *Morbidity and Mortality Weekly Report*, **41**, 787–90.

Doyle, R. J. & Lee, N. C. (1986). Microbes, warfare, religion and human institutions. *Canadian Journal of Microbiology*, **32**, 193–200.

Dryden, F. D. & Chen, C.-L. (1978). Groundwater recharge with reclaimed waters from the Pomona, San Jose Creek, and Whittier Narrows plants. In *State of Knowledge in Land Treatment of Wastewater*, vol. 1, coordinator, H. L. McKim, pp. 241–51. Hanover, New Hampshire: The United States Army Corps of Engineers.

Echeverria, P., Harrison, B. A., Tirapat, C. & McFarland, A. (1983). Flies as a source of enteric pathogens in a rural village in Thailand. *Applied and Environmental Microbiology*, **46**, 32–6.

Edmonds, R. L. (1976). Survival of coliform bacteria in sewage sludge applied to a forest clearcut and potential movement into groundwater. *Applied and Environmental Microbiology*, **32**, 537–46.

Edmonds, R. L. & Littke, W. (1978). Coliform aerosols generated from the surface of dewatered sewage applied to a forest clearcut. *Applied and Environmental Microbiology*, **36**, 972–4.

Eichold, B. H., Williamson, J. R., Woernle, C. H. & McPhearson, R. M. (1993). Isolation of *Vibrio cholerae* 01 from oysters – Mobile Bay, 1991–1992. *Morbidity and Mortality Weekly Report*, **42**, 91–3.

Ellis, K. V., Cotton, A. P. & Khowaja, M. A. (1993). Iodine disinfection of poor quality waters. *Water Research*, **27**, 369–75.

Emory University & Centers for Disease Control (1993). Update: respiratory syncytial virus activity – United States, 1993. *Morbidity and Mortality Weekly Report*, **42**, 971–3.

Engleberg, N. C., Holburt, E. N., Barrett, T. J., Gary, G. W. Jr, Trujillo, M. H., Feldman, R. A. & Hughes, J. M. (1982). Epidemiology of diarrhea due to rotavirus on an indian reservation: risk factors in the home environment. *Journal of Infectious Diseases*, **145**, 894–8.

Erlandsen, S. L., Sherlock, L. A., Januschka, M., Schupp, D. G., Schaefer, F. W. III, Jakubowski, W. & Bemrick, W. J. (1988). Cross-species transmission of *Giardia* spp.: inoculation of beavers and muskrats with cysts of human, beaver, mouse, and muskrat origin. *Applied and Environmental Microbiology*, **54**, 2777–85.

Erlandsen, S. L., Sherlock, L. A., Bemrick, W. J., Ghobrial, H. & Jakubowski,

W. (1990). Prevalence of *Giardia* spp. in beaver and muskrat populations in Northeastern states and Minnesota: detection of intestinal trophozoites at necropsy provides greater sensitivity than detection of cysts in fecal samples. *Applied and Environmental Microbiology*, **56**, 31–6.

Ewert, D., Bendaña, N., Tormey, M., Kilman, L., Mascola, L., Gresham, L. S., Ginsberg, M. M., Tanner, P. A., Bartzen, M. E., Hunt, S., Marks, R. S., Peter, C. R., Mohle-Boetani, J., Fenstersheib, M., Gans, J., Coy, K., Liska, S., Abbott, S., Bryant, R., Barrett, L., Reilly, K., Wang, M., Werner, S. B., Jackson, R. J. & Rutherford, G. W. (1993). Outbreaks of *Salmonella enteritidis* gastroenteritis – California, 1993. *Morbidity and Mortality Weekly Report*, **42**, 793–7.

Falsey, A. R. & Walsh, E. E. (1993). Transmission of *Chlamydia pneumoniae*. *Journal of Infectious Diseases*, **168**, 493–6.

Farley, T. A., Wilson, S. A., Mahoney, F., Kelso, K. Y., Johnson, D. R. & Kaplan, E. L. (1993). Direct inoculation of food as the cause of an outbreak of group A streptococcal pharyngitis. *Journal of Infectious Diseases*, **167**, 1232–5.

Farooq, S., Kurucz, C. N., Waite, T. D. & Cooper, W. J. (1993). Disinfection of waste-waters: high-energy electron vs gamma irradiation. *Water Research*, **27**, 1177–84.

Farzadegan, H., Vlahov, D., Solomon, L., Muñoz, A., Astemborski, J., Taylor, E., Burnley, A. & Nelson, K. E. (1993). Detection of human immunodeficiency virus type 1 infection by polymerase chain reaction in a cohort of seronegative intravenous drug users. *Journal of Infectious Diseases*, **168**, 327–31.

Fattal, B., Peleg-Olevsky, E. & Cabelli, V. J. (1991). Bathers as a possible source of contamination for swimming-associated illness at marine bathing beaches. *International Journal of Environmental Health Research*, **1**, 204–14.

Fayer, R. (1994). Effect of high temperature on infectivity of *Cryptosporidium parvum* oocysts in water. *Applied and Environmental Microbiology*, **60**, 2732–5.

Feachem, R. G., Hogan, R. C. & Merson, M. H. (1983). Diarrhoeal disease control: reviews of potential interventions. *Bulletin of the World Health Organization*, **61**, 637–40.

Fields, B. S. (1993). *Legionella* and protozoa: interaction of a pathogen and its natural host. In *Legionella: Current Status and Emerging Perspectives*, ed. J. M. Barbaree, R. F. Breiman & A. P. Dufour, pp. 129–36. Washington: The American Society for Microbiology.

Fiksdal, L., Pommepuy, M., Caprais, M.-P. & Midttun, I. (1994). Monitoring of fecal pollution in coastal waters by use of rapid enzymatic techniques. *Applied and Environmental Microbiology*, **60**, 1581–4.

Fujioka, R. S. & Narikawa, O. T. (1982). Effect of sunlight on enumeration of indicator bacteria under field conditions. *Applied and Environmental Microbiology*, **44**, 395–401.

Glass, R. I. & Black, R. E. (1992). The epidemiology of cholera. In *Cholera*, ed. D. Barua & W. B. Greenough III, pp. 129–54. New York: Plenum Medical Book Company.

Glass, R. I., Libel, M. & Brandling-Bennett, A. D. (1992). Epidemic cholera in the Americas. *Science*, **256**, 1524–5.

Gloster, J. (1983). Forecasting the airborne spread of foot-and-mouth disease and Newcastle disease. *Philosophical Transactions of the Royal Society of London*, B, **302**, 535–41.

Gode, G. R. & Bhide, N. K. (1988). Two rabies deaths after corneal grafts from one donor. *Lancet*, **ii**, 791.

Gordon, V., Parry, S., Bellamy, K. & Osborne, R. (1993). Assessment of chemical disinfectants against human immunodeficiency virus: overcoming the problem of cytotoxicity and the evaluation of selected actives. *Journal of Virological Methods*, **45**, 247–57.

Goswami, B. B., Koch, W. H. & Cebula, T. A. (1993). Detection of hepatitis A virus in *Mercenaria mercenaria* by coupled reverse transcription and polymerase chain reaction. *Applied and Environmental Microbiology*, **59**, 2765–70.

Grabow, W. O. K. (1991). New trends in infections associated with swimming pools. *Water SA* (South Africa), **17**, 173–7.

Grabowski, D. J., Tiggs, K. J., Senke, H. W., Salas, A. J., Powers, C. M., Knott, J. A. & Sewell, C. M. (1989). Common-source outbreak of giardiasis – New Mexico. *Morbidity and Mortality Weekly Report*, **38**, 405–7.

Grayston, J. T., Aldous, M. B., Easton, A., Wang, S.-P., Kuo, C.-C., Campbell, L. A. & Altman, J. (1993). Evidence that *Chlamydia pneumoniae* causes pneumonia and bronchitis. *Journal of Infectious Diseases*, **168**, 1231–5.

Grimes, D. J. (1991). Ecology of estuarine bacteria capable of causing human disease: a review. *Estuaries*, **14**, 345–60.

Grundmann, H., Kropec, A., Hartung, D., Berner, R. & Daschner, F. (1993). *Pseudomonas aeruginosa* in a neonatal intensive care unit: reservoirs and ecology of the nosocomial pathogen. *Journal of Infectious Diseases*, **168**, 943–7.

Haas, C. N. (1986). Wastewater disinfection and infectious disease risks. *CRC Critical Reviews in Environmental Control*, **17**, 1–20.

Hamed, K. A., Winters, M. A., Holodniy, M., Katzenstein, D. A. & Merigan, T. C. (1993). Detection of human immunodeficiency virus type 1 in semen: effects of disease stage and nucleoside therapy. *Journal of Infectious Diseases*, **167**, 798–802.

Harewood, P., Rippey, S. & Montesalvo, M. (1994). Effect of gamma irradiation on shelf life and bacterial and viral loads in hard-shelled clams (*Mercenaria mercenaria*). *Applied and Environmental Microbiology*, **60**, 2666–70.

Harper, G. J. (1961). Airborne microorganisms: survival tests with four viruses. *Journal of Hygiene, Cambridge*, **59**, 479–86.

Hayes, E. B., Matte, T. D., O'Brien, T. R., McKinley, T. W., Logsdon, G. S., Rose, J. B., Ungar, B. L. P., Word, D. M., Pinsky, P. F., Cummings, M. L., Wilson, M. A., Long, E. G., Hurwitz, E. S. & Juranek, D. D. (1989). Large community outbreak of cryptosporidiosis due to contamination of a filtered public water supply. *New England Journal of Medicine*, **320**, 1372–6.

Healy, G. R. (1990). Giardiasis in perspective: the evidence of animals as a source of human *Giardia* infections. In *Giardiasis*, ed. E. A. Meyer, pp. 305–13. Amsterdam: Elsevier Science Publishers.

Hedberg, C. W., David, M. J., White, K. E., MacDonald, K. L. & Osterholm, M. T. (1993). Role of egg consumption in sporadic *Salmonella enteritidis* and *Salmonella typhimurium* infections in Minnesota. *Journal of Infectious Diseases*, **167**, 107–11.

Hejkal, T. W., Keswick, B., LaBelle, R. L., Gerba, C. P., Sanchez, Y., Dreesman, G., Hafkin, B. & Melnick J. L. (1982). Viruses in a community water supply associated with an outbreak of gastroenteritis and infectious hepatitis. *Journal, American Water Works Association*, **74** (6), 318–21.

Herwaldt, B. L., Lew, J. F., Moe, C. L., Lewis, D. C., Humphrey, C. D., Monroe, S. S., Pon, E. W. & Glass, R. I. (1994). Characterization of a variant strain of Norwalk virus from a food-borne outbreak of gastroenteritis on a cruise ship in Hawaii. *Journal of Clinical Microbiology*, **32**, 861–6.

Hewlett, D., Franchini, D., Horn, D., Alfalla, C., Yap, R., DiPietro, D., Peterson, S., Eisenberg, H., Lue, Y., Rodriguez, M., Roberto, M., Alland, D. & Opal, S. (1993). Outbreak of multidrug-resistant tuberculosis at a hospital – New York City, 1991. *Morbidity and Mortality Weekly Report*, **42**, 427–34.

Higashihara, S., Kanenaka, K., Ching-Lee, M., Effler, P., Akiyama, D., Sugi, M., Pon, E., Sharifzadeh, K., Waskiewicz, R., Ridley, N., Hohmann, W., Higson, W., Sobsey, M. & Wait, D. (1991). Gastroenteritis associated with consumption of raw shellfish – Hawaii, 1991. *Morbidity and Mortality Weekly Report*, **40**, 303–5.

Hlady, W. G. (1993). Arboviral diseases – United States, 1992. *Morbidity and Mortality Weekly Report*, **42**, 467–8.

Hlady, W. G., Hopkins, R. S., Ogilby, T. E. & Allen, S. T. (1993). Patient-to-patient transmission of hepatitis B in a dermatology practice. *American Journal of Public Health*, **83**, 1689–93.

Holmes, S. E., Kinde, M. R., Pearson, J. L. & Hennes, R. F. (1989). Gastroenteritis outbreak disease linked to swimming pool and spa use. *Journal of Environmental Health*, **51** (5), 286–8.

Holmquist, C. A. (1924). Relation of improvements in water supplies to typhoid fever and other intestinal diseases. *Journal of the New England Water Works Association*, **38**, 237–47.

Hopkins, R. J., Vial, P. A., Ferreccio, C., Ovalle, J., Prado, P., Sotomayor, V., Russell, R. G., Wasserman, S. S. & Morris, J. G. Jr (1993). Seroprevalence of *Helicobacter pylori* in Chile: vegetables may serve as one route of transmission. *Journal of Infectious Diseases*, **168**, 222–6.

Hou, M.-C., Wu, J.-C., Kuo, B. I.-T., Sheng, W.-Y., Chen, T.-Z., Lee, S.-D. & Lo, K.-J. (1993). Heterosexual transmission as the most common route of acute hepatitis B virus infection among adults in Taiwan – the importance of extending vaccination to susceptible adults. *Journal of Infectious Diseases*, **167**, 938–41.

Hurst, C. J. (1991). Using linear and polynomial models to examine the environmental stability of viruses. In *Modeling the Environmental Fate of Microorganisms*, ed. C. J. Hurst, pp. 137–59. Washington: American Society for Microbiology.

Hurst, C. J., Wild, D. K. & Clark, R. M. (1992) Comparing the accuracy of equation formats for modeling microbial population decay rates. In *Modeling the Metabolic and Physiologic Activities of Microorganisms*, ed. C. J. Hurst, pp. 149–75. New York: John Wiley & Sons.

Isaac-Renton, J. L., Cordeiro, C., Sarafis, K. & Shahriari, H. (1993). Characterization of *Giardia duodenalis* isolates from a waterborne outbreak. *Journal of Infectious Diseases*, **167**, 431–40.

Isenberg, H. D. & D'Amato, R. F. (1991). Indigenous and pathogenic microorganisms of humans. In *Manual of Clinical Microbiology*, 5th edn, ed. A. Balows, W. J. Hausler Jr, K. L. Herrmann, H. D. Isenberg & H. J. Shadomy, pp. 2–14. Washington: The American Society for Microbiology.

James, A. D. & Rossiter, P. B. (1989). An epidemiological model of rinderpest. I. Description of the model. *Tropical Animal Health and Production*, **21**, 59–68.

Jennings, L. C., Dick, E. C., Mink, K. A., Wartgow, C. D. & Inhorn, S. L. (1988). Near disappearance of rhinovirus along a fomite transmission chain. *Journal of Infectious Diseases*, **158**, 888–92.

Jereb, J. A., Burwen D. R., Dooley, S. W., Haas, W. H., Crawford, J. T., Geiter, L. J., Edmond, M. B., Dowling J. N., Shapiro, R., Pasculle, A. W., Shanahan, S. L. & Jarvis, W. R. (1993). Nosocomial outbreak of tuberculosis in a renal transplant unit: application of a new technique for restriction fragment length polymorphism analysis of *Mycobacterium tuberculosis* isolates. *Journal of Infectious Diseases*, **168**, 1219–24.

Jernigan, J. A., Lowrey, B. S., Hayden, F. G., Kyger, S. A., Conway, B. P., Gröschel, D. H. M. & Farr, B. M. (1993). Adenovirus type 8 epidemic keratoconjunctivitis in an eye clinic: risk factors and control. *Journal of Infectious Diseases*, **167**, 1307–13.

Joce, R. E., Bruce, J., Kiely, D., Noah, N. D., Dempster, W. B., Stalker, R., Gumsley, P., Chapman, P. A., Norman, P., Watkins, J., Smith, H. V., Price, T. J. & Watts, D. (1991). An outbreak of cryptosporidiosis associated with a swimming pool. *Epidemiology and Infection*, **107**, 497–508.

Johnson, G. A. (1916). The typhoid toll. *Journal, American Water Works Association*, **3** (2), 249–326.

Joly, J. R. (1993). Monitoring for the presence of *Legionella*: where, when, and how? In *Legionella: Current Status and Emerging Perspectives*, ed. J. M. Barbaree, R. F. Breiman & A. P. Dufour, pp. 211–16. Washington: The American Society for Microbiology.

Kaspar, C. W. & Tamplin, M. L. (1993). Effects of temperature and salinity on the survival of *Vibrio vulnificus* in seawater and shellfish. *Applied and Environmental Microbiology*, **59**, 2425–9.

Katzenelson, E., Buium, I. & Shuval, H. I. (1976). Risk of communicable disease infection associated with wastewater irrigation in agricultural settlements. *Science*, **194**, 944–6.

Kehr, R. W. & Butterfield, C. T. (1943). Notes on the relation between coliforms and enteric pathogens. *Public Health Reports*, **58**, 589–607.

Keswick, B. H., Pickering, L. K., DuPont, H. L. & Woodward, W. E. (1983). Survival and detection of rotaviruses on environmental surfaces in day-care centers. *Applied and Environmental Microbiology*, **46**, 813–16.

Khan, A. S., Moe, C. L., Glass, R. I., Monroe, S. S., Estes, M. K., Chapman, L. E., Jiang, X., Humphrey, C., Pon, E., Iskander, J. K. & Schonberger, L. B. (1994). Norwalk virus-associated gastroenteritis traced to ice consumption aboard a cruise ship in Hawaii: comparison and application of molecular method-based assays. *Journal of Clinical Microbiogy*, **32**, 318–22.

Kim-Farley, R. J., Bart, K. J., Schonberger, L. B., Orenstein, W. A., Nkowane, B. M., Hinman, A. R., Kew, O. M., Hatch, M. H. & Kaplan, J. E. (1984). Poliomyelitis in the USA: virtual elimination of disease caused by wild virus. *Lancet*, **ii**, 1315–17.

Kladivko, E. J. & Nelson, D. W. (1979*a*). Surface runoff from sludge-amended soils. *Journal, Water Pollution Control Federation*, **51**, 100–10.

Kladivko, E. J. & Nelson, D. W. (1979*b*). Changes in soil properties from application of anaerobic sludge. *Journal, Water Pollution Control Federation*, **51**, 325–32.

Koopman, J. S., Eckert, E. A., Greenberg, H. B., Strohm, B. C., Isaacson, R. E. & Monto, A. S. (1982). Norwalk virus enteric illness acquired by swimming exposure. *American Journal of Epidemiology*, **115**, 173–7.

Lawson, H. W., Braun, M. M., Glass, R. I. M., Stine, S. E., Monroe, S. S., Atrash, H. K., Lee, L. E. & Englender, S. J. (1991). Waterborne outbreak of Norwalk virus gastroenteritis at a southwest US resort: role of geological formations in contamination of well water. *Lancet*, **337**, 1200–4.

Le Saux, N., Spika, J. S., Friesen, B., Johnson, I., Melnychuck, D., Anderson, C., Dion, R., Rahman, M. & Tostowaryk, W. (1993). Ground beef consumption in noncommercial settings is a risk factor for sporadic *Escherichia coli* 0157:H7 infection in Canada. *Journal of Infectious Diseases*, **167**, 500–2.

Levy, M., Moll, M. & Jones, B. R. (1993). Improper infection-control practices during employee vaccination programs – District of Columbia and Pennsylvania, 1993. *Morbidity and Mortality Weekly Report*, **42**, 969–71.

Lindenauer, K. G. & Darby, J. L. (1994). Ultraviolet disinfection of wastewater: effect of dose on subsequent photoreactivation. *Water Research*, **28**, 805–17.

Lotka, A. J. (1956). *Elements of Mathematical Biology*. New York: Dover Publications.

Lowry, P. W., Blankenship, R. J., Gridley, W., Troup, N. J. & Tompkins, L. S. (1991). A cluster of legionella sternal-wound infections due to postoperative topical exposure to contaminated tap water. *New England Journal of Medicine*, **324** (2), 109–13.

Luby, S., Jones, J., Dowda, H., Kramer, J. & Horan, J. (1993). A large outbreak of gastroenteritis caused by diarrheal toxin-producing *Bacillus cereus*. *Journal of Infectious Diseases*, **167**, 1452–5.

Lund, E. & Nissen, B. (1986). Low technology water purification by bentonite clay flocculation as performed in Sudanese villages virological examinations. *Water Research*, **20**, 37–43.

MacKenzie, W. R., Hoxie, N. J., Proctor, M. E., Gradus, M. S., Blair, K. A., Peterson, D. E., Kazmierczak, J. J., Addiss, D. G., Fox, K. R., Rose, J. B. & Davis, J. P. (1994). A massive outbreak in Milwaukee of cryptosporidium infection transmitted through the public water supply. *New England Journal of Medicine*, **331**, 161–7.

Martone, W. J., Hierholzer, J. C., Keenlyside, R. A., Fraser, D. W., D'Angelo, L. J. & Winkler, W. G. (1980). An outbreak of adenovirus type 3 disease at a private recreation center swimming pool. *American Journal of Epidemiology*, **111** (2), 229–37.

Mbithi, J. N., Springthorpe, V. S. & Sattar, S. A. (1993). Comparative *in vivo* efficiencies of hand-washing agents against hepatitis A virus (HM-175) and poliovirus type 1 (Sabin). *Applied and Environmental Microbiology*, **59**, 3463–9.

McCarthy, S. A. & Khambaty, F. M. (1994). International dissemination of epidemic *Vibrio cholerae* by cargo ship ballast and other nonpotable waters. *Applied and Environmental Microbiology*, **60**, 2597–601.

McFeters, G. A., Barry, J. P. & Howington, J. P. (1993). Distribution of enteric bacteria in antarctic seawater surrounding a sewage outfall. *Water Research*, **27**, 645–50.

McMillan, N. S., Martin, S. A., Sobsey, M. D., Wait, D. A., Meriwether, R. A. & MacCormack, J. N. (1992). Outbreak of pharyngoconjunctival fever at a summer camp – North Carolina, 1991. *Morbidity and Mortality Weekly Report*, **41**, 342–4.

McPherson, J. B. (1979). Land treatment of wastewater at Werribee past, present and future. *Progress in Water Technology*, **11** (4/5), 15–31.

Melnick, J. L. (1947). Poliomyelitis virus in urban sewage in epidemic and in nonepidemic times. *American Journal of Hygiene*, **45**, 240–53.

Melnick, J. L., Emmons, J., Coffey, J. H. & Schoop, H. (1954a). Seasonal distribution of coxsackie viruses in urban sewage and flies. *American Journal of Hygiene*, **59**, 164–84.

Melnick, J. L., Emmons, J., Opton, E. M. & Coffey, J. H. (1954b). Coxsackie viruses from sewage methodology including an evaluation of the grab sample and gauze pad collection procedures. *American Journal of Hygiene*, **59**, 185–95.

Miller, W. M. (1979). A state-transition model of epidemic foot-and-mouth disease. *Technical Bulletin 1597: A Study of the Potential Economic Impact of Foot-and-Mouth Disease in the United States*, pp. 113–31. Washington: United States Department of Agriculture.

Millner, P. D., Bassett, D. A. & Marsh, P. B. (1980). Dispersal of *Aspergillus fumigatus* from sewage sludge compost piles subjected to mechanical agitation in open air. *Applied and Environmental Microbiology*, **39**, 1000–9.

Mims, C. A. (1981). Vertical transmission of viruses. *Microbiological Reviews*, **45**, 267–86.

Mintz, E. D., Hudson-Wragg, M., Mshar, P., Cartter, M. L. & Hadler, J. L. (1993). Foodborne giardiasis in a corporate office setting. *Journal of Infectious Diseases*, **167**, 250–3.

Monto, A. S. & Koopman, J. S. (1980). The Tecumseh Study XI. Occurrence of acute enteric illness in the community. *American Journal of Epidemiology*, **112** (3), 323–33.

Moore, A. C., Herwaldt, B. L., Craun, G. F., Calderon, R. L., Highsmith, A. K. & Juranek, D. D. (1993). Surveillance for waterborne disease outbreaks – United States, 1991–1992. *Morbidity and Mortality Weekly Report*, **42** (SS-5), 1–22.

Moore, B. E. (1993). Survival of human immunodeficiency virus (HIV), HIV-infected lymphocytes, and poliovirus in water. *Applied and Environmental Microbiology*, **59**, 1437–43.

Mosley, J. W. (1966). Transmission of viral diseases by drinking water. In *Transmission of Viruses by the Water Route*, ed. G. Berg, pp. 5–23. New York: Interscience Publishers, John Wiley & Sons.

Murphy, A. M., Grohmann, G. S. & Sexton, M. F. H. (1983). Infectious gastroenteritis in Norfolk Island and recovery of viruses from drinking water. *Journal of Hygiene, Cambridge*, **91**, 139–46.

O'Neill, K. R., Jones, S. H. & Grimes, D. J. (1990). Incidence of *Vibrio vulnificus* in northern New England water and shellfish. *Federation of European Microbiological Societies Microbiology Letters*, **72**, 163–8.

Opulski, A., MacNeill, E., Rosales, C., Hartsough, A., Doll, J., Levy, C., Fink, M., Erickson, B., Slanta, W., Cage, G., Sands, L., Lofgren, J., Gentry, G., Davis, T., Pape, J. & Hoffman, R. E. (1992). Pneumonic plague – Arizona, 1992. *Morbidity and Mortality Weekly Report*, **41**, 737–9.

Ormsby, H. L. & Aitchison, W. S. (1955). The role of the swimming pool in the transmission of pharyngeal-conjunctival fever. *Canadian Medical Association Journal*, **73**, 864–6.

Ozanne, G., Huot, R. & Montpetit, C. (1993). Influence of packaging and processing conditions on the decontamination of laboratory biomedical waste by steam sterilization. *Applied and Environmental Microbiology*, **59**, 4335–7.

Pacha, R. E., Clark, G. W. & Williams, E. A. (1985). Occurrence of *Campylobacter jejuni* and *Giardia* species in muskrat (*Ondatra zibethica*). *Applied and Environmental Microbiology*, **50**, 177–8.

Pacha, R. E., Clark, G. W., Williams, E. A. & Carter, A. M. (1988). Migratory birds of central Washington as reservoirs of *Campylobacter jejuni*. *Canadian Journal of Microbiology*, **34**, 80–2.

Pacha, R. E., Clark, G. W., Williams, E. A., Carter, A. M., Scheffelmaier, J. J. & Debusschere, P. (1987). Small rodents and other mammals associated with mountain meadows as reservoirs of *Giardia* spp. and *Campylobacter* spp. *Applied and Environmental Microbiology*, **53**, 1574–9.

Payment, P., Richardson, L., Siemiatycki, J., Dewar, R., Edwardes, M. & Franco, E. (1991). A randomized trial to evaluate the risk of gastrointestinal disease due to consumption of drinking water meeting current microbiological standards. *American Journal of Public Health*, **81**, 703–8.

Pearson, A. D., Greenwood, M., Healing, T. D., Rollins, D., Shahamat, M., Donaldson, J. & Colwell, R. R. (1993). Colonization of broiler chickens by waterborne *Campylobacter jejuni*. *Applied and Environmental Microbiology*, **59**, 987–96.

Pinfold, J. V., Horan, N. J., Wirojanagud, W. & Mara, D. (1993). The bacteriological quality of rainjar water in rural northeast Thailand. *Water Research*, **27**, 297–302.

Porter, J. D., Ragazzoni, H. P., Buchanon, J. D., Waskin, H. A., Juranek, D. D. & Parkin, W. E. (1988). *Giardia* transmission in a swimming pool. *American Journal of Public Health*, **78**, 659–62.

Poupard, J. A., Miller, L. A. & Granshaw, L. (1989). The use of smallpox as a biological weapon in the French and Indian war of 1763. *American Society for Microbiology News*, **55**, 122–4.

Powelson, D. K., Gerba, C. P. & Yahya, M. T. (1993). Virus transport and removal in wastewater during aquifer recharge. *Water Research*, **27**, 583–90.

Powers, E. M., Hernandez, C., Boutros, S. N. & Harper, B. G. (1994). Biocidal efficacy of a flocculating emergency water purification tablet. *Applied and Environmental Microbiology*, **60**, 2316–23.

Pratt, H. D. & Smith, J. W. (1991). Arthropods of medical importance. In *Manual of Clinical Microbiology*, 5th edn, ed. A. Balows, W. J. Hausler Jr, K. L. Herrmann, H. D. Isenberg & H. J. Shadomy, pp. 796–810. Washington: The American Society for Microbiology.

Preston, D. R., Farrah, S. R., Bitton, G. & Chaudhry, G. R. (1991). Detection of nucleic acids homologous to human immunodeficiency virus in wastewater. *Journal of Virological Methods*, **33**, 383–90.

Raad, I., Costerton, W., Sabharwal, U., Sacilowski, M., Anaissie, E. & Bodey, G. P. (1993). Ultrastructural analysis of indwelling vascular catheters: a quantitative relationship between luminal colonization and duration of placement. *Journal of Infectious Diseases*, **168**, 400–7.

Ramelow, C., Süss, J., Berndt, D., Roggendorf, M. & Schreier, E. (1993). Detection of tick-borne encephalitis virus RNA in ticks (*Ixodes ricinus*) by the polymerase chain reaction. *Journal of Virological Methods*, **45**, 115–19.

Rehle, T. M. (1989). Classification, distribution and importance of arboviruses. *Tropical Medicine and Parasitology*, **40**, 391–5.

Rice, E. W. & Johnson, C. H. (1991). Cholera in Peru. *Lancet*, **338**, 455.

Roman, R., Middleton, M., Cato, S., Bell, E., Ong, K. R., Gruenewald, R., Ferguson, A. & Ramon, A. (1991). Cholera – New York, 1991. *Morbidity and Mortality Weekly Report*, **40**, 516–18.

Rosas, I., Báez, A. & Coutiño, M. (1984). Bacteriological quality of crops irrigated with wastewater in the Xochimilco plots, Mexico City, Mexico. *Applied and Environmental Microbiology*, **47**, 1074–9.

Rose, J. B. (1988). Occurrence and significance of cryptosporidium in water. *Journal, American Water Works Association*, **80** (2), 53–8.

Rossiter, P. B. & James, A. D. (1989). An epidemiological model of rinderpest. II. Simulations of the behavior of rinderpest virus in populations. *Tropical Animal Health and Production*, **21**, 69–84.

Salzman, M. B., Isenberg, H. D., Shapiro, J. F., Lipsitz, P. J. & Rubin, L. G. (1993). A prospective study of the catheter hub as the portal of entry for microorganisms causing catheter-related sepsis in neonates. *Journal of Infectious Diseases*, **167**, 487–90.

Sattar, S. A., Jacobsen, H., Springthorpe, V. S., Cusack, T. M. & Rubino, J. R. (1993). Chemical disinfection to interrupt transfer of rhinovirus type 14 from environmental surfaces to hands. *Applied and Environmental Microbiology*, **59**, 1579–85.

Sattar, S. A., Lloyd-Evans, N., Springthorpe, V. S. & Nair, R. C. (1986). Institutional outbreaks of rotavirus diarrhoea: potential role of fomites and environmental surfaces as vehicles for virus transmission. *Journal of Hygiene, Cambridge*, **96**, 277–89.

Sawyer, B., Elenbogen, G., Rao, K. C., O'Brien, P., Zenz, D. R. & Lue-Hing, C. (1993). Bacterial aerosol emission rates from municipal wastewater aeration tanks. *Applied and Environmental Microbiology*, **59**, 3183–6.

Sawyer, L. A., Murphy, J. J., Kaplan, J. E., Pinsky, P. F., Chacon, D., Walmsley, S., Schonberger, L. B., Phillips, A., Forward, K., Goldman, C., Brunton, J., Fralick, R. A., Carter, A. O., Gary, W. G. Jr, Glass, R. I. & Low, D. E. (1988). 25- to 30-nM virus particle associated with a hospital outbreak of acute gastro-enteritis with evidence for airborne transmission. *American Journal of Epidemiology*, **127**, 1261–71.

Sawyer, M. H., Chamberlin, C. J., Wu, Y. N., Aintablian, N. & Wallace, M. R. (1994). Detection of varicella-zoster virus DNA in air samples from hospital rooms. *Journal of Infectious Diseases*, **169**, 91–4.

Schürmann, W. & Eggers, H. J. (1985). An experimental study on the epidemiology of enteroviruses: water and soap washing of poliovirus 1-contaminated hands, its effectiveness and kinetics. *Medical Microbiology and Immunology*, **174**, 221–36.

Schwartzbrod, L., Finance, C., Aymard, M., Brigaud, M. & Lucena, F. (1985). Recovery of reoviruses from tap water. *Zentralblatt für Bakteriologie und Hygiene, 1. Abteilung Originale Reihe B*, **181**, 383–9.

Service, R. F. (1994). Triggering the first line of defense. *Science*, **265**, 1522–4.

Seyfried, P. L., Tobin, R. S., Brown, N. E. & Ness, P. F. (1985a). A prospective study of swimming-related illness I. Swimming-associated health risk. *American Journal of Public Health*, **75**, 1068–70.

Seyfried, P. L., Tobin, R. S., Brown, N. E. & Ness, P. F. (1985b). A prospective study of swimming-related illness II. Morbidity and the microbiological quality of water. *American Journal of Public Health*, **75**, 1071–5.

Sims, C., Hill, J., Rizer, M., Miller, T., Beard, D., Kerr, M., O'Cain, M., Williamson, D. & Woernle, C. (1992). Epidemic early syphilis – Montgomery County, Alabama, 1990–1991. *Morbidity and Mortality Weekly Report*, **41**, 790–4.

Smith, G. & Grenfell, B. T. (1990). Population biology of pseudorabies in swine. *American Journal of Veterinary Research*, **51**, 148–55.

Sonzogni, W. C., Chesters, G., Coote, D. R., Jeffs, D. N., Konrad, J. C., Ostry, R. C. & Robinson, J. B. (1980). Pollution from land runoff. *Environmental Science and Technology*, **14** (2), 148–53.

Sorber, C. A., Moore, B. E., Johnson, D. E., Harding, H. J. & Thomas, R. E. (1984). Microbiological aerosols from the application of liquid sludge to land. *Journal, Water Pollution Control Federation*, **56**, 830–6.

Sorvillo, F. J., Fujioka, K., Nahlen, B., Tormey, M. P., Kebabjian, R. & Mascola, L. (1992). Swimming-associated cryptosporidiosis. *American Journal of Public Health*, **82**, 742–4.

Srikanth, S. & Berk, S. G. (1993). Stimulatory effect of cooling tower biocides on amoebae. *Applied and Environmental Microbiology*, **59**, 3245–9.

Sternberg, S. (1994). The emerging fungal threat. *Science*, **266**, 1632–4.

Stetler, R. E., Waltrip, S. C. & Hurst, C. J. (1992). Virus removal and recovery in the drinking water treatment train. *Water Research*, **26**, 727–31.

Stewart, S. J. (1991). *Francisella*. In *Manual of Clinical Microbiology*, 5th edn, ed. A. Balows, W. J. Hausler Jr, K. L. Herrmann, H. D. Isenberg & H. J. Shadomy, pp. 454–6. Washington: The American Society for Microbiology.

St Louis, M. E. (1988). Water-related disease outbreaks, 1985. *Morbidity and Mortality Weekly Report* **37** (SS-2), 15–24.

Svitlik, C., Cartter, M., McCarter, Y., Hadler, J. L., Goeller, D., Groves, C., Dwyer, D., Tilghman, D., Israel, E., Housenecht, R., Yeager, S. & Tavris, D. R. (1992). *Salmonella hadar* associated with pet ducklings – Connecticut, Maryland and Pennsylvania, 1991. *Morbidity and Mortality Weekly Report*, **41**, 185–7.

Swerdlow, D. L. & Ries, A. A. (1992). Cholera in the Americas. Guidelines for the clinician. *Journal of the American Medical Association*, **267**, 1495–9.

Tabet, S. R., Goldbaum, G. M., Hooton, T. M., Eisenach, K. D., Cave, M. D. & Nolan, C. M. (1994). Restriction fragment length polymorphism analysis detecting a community-based tuberculosis outbreak among persons infected with human immunodeficiency virus. *Journal of Infectious Diseases*, **169**, 189–92.

Tao, H., Guangmu, C., Changan, W., Henli, Y., Zhaoying, F., Tungxin, C., Zinyi, C., Weiwe, Y., Xuejian, C., Shuasen, D., Xiaoquang, L. & Weicheng, C. (1984). Waterborne outbreak of rotavirus diarrhoea in adults in China caused by a novel rotavirus. *Lancet*, **i**, 1139–42.

Task Committee on the Status of Irrigation and Drainage Research (1984). Status of irrigation and drainage research in the United States. *Journal of Irrigation and Drainage Engineering*, **110**, 55–74.

Tauxe, R. V., Hargrett-Bean, N., Patton, C. M. & Wachsmuth, I. K. (1988). *Campylobacter* isolates in the United States, 1982–1986. *Morbidity and Mortality Weekly Report*, **37** (SS-2), 1–13.

Taylor, F. S. (1942). *The Conquest of Bacteria*. New York: The Philosophical Library.

Taylor, J. L., Tuttle, J., Pramukul, T., O'Brien, K., Barrett, T. J., Jolbitado, B., Lim, Y. L., Vugia, D., Morris, J. G. Jr, Tauxe, R. V. & Dwyer, D. M. (1993). An outbreak of cholera in Maryland associated with imported commercial frozen fresh coconut milk. *Journal of Infectious Diseases*, **167**, 1330–5.

Theunissen, H. J. H., Lemmens-den Toom, N. A., Burggraaf, A., Stolz, E. & Michel, M. F. (1993). Influence of temperature and relative humidity on the survival of *Chlamydia pneumoniae* in aerosols. *Applied and Environmental Microbiology*, **59**, 2589–93.

Tulchinsky, T. H., Levine, I., Abrookin, R. & Halperin, R. (1988). Waterborne enteric disease outbreaks in Israel, 1976–1985. *Israeli Journal of Medical Sciences*, **24**, 644–51.

Twartz, J. C., Shiral, A., Selvaraju, G., Saunders, J. P., Huxsoll, D. L. & Groves, M. G. (1982). Doxycycline prophylaxis for human scrub typhus. *Journal of Infectious Diseases*, **146**, 811–18.

Van Steenderen, R. A. (1977). Studies on the efficiency of a solar distillation still for supplementing drinking water supplies in South West Africa. *Water SA* (South Africa), **3**, 1–14.

Vazquez, J. A., Sanchez, V., Dmuchowski, C., Dembry, L. M., Sobel, J. D. & Zervos, M. J. (1993). Nosocomial acquisition of *Candida albicans*: an epidemiologic study. *Journal of Infectious Diseases*, **168**, 195–201.

Viscidi, R. P., Bobo, L., Hook, E. W. III & Quinn, T. C. (1993). Transmission of *Chlamydia trachomatis* among sex partners assessed by polymerase chain reaction. *Journal of Infectious Diseases*, **168**, 488–92.

Viswanathan, R. (1957). Epidemiology. *Indian Journal of Medical Research*, **45** (Supplement), 1–29.

Walker, J. M. & Demirjian, Y. A. (1978). Muskegon County, Michigan's own land wastewater treatment system. In *State of Knowledge in Land Treatment of Wastewater*, vol. 1, coordinator H. L. McKim, pp. 417–27. Hanover, New Hampshire: The United States Army Corps of Engineers.

Wallach, F. R., Forni, A. L., Hariprashad, J., Stoeckle, M. Y., Steinberg, C. R., Fisher, L., Malawista, S. E. & Murray, H. W. (1993). Circulating *Borrelia burgdorferi* in patients with acute Lyme disease: results of blood cultures and serum DNA analysis. *Journal of Infectious Diseases*, **168**, 1541–3.

Ward, R. L., Knowlton, D. R., Stober, J., Jakubowski, W., Mills, T., Graham, P. & Camann, D. E. (1989). Effect of wastewater spray irrigation on rotavirus infection rates in an exposed population. *Water Research*, **23**, 1503–9.

Watson, J. C., Fleming, D. W., Borella, A. J., Olcott, E. S., Conrad, R. E. & Baron, R. C. (1993). Vertical transmission of hepatitis A resulting in an outbreak in a neonatal intensive care unit. *Journal of Infectious Diseases*, **167**, 567–71.

Wegmüller, B., Lüthy, J. & Candrian, U. (1993). Direct polymerase chain reaction detection of *Campylobacter jejuni* and *Campylobacter coli* in raw milk and dairy products. *Applied and Environmental Microbiology*, **59**, 2161–5.

Welt, B. A., Tong, C. H., Rossen, J. L. & Lund, D. B. (1994). Effect of microwave radiation on inactivation of *Clostridium sporogenes* (PA 3679) spores. *Applied and Environmental Microbiology*, **60**, 482–8.

Westwood, J. C. N. & Sattar, S. A. (1976). The minimal infective dose. In *Viruses in Water*, eds. G. Berg, H. L. Bodily, E. H. Lennette, J. L. Melnick & T. G. Metcalf, pp. 61–9. Washington: American Public Health Association.

Wolfe, M., Parenti, D., Pollner, J., Kobrine, A. & Schwartz, A. (1993). Schisto-somiasis in US Peace Corps volunteers – Malawi, 1992. *Morbidity and Mortality Weekly Report*, **42**, 565–70.

Yeung, S. C. H., Kazazi, F., Randle, C. G. M., Howard, R. C., Rizvi, N., Downie, J. C., Donovan, B. J., Cooper, D. A., Sekine, H., Dwyer, D. E. & Cunningham, A. L. (1993). Patients infected with human immunodeficiency virus type 1 have low levels of virus in saliva even in the presence of periodontal disease. *Journal of Infectious Diseases*, **167**, 803–9.

Chapter 2

Strategies for modeling microbial colonization of the human body in health and disease

ROBIN A. ROSS and MEI-LING T. LEE

Introduction

The human body is colonized by an array of microorganisms. For a particular anatomical site, there is a microflora that is normally present and considered nonpathogenic. We have come to understand that this normal microflora is part of a synergistic host–microbe relationship in which the microflora contributes to the maintenance of a healthy host. By preventing the colonization or overgrowth of pathogens, the normal microflora of the human body acts as a barrier to disease. It is this barrier that should be a part of the focus of any model of microbial colonization of the human body.

The following is a discussion of several strategies for modeling human host–microbe ecosystems. It includes a presentation of the various microbial ecosystems of the human body and methods for the modeling of data obtained from these ecosystems. Two types of models that are useful in the modeling of microbial colonization are analytic and simulation models. Examples of both, that have applications in the modeling of human host–microbe ecosystems, are given.

Host–microbe ecosystems of the human body

The human body contains a number of ecological niches, the major host–microbe ecosystems being the oronasal tract, skin, intestinal tract and urogenital tract. Each of these ecosystems has its own microflora. At the generic level these various microflora can appear similar, but, upon closer examination (identification to the species level), the microflora of each ecological niche is seen to be unique to that niche.

The microflora of a human develops over time, but initial colonization occurs rapidly after birth. Under normal conditions, the body is sterile until birth. At birth the infant is immediately exposed to the microorganisms of the birth canal and subsequently the external environment, some of which permanently establish at various anatomic sites of the body. For example, the oral cavity is colonized within 72 h after birth, as is the rectum (Bhatia *et al.*, 1988). The skin is colonized within 2 h at almost all sites (Leyden, 1982) and the vagina within four days after birth (Kistner, 1978).

The microflora of each ecosystem is dynamic. Changes in the number of individual components of the microflora and the population as a whole occur on both a regular and irregular basis. There is an overall flexibility inherent in the microflora of the various ecological niches of the human body. It is the combined flexibility of the microflora that allows it to change in response to external conditions and to return to a normal state (Onderdonk & Wissemann, 1993).

The following are descriptions of four major microbial ecosystems of the human body – skin, mouth, intestinal tract and vagina – and include brief descriptions of the ecosystem environment and a description of the normal microflora and their prevalence.

Cutaneous ecosystem

The skin, for the most part, is a tough dry exterior that protects the body from the external environment (McBride, 1993). The skin provides an environment of low humidity on exposed surfaces and of higher humidity in occluded sites such as the groin, axilla and toe web (Noble & Pitcher, 1978; McBride, 1993). It is predominately acidic, but anatomical site variation does exist (Behrendt & Green, 1971; Noble, 1981). The temperature of the skin is also site dependent and can range from 26 to 35 °C (Dubois, 1939).

The structure of the skin is of two layers, the inner dermis and the outer epidermis (Montagna & Parakaal, 1974; Holbrook & Wolff, 1987). The dermal layer is composed of connective tissue, tendons, elastic fibers, fat and collagen (Sengel, 1985). The epidermis consists of epithelial cells continuously differentiating towards the skin surface to form its outermost layer, the stratum corneum. The cutaneous ecosystem also includes hair follicles distributed over the entire body except on the palms of the hands and soles and dorsa of the feet (Montagna & Parakaal, 1974).

The cutaneous ecosystem has the capacity to maintain a stable microbial population (Evans, 1975; Evans & Strom, 1982). The majority of the microflora are found within the outermost layers of the stratum corneum and the hair follicles (Price, 1938; Updegraff, 1964; Montes & Wilborn, 1969). Most of the bacteria exist as microcolonies (Noble & Pitcher, 1978), with the largest number of microorganisms at the hair follicle (Montes & Wilborn, 1969). The actual concentration of bacteria on the skin varies with anatomical site, from a mean value of 10 colony forming units (CFU)/cm^2 of skin (back) to 1×10^5 CFU/cm^2 of skin (feet) (McBride, Duncan & Knox, 1977).

The surface of the skin on the human body is colonized by a variety of microorganisms in varying densities and locations (McBride, 1993). The microflora of the skin is composed predominantly of coagulase-negative *Staphylococcus* (Kloos & Musselwhite, 1975) and *Propionibacterium acnes* (McGinley, Webster & Leyden, 1978). *Micrococcus* (Kloos & Musselwhite, 1975; Kloos, 1981), lipophilic fungi (Mok & Barreto da Silva, 1984), *Lactobacillus*, *Corynebacterium* and *Brevibacterium* (Pitcher, 1981) are frequently isolated from the skin. Also common to the skin microflora are *Acinetobacter* (Retailliau *et al.*, 1977).

Oral ecosystem

The oral environment is a combination of mucosal surfaces continually shedding layers of cells by desquamation and hard 'non-shedding' surfaces such as teeth or fillings (Marsh, 1991). The oral ecosystem, as with all the microbial ecosystems of the human body, can be viewed as a group of subecosystems each containing a characteristic microflora (Milton-Thompson *et al.*, 1982; Marsh & Martin, 1982; Theilade, 1990).

The oral cavity is a moist environment that ranges in temperature from 34 to 36 °C (Theilade, 1990). All surfaces of the mouth are constantly washed by saliva, which maintains a relatively neutral pH except when high-sucrose foods are eaten, at which time the oral pH can drop to 4–5 (Theilade, 1990). The level of oxygen in the mouth varies, as does the redox potential, depending on where in the mouth the measurements are taken (Theilade, 1990).

The microflora of the oral cavity consists of > 200 bacterial species, some protozoa, yeasts and mycoplasmas (Theilade, 1990). The mucosal surfaces of the mouth are sparsely colonized, whereas the teeth harbor a much more diverse microflora (Marsh & Martin, 1982; Theilade, 1990). The microflora of the oral ecosystem is in large part made up of alpha- and nonhemolytic streptococci and group D *Streptococcus*, with species specificity for each oral niche (Bowden, Ellwood & Hamilton, 1979; Nisengard & Newman, 1994). Other organisms frequently isolated are *Neisseria* and *Branhamella catarrhalis* (Nisengard & Newman, 1994). Also characteristic of this ecosystem are *Corynebacterium* and *Staphylococcus* (Theilade, 1990; Nisengard & Newman, 1994). Less frequently recovered are *Peptotostreptococcus*, *Propionibacterium acnes*, *Hemophilus*, *Lactobacillus*, *Bacteroides* and *Candida albicans* (Nisengard & Newman, 1994; Theilade, 1990).

Intestinal ecosystem

The intestinal ecosystem is an environment of low pH and low redox potential (Simon & Gorbach, 1986). The intestine is lined with an epithelial layer that is coated with mucin, which fills the lumen of the intestine (Bollard et al., 1986). Bacteria are associated with the mucin and with the epithelial-mucin interface (Savage, McAllister & Davis, 1971; Croucher et al., 1983; Bollard et al., 1986). It can be divided into two subecosystems, the small intestine and the large intestine.

The concentration of microorganisms in the contents of the intestinal tract increases from a low level of 10^2 CFU/ml at the upper small intestine to 10^{12} CFU/ml at the lower large intestine. The composition of the microflora also changes from primarily alpha-hemolytic *Streptococcus* and *Lactobacillus* high in the intestinal tract to primarily *Bacteroides* and *Clostridium* low in the intestinal tract (Feingold, Attebery & Sutter, 1974; Savage, 1977; Simon & Gorbach, 1986). The microflora of the large intestine also consists of yeasts, spirochetes, protozoa, and over 400 species of bacteria (Savage, 1977). A total of 30–40 species of bacteria make up 99% of the microbial mass of the large intestine (Drasar & Roberts, 1990), with nonsporulating obligately anaerobic bacteria predominating. Numerically the most important genus of the intestinal bacteria is *Bacteroides* (Drasar & Roberts, 1990). *Clostridium* is ubiquitous in the gut, although not present in high numbers (Drasar & Roberts, 1990). The numerically important nonsporulating gram-positive organisms include *Bifidobacterium*, *Lactobacillus* and *Eubacterium* (Drasar & Roberts, 1990). Facultative *Streptococcus* species are also dominant among the intestinal microflora. Other prevalent bacteria are those of the genera *Fusobacterium*, *Peptostreptococcus* and *Bacillus*, and Enterobacteriaceae such as *Escherichia coli*. Less frequently isolated are anaerobic *Streptococcus*, *Ruminococcus*, *Veillonella*, *Klebsiella pneumoniae* and *Staphylococcus epidermidis* (Savage, 1977).

Vaginal ecosystem

The vagina is the final portion of the internal female genital and parturient canal. Its walls are lined with stratified squamous epithelium over a bed of connective tissue (Kistner, 1978) and bathed in a fine layer of moisture from vaginal secretions (Wagner & Levin, 1978). This multi-layered structure lacks hair follicles or glands (Kistner, 1978). Depending on the stage of the menstrual cycle, the pH of the vagina is mostly acidic, ranging from 4.0 to 6.0 (Onderdonk & Wissemann, 1993).

The microflora of the vaginal environment is present at a level of 10^8–10^9 organisms/g of secretion (Onderdonk et al., 1977; Lindner & Plantema, 1978; Bartlett & Polk, 1989) and consists predominantly of gram-positive organisms (Onderdonk & Wissemann, 1993). The vaginal microflora is a heterogeneous population of facultative and obligately anaerobic micro-organisms, with anaerobes outnumbering the facultative organisms 10:1 (Onderdonk *et al.*, 1977; Burgos & Roig de Vargas-Linares, 1978; Walz, Metzger & Ludwig, 1978; Bartlett & Polk, 1989). The numerically domi-nant, frequently isolated members of the vaginal microflora are species of *Lactobacillus, Corynebacterium* and *Streptococcus*; *Staphylococcus epidem-idis* and other coagulase-negative *Staphylococcus*; *Bacteroides, Eubacter-ium, Mycoplasma hominis* and *Peptostreptococcus*. Members frequently isolated, but in low concentrations, include species of *Micrococcus, Pro-pionibacterium* and *Veillonella*. Microorganisms isolated from only some women include species of *Clostridium, Ureoplasma urealyticum* and *Fuso-bacterium*. Rarely isolated organisms include *Staphylococcus aureus, Neisseria*, and *Gardnerella vaginalis* (Ohm & Galask, 1975; Bartlett *et al.*, 1977; Levinson *et al.*, 1977; Onderdonk *et al.*, 1977; Lindner & Plantema, 1978; Larsen & Galask, 1982; Hill, Eschenbach & Holmes, 1984; Bartlett & Polk, 1989). These do not include all organisms isolated from the vagina (Onderdonk & Wissemann, 1993).

The characteristics of the vaginal microflora change not only throughout the reproductive life of a woman (i.e. premenarchial, menarchial and post-menopausal stages), but also within the menstrual cycle. For an individual, the genotypes of anaerobic and aerobic bacteria and their concentrations can vary significantly during a single menstrual cycle (Bartlett *et al.*, 1977; Sauter & Brown, 1980; Larsen & Galask, 1982; Johnson, Petzold & Galask, 1985; Onderdonk *et al.*, 1986; Bartlett & Polk, 1989). The concentration of anaerobes remains constant throughout the menstrual cycle, but the total aerobic and facultative organism counts decrease 100-fold during the week preceding menstrual flow (Bartlett *et al.*, 1977). During the menstrual phase there is a decrease in the total bacterial count, while an increase in various individual anaerobic organisms is observed (Onderdonk *et al.*, 1986, 1987*a,b*). A decrease in *Lactobacillus* concentrations and an increase in other gram-positive organisms, such as *Staphylococcus, Corynebacterium* and *Streptococcus*, has also been reported to occur during menstrual flow (Sauter & Brown, 1980; Larsen & Galask, 1982; Onderdonk *et al.*, 1986). It has also been noted that a greater variety of phenotypes of facultative and obligately anaerobic organisms are isolated during the menstrual phase (Larsen & Galask, 1982; Johnson *et al.*, 1985).

Modeling host–microbe ecosystems

The goal of any model is to put knowledge into a more useful form. Models are essentially simplifications of reality. Modeling begins with the identification of the important components and interactions that define an ecosystem and their values. Any particular model must be complex enough to reflect the microbial ecosystem accurately, but not so complex that it becomes useless. An important factor to keep in mind when modeling ecosystems is that we do not know all the details of a system or their importance. We make assumptions (e.g. including only numerically dominant isolates in growth system models) with the hope that the smaller unknown interactions are included.

From the pool of data available describing a particular ecosystem, relevant information must be identified and organized in a manner suitable for the model being used. The skill is in knowing what information to exclude and how the components of the ecosystem are linked (Overton, 1990). Success lies in identifying the essence of the ecosystem and including it in the model.

Studies of the normal microflora of the human body tend to be akinetic cross sections of the specific microbial ecosystem. These types of studies fail to reflect the true complexity of the host–microbe ecosystem (Onderdonk & Wissemann, 1993). If the same location is sampled over a period of time, a dimension is added to the database generated. This added dimension is important in the modeling of the microflora ecosystems of the human body (Onderdonk & Wissemann, 1993).

The two major types of models that are used in modeling host–microbe ecosystems can be classified as analytic and simulation models. Analytic modeling is mathematical, whereas simulation modeling involves the development of an *in vitro* growth system. Both methods are directed at increasing our understanding of and ability to predict a microbial ecosystem.

Analytic modeling

Analytic or mathematical models generally involve the comparison of one group to another or attempt to predict outcomes based on previous observations. The various strategies for making comparisons or predictions are discussed in this section.

Comparing treatment effects

Continuous data

In experimental studies we often want to compare the effects of two or more treatments. When two treatments are applied to independent samples, a simple two-sample t-test can be used to test the difference of effects, if any. When more than two treatments are applied to independent samples, one can use the method of one-way analysis of variance to compare the differences.

When an experiment is designed to study effects classified by two characteristics, a two-way analysis of variance should be used to analyze the data.
For example, if we want to take into account the possibility of an interaction
of effects among four different pH values and three treatments, we can
design the experiment so that equal numbers of observations can be made
for all possible combinations of pH levels and treatments. The method of
two-way analysis of variance can be used to answer questions such as: is
there a difference due to pH levels?; is there a difference due to treatments?; and is there an interaction effect between pH and treatments?
The analysis of variance can be carried out by most statistical software.

Discrete ordinal data

When similar questions are addressed with the use of data of a discrete
and ordinal nature instead of a numerical nature, nonparametric methods
can be applied to test the differences between treatment effects.

Finding groups in data

A host–microbe ecosystem is a dynamic system involving many microbial
interactions. Cluster analysis is the art of finding groups of similar observations within a complex data set. The classification of similar objects into
groups can lead to important correlations that may not be apparent when
other analytic methods are used.

Factor analysis

When variables forming the data set are mostly continuous in nature, a
factor analysis can be used to calculate and extract principal factors that
account for most of the variation within the data set. The factor analysis
typically begins with a matrix of correlation coefficients between the data
variables that are being studied. Factor analysis provides a way of thinking about interrelationships by determining the existence of underlying
'factors' that account for the values appearing in the correlation matrix.
One common objective of factor analysis is to provide a relatively small
number of factor constructs that will serve as satisfactory substitutes for
a much larger number of variables. These factor constructs are linear
combinations of variables that may prove to be more useful than the
original variables from which they were derived.

Regression trees

When variables are mostly categorical and ordinal but not normally distributed, the method of the regression tree can be applied. Tree-based
modeling is an explanatory technique for uncovering structure in data. In
order to construct a regression tree, four steps are required:

(1) Form a region of interest, say region A, and then create a set of binary conditions of the form 'if predictor variables X satisfy a certain classification, then response Y belongs to region A'. Such a condition induces a split of the predictor space creating subsets of sample data. That is, the cases for which the answer is 'yes' are associated with response region A and those for which the answer is 'no' are associated with the complement of A; the subsets formed in this way are called nodes.

(2) Design a goodness-of-split criterion that can be evaluated for any split of any node t. The criterion used will help to assess the worth of the competing splits.

(3) Provide a means for determining the appropriate tree size.

(4) Establish statistical summaries for the terminal nodes of the selected tree. These can be as simple as calculating an average value, depending on the context of the study.

Regression trees can be computed using commercially available statistical programs such as CART (California Statistical Software, Inc.).

Regression analysis for longitudinal data

Longitudinal data often occur in clinical studies when repeated measurements are collected on the same subject over time. For example, vaginal swab samples may be obtained from women for several menstrual cycles (Onderdonk et al., 1986). In one particular study, at each cycle, swab samples were obtained on days 2, 4 and 21 after the start of menstrual flow. For each woman in the study, therefore, the repeated observations collected over time were highly correlated. This within-subject correlation must be accounted for in order to derive a proper statistical analysis.

For balanced data, the simplest way to analyze longitudinal observations is to calculate summary statistics for each subject and then analyze the summary statistics across subjects. Since data from different subjects are assumed to be independent, the summary statistics will be independent, and thus general methods such as t-test or Wilcoxon rank sum test can be applied in comparing results obtained from different groups.

However, in reality, there are many possible complications in obtaining data. One cannot always obtain balanced data. It is not always possible to obtain measurements on subjects at equally spaced intervals, because they may not return at prespecified times. Also, there can be the problem of missing observations for both outcome and covariate measurements.

In this section, three general methods for longitudinal regression models are considered. Assume that, except for missing values, repeated measurements are equally spaced in time and that measurements at n_i time points are taken from subject i. If, for any subject in the study, measurements taken at different time points follow a multivariate normal distribution, then longitudinal measurements can be completely modeled by the mean vector and the corresponding covariance matrix.

Let $Y = (Y_1, \ldots, Y_n)$ denote the observed longitudinal vector from subject i and $X = (X_1, \ldots, X_P)$ denote the corresponding time-dependent matrix of P covariates, where X_P is an n-dimensional vector that denotes the vector of n-repeated measurements taken over time on the Pth covariate from the i th subject. For any given subject i, pairs of repeated measurements (Y_{t_1}, Y_{t_2}), taken at time points t_1, and t_2, follow a bivariate normal distribution with means (μ_{t_1}, μ_{t_2}), standard deviations $(\sigma_{t_1}, \sigma_{t_2})$, and correlation $r_{1,2}$. In this section, three longitudinal regression models are reviewed that can be used to discuss the goodness-of-fit procedures.

Mixed-effects models

Random-effects models have been investigated by Laird & Ware (1982) and later modified by Jennrich & Schluchter (1986) and Lindstrom & Bates (1988). The following is a discussion of the program implemented by Lindstrom & Bates (1988). The mixed-effects model (MIXMOD) takes into account the heterogeneity of regression parameter estimates that may exist in the sample.

The mixed-effects regression model has the general form shown in Equation 2.1:

$$Y_t = (\beta_0 + b_0) + (\beta_1 + b_1) X_{1t} + \ldots + (\beta_P + b_P) X_{Pt} + e_t \quad (2.1)$$

where $(X_{1t}, \ldots, X_{Pt})$ is the observed covariate vector at time t; $\beta = (\beta_0, \beta_1, \ldots, \beta_P)$ the vector of fixed-effects population parameters; $b = (b_0, b_1, \ldots, b_P)$ the corresponding vector of random effects for individual deviation (b has the normal distribution with mean 0 and variance $\sigma_m^2 D$); and e_t is the error random variable distributed as normal with mean 0 and variance σ^2.

Ross *et al.* (1994) found that for their data set, containing repeated measurements for a subject, the MIXMOD was the best fit for predicting microbial interactions within the vaginal ecosystem.

Regression with damped exponential correlations

If pairwise correlations between variable measurements have a tendency to decrease as the time between measurements increases, then we can consider the regression model with damped exponential correlations (RDEC; Muñoz *et al.*, 1992).

For any subject, repeated measurements taken at time t_k and t_j have a multivariate normal distribution with mean $X\beta$ and variance–covariance matrix $\sigma^2 V$, where the matrix V is the correlation coefficient defined by Equation 2.2:

$$\text{Correlation } (\mathbf{Y}_{t_k}, \mathbf{Y}_{t_j}) = r^{|k-j|^\theta} \tag{2.2}$$

where $t_1 = 0$ is the time at first measurement, t_k is the number of time intervals from baseline to the kth follow-up, $k = 1, 2, \ldots$ n, t_n is the number of time intervals from baseline to the last follow-up, and θ is the damping parameter. One of the advantages of using the RDEC model is that it includes several well-known models as special cases: (a) when $\theta = 0$, it is equivalent to the compound symmetry model, i.e. all pairs of measurements at different times are assumed to have the same correlation; (b) when $\theta = 1$, it is equivalent to the first-order autoregressive model of order 1, i.e. Correlation $(\mathbf{Y}_{t_k}, \mathbf{Y}_{t_j}) = r^{|k-j|}$; (c) when $\theta = $ infinity, it is the first-order moving average model.

Predictive probabilities

The regression methods discussed above are applicable when the outcome variable is continuous and normally distributed. When the outcome variable for an experimental study is binary or dichotomous, either a logistic regression or a generalized estimation equation method can be applied.

Logistic regression

In a logistic regression equation, we can link the logit with explanatory regressor variables as shown in Equation 2.3, where p(t) is the probability that the specified event occurs at time t.

$$\log_e[\text{p}(t)/1 - \text{p}(t)] = \beta_0 + \beta_1 X_1 + \ldots + \beta_P X_P \tag{2.3}$$

That is, the logit is modeled as a linear function of the independent variables with regression parameters $\beta_0, \ldots, \beta_P$ and explanatory covariates $X_1, \ldots, X_P$.

Generalized estimation equation

When the data set is longitudinal in nature, the general logistic regression methods do not apply, since they require the assumption of independent observations. Using the methodology of quasi-likelihood inference, Liang & Zeger (1986) introduced a model based on generalized estimating equations (GEE). To use the GEE model, one need only specify the relationships between the outcome mean and covariates, and between the mean and variance. The expected value of $\mathbf{Y}$, E[$\mathbf{Y}$], is modeled by E[$\mathbf{Y}$] = $h(\mathbf{X}\beta)$, where the inverse of h is the link function, and the variance of $\mathbf{Y}$, Var[$\mathbf{Y}$], is modeled by Var[$\mathbf{Y}$] = $g(\text{E}[\mathbf{Y}])$, where g is a known function. The GEE method can be used in longitudinal data analysis for discrete and continous outcomes and for both Gaussian and non-Gaussian data.

Lee *et al.* (1994) investigated a predictive logit model to identify abnormal microbial population levels in the vaginal ecosystem using the GEE method. They expressed the logit of having abnormal microflora concentrations at time *t* as a linear function of the corresponding pH, *Staphylococcus* concentration, *Lactobacillus* concentration, total aerobic bacterial concentration and *Prevotella* concentration.

Simulation modeling

The microfloras of human microbial ecosystems significantly affect the host's general state of health by acting as a biological barrier to infectious agents. Interactions of the microflora with its environment and interactions among the members of the microflora define the status of this barrier. By examining these relationships, one can begin to determine which factors contribute to an increased risk of disease and which contribute to the prevention of disease or disease transmission. Identification of alterations in the microflora that lead to barrier disruptions would make it possible to optimize treatments for a particular infectious process or to prevent such infection altogether. To conduct studies that determine the effectiveness of various techniques for preventing disease and disease transmission, it is necessary to develop an *in vitro* model that uses a growth system simulating the *in vivo* environment.

A number of strategies are useful in the successful modeling of human microbial ecosystems. First, it is important to use a chemically defined medium that simulates the growth milieu, since components of the growth medium could be very important with regard to interactions of the microflora. The microflora selected for use as an *in vitro* model should be selected on the basis of *in vivo* data with the results of *in vivo* studies used to guide microorganism choice. Mixed cultures of microorganisms are more useful and realistic than are pure cultures; microorganisms should originate from *in vivo* isolation. Finally, the selection of a growth system should be based on evaluation of the growth characteristics of each system that are important in the ecosystem to be modeled. An open, flowing ecosystem should not be modeled with a closed growth system such as batch culture. In developing an *in vitro* model, it is necessary to use a growth system that allows for the cultivation of the microflora under controlled conditions and thus allows for the manipulation of growth parameters.

Batch culture

Most *in vitro* studies of host–microbe ecosystems employed for the study of microbial colonization use batch culture techniques, mainly because of the convenience of such techniques. Such systems are closed to the addition of nutrients and microorganisms, which results in a continual change in cell numbers and the accumulation of metabolic products and, most important, bear no resemblance to any host–microbe ecosystems in living human or animal hosts, which are characteristically open systems with nutrients entering and products of metabolism and cells leaving (Chesbro, Arbige & Eifert, 1990). For these reasons, more sophisticated methods are necessary for the study of microbial colonization.

Fed-batch culture

Fed-batch culturing is similar to batch culturing in that it is a closed system with regard to biomass, but it is open to the addition of nutrients (Chesbro et al., 1990). Therefore, the volume of a fed-batch culture continually increases, which effectively extends the growth phase of the culture while a partially closed system is maintained. The rate of growth of microorganisms in this system is constantly decreasing (Weusthuis et al., 1994). As is the case in batch culture, by-products of metabolism accumulate, which is most probably not the case for in vivo systems using living hosts. Ultimately, the accumulation of biomass in a fed-batch culture leads to the disruption of maintained growth conditions, usually within hours of experiment initiation (Weusthuis et al., 1994).

Cell-recycle culture

Cell-recycle culture is similar to fed-batch culture in that it is a nutrient-open, biomass-closed growth system (Chesbro et al., 1990). The major difference between the two types of culture systems is that in cell recycle, the volume of the culture remains constant. Cells are retained through filtration and recycled into the culture vessel, and only spent medium is removed from the system. As with the fed-batch culture technique, the growth rate of the microflora in cell-recycle culture systems is continually slowing as cell concentration increases, while the feeding of fresh nutrients remains constant, which results in the concentration of nutrients available to each cell decreasing as the cell concentration increases. Thus, the culture is constantly changing.

Continuous culture

Continuous culture is an open system for cells, nutrients and any metabolic products (Chesbro et al., 1990). Nutrients are continuously fed into the culture while spent medium and cells are continuously removed. No change in culture volume occurs. It is also possible to select and maintain the growth rate of the microflora in this type of growth system, which is not possible in the other systems mentioned above. A limitation of this system is that the range of growth rates that produce a homogeneous population of cells is restricted to doubling times of approximately 14 h or faster.

The strongest feature of the continuous culture growth system is its ability to maintain constant microbial growth conditions. The microflora growing in this system constantly exhibits a physiological response to the environment of the model. After an initial period of adjustment, the microflora enters a steady state of growth and metabolic activity. The concentration of nutrients in the medium likewise remains constant. The combination of these factors results in a highly reproducible system.

Because constant growth conditions can be maintained, continuous culture allows the effects of specific environmental parameters on an ecosystem's microflora to be evaluated. All parameters can be held constant while nutrients and a single compound of interest are added to the growth system. Thus, only the effect of the added compound is evaluated at any given time, independent of a change in growth rate, which in other types of culture systems could be affected by the addition of the compound. In the other growth systems this maintenance of growth rate is not possible.

Summary

The power of a growth system model and a mathematical model to describe microbial interactions in a human ecosystem is greatest when the two types of models are combined. The combination of these two approaches allows for the data generated by the growth system model to be related to the *in vivo* situation. Bacteriological measurements indicating changes in the concentration of different types of microorganisms can then be further analyzed with the use of predictive mathematical formulas developed from *in vivo* data to determine the impact of certain manipulations (e.g. drug treatment) on the health of the ecosystem and whether this manipulation may increase or decrease the risk of successful disease transmission.

In order to understand disease transmission and prevention, it is necessary to understand how the normal microflora of the human body functions. Models of microbial ecosystems are developed to explain and predict the behavior of the system as a whole by quantitatively describing interactions between parts of the ecosystem (Shoemaker, 1990). It is understood that such models cannot be complete and must ignore some of the information that is a part of the system of interest. The ultimate goal is to take a complex system such as a host–microbe ecosystem of the human body and reduce and refine the *in vivo* data into a form that portrays the essential features of the system.

68   R. A. ROSS AND M.-L. T. LEE

References

Bartlett, J. G., Onderdonk, A. B., Drude, E., Goldstein, C., Anderka, M., Alpert, S. & McCormack, W. M. (1977). Quantitative bacteriology of the vaginal flora. *Journal of Infectious Diseases*, **136**, 271–7.

Bartlett, J. G. & Polk, B. F. (1989). Normal vaginal flora in relation to vaginitis. *Obstetrics and Gynecology Clinics of North America*, **16**, 329–36.

Behrendt, H. & Green, M. (1971). *Patterns of Skin pH From Birth Through Adolescence*. Springfield: Thomas.

Bhatia, B. D., Chug, S., Narang, P. & Singh, M. N. (1988). Bacterial flora of newborns at birth and 72 hours of age. *Indian Pediatrics*, **25**, 1058–65.

Bollard, J. E., Vanderwee, M. A., Smith, G. W., Tasman-Jones, C., Gavin, J. B. & Lee, S. P. (1986). Location of bacteria in the mid-colon of the rat. *Applied and Environmental Microbiology*, **51**, 604–8.

Bowden, G. H., Ellwood, D. C. & Hamilton, I. R. (1979). Microbial ecology of the oral cavity. In *Advances in Microbial Ecology*, vol. 3, ed. M. Alexander, pp. 135–77. New York: Plenum Press.

Burgos, M. H. & Roig de Vargas-Linares, C. E. (1978). Ultrastructure of the vaginal mucosa. In *Human Reproductive Medicine: The Human Vagina*, vol. 2, ed. E. S. Hafez & T. N. Evans, pp. 63–93. Amsterdam: Elsevier/North Holland Biomedical Press.

Chesbro, W., Arbige, M. & Eifert, R. (1990). When nutrient limitation places bacteria in the domains of slow growth: metabolic, morphologic and cell cycle behavior. *FEMS Microbiology and Ecology*, **74**, 103–20.

Croucher, S. C., Houston, A. P., Bayliss, C. E. & Turner, R. J. (1983). Bacterial populations associated with different regions of the human colon wall. *Applied and Environmental Microbiology*, **45**, 1025–33.

Drasar, B. S. & Roberts, A. K. (1990). Control of the large bowel microflora. In *Human Microbial Ecology*, ed. M. J. Hill & P. D. Marsh, pp. 87–110. Boca Raton: CRC Press.

Dubois, E. F. (1939). Heat loss from the body. *Bulletin of the New York Academy of Medicine*, **15**, 143–73.

Evans, C. A. (1975). Persistent individual differences in bacterial flora of the skin of the forehead: numbers of propionibacteria. *Journal of Investigative Dermatology*, **64**, 42–6.

Evans, C. A. & Strom, M. S. (1982). Eight year persistence of individual differences in the bacterial flora of the forehead. *Journal of Investigative Dermatology*, **79**, 51–2.

Feingold, S. M., Attebery, H. R. & Sutter, V. L. (1974). Effect of diet on human fecal flora: comparison of Japanese and American diets. *American Journal Clinical Nutrition*, **27**, 1456–69.

Hill, G. B., Eschenbach, D. A. & Holmes, K. K. (1984). Bacteriology of the vagina. *Scandinavian Journal of Urology and Nephrology* (Supplement), **86**, 23–39.

Holbrook, K. A. & Wolff, K. (1987). Structure and development of skin. In *Dermatology in General Medicine*, 3rd edn, ed. T. B. Fitzpatrick, A. Z. Eisen, K. Wolff, I. M. Feedberg & K. F. Austen, pp. 93–130. New York: McGraw-Hill.

Jennrich, R. I. & Schluchter, M. D. (1986). Unbalanced repeated-measures models with structured covariance matrices. *Biometrics*, **42**, 805–20.

Johnson, S. R., Petzold, C. R. & Galask, R. P. (1985). Qualitative and quantitative changes of the vaginal microbial flora during the menstrual cycle. *American Journal of Reproductive Immunology and Microbiology*, **9**, 1–5.

Kistner, R. W. (1978). Physiology of the vagina. In *Human Reproductive Medicine: The Human Vagina*, vol. 2, ed. E. S. Hafez & T. N. Evans, pp. 109–20. Amsterdam: Elsevier/North Holland Biomedical Press.

Kloos, W. E. (1981). The identification of *Staphylococcus* and *Micrococcus* species isolated from human skin. In *Skin Microbiology: Relevance to Clinical Infection*, ed. H. Maibach & R. Aly, pp. 3–12. New York: Springer-Verlag.

Kloos, W. E. & Musselwhite, M. S. (1975). Distribution and persistence of *Staphylococcus* and *Micrococcus* species and other anaerobic bacteria on human skin. *Applied Microbiology*, **30**, 381–95.

Laird, N. M. & Ware, J. H. (1982). Random-effects models for longitudinal data. *Biometrics*, **38**, 963–74.

Larsen, B. & Galask, R. P. (1982). Vaginal microbial flora: composition and influences of host physiology. *Annals of Internal Medicine*, **92**, 926–30.

Lee, M.-L. T., Ross, R. A., Delaney, M. L. & Onderdonk, A. B. (1994). Predicting abnormal microbial population levels in the vaginal ecosystem. *Microbial Ecology in Health and Disease*, **7**, 235–40.

Levinson, M. E., Corman, L. C., Carrington, E. R. & Kaye, D. (1977). Quantitative microflora of the vagina. *American Journal of Obstetrics and Gynecology*, **127**, 80–5.

Leyden, J. J. (1982). Bacteriology of newborn skin. In *Neonatal Skin: Structure and Function*, ed. H. I. Maibach & T. V. Boisits, pp. 167–81. New York: Marcel Dekker.

Liang, K.-Y. & Zeger, S. L. (1986). Longitudinal data analysis using generalized linear models. *Biometrika*, **73**, 13–22.

Lindner, J. G. E. M. & Plantema, F. H. F. (1978). Quantitative studies of the vaginal flora of healthy women and of obstetric and gynecological patients. *The Journal of Medical Microbiology*, **11**, 233–41.

Lindstrom, M. J. and Bates, D. M. (1988). Newton-Raphson and EM algorithms for linear mixed-effects models for repeated-measures data. *Journal of The American Statistical Association*, **83**, 1014–22.

Marsh, P. D. (1991). The significance of maintaining the stability of the natural microflora of the mouth. *British Dental Journal*, **171**, 174–7.

Marsh, P. D. & Martin, M. V. (1982). *Oral Microbiology*, 2nd edn. Wokingham: van Nostrand Rheinhold.

McBride, M. E. (1993). Physical factors affecting skin flora and disease. In *The Skin Microflora and Skin Microbial Skin Diseases*, ed. W. C. Noble, pp. 73–101. Cambridge: Cambridge University Press.

McBride, M. E., Duncan, W. C. & Knox, J. M. (1977). The environment and the microbial ecology of the human skin. *Applied and Environmental Microbiology*, **33**, 603–8.

McGinley, K. J., Webster, G. F. & Leyden J. J. (1978). Regional variations of cutaneous propionibacteria. *Applied and Environmental Microbiology*, **35**, 62–6.

Milton-Thompson, G. J., Lightfoot, N. F., Ahmet, Z., Hunt, R. H., Barnard, J., Bavin, P. M., Brimblecombe, R. W., Darkin, D. W., Moore, P. J. & Viney, N.

(1982). Intragastric acidity, bacteria, nitrate, and N-nitroso compounds before, during, and after treatment with cimetidine. *Lancet*, **1**, 1091–5.

Mok, W. Y. & Barreto da Silva, M. S. (1984). Mycoflora of the human dermal surfaces. *Canadian Journal of Microbiology*, **30**, 1205–9.

Montagna, W. & Parakaal, P. F. (1974). *The Structure and Function of Skin*. New York: Academic Press.

Montes, L. F. & Wilborn, W. H. (1969). Location of bacterial skin flora. *British Journal of Dermatology*, **81** (Supplement 1), 23–6.

Muñoz, A., Carey, V., Schouten, J., Segal, M. & Rosner, B. (1992). A parametric family of correlation structures for the analysis of longitudinal data. *Biometrics*, **48**, 733–42.

Nisengard, R. J. & Newman, M. G. (1994). *Oral Microbiology and Immunity*, 2nd edn. Philadelphia: W. B. Saunders Company.

Noble, W. C. (1981). *Microbiology of Human Skin*, 2nd edn, pp. 339–57. London: Lloyd-Luke.

Noble, W. C. & Pitcher, D. G. (1978). Microbial ecology of human skin. In *Advances in Microbial Ecology*, vol. 2, ed. M. Alexander, pp. 245–89. New York: Plenum Press.

Ohm, M. J. & Galask, R. P. (1975). Bacterial flora of the cervix from 100 pre-hysterectomy patients. *American Journal of Obstetrics and Gynecology*, **122**, 683–7.

Onderdonk, A. B., Polk, B. F., Moon, N. E., Goren, B. & Bartlett, J. G. (1977). Methods for quantitative vaginal flora studies. *American Journal of Obstetrics and Gynecology*, **128**, 777–81.

Onderdonk, A. B. & Wissemann, K. W. (1993). Normal vaginal microflora. In *Vulvovaginitis*, ed. P. Elsner & J. Martius, pp. 285–304. New York: Marcel Dekker.

Onderdonk, A. B., Zamarchi, G. R., Rodriguez, M. L., Hirsch, M. L., Munoz, A. & Kass, E. H. (1987a). Quantitative assessment of vaginal microflora during use of tampons of various compositions. *Applied and Environmental Microbiology*, **53**, 2774–8.

Onderdonk, A. B., Zamarchi, G. R., Rodriguez, M. L., Hirsch, M. L., Munoz, A. & Kass, E. H. (1987b). Qualitative assessment of vaginal microflora during use of tampons of various compositions. *Applied and Environmental Microbiology*, **53**, 2779–84.

Onderdonk, A. B., Zamarchi, G. R., Walsh, J. A., Mellor, R. D., Munoz, A. & Kass, E. H. (1986). Methods for quantitative and qualitative evaluation of vaginal microflora during menstruation. *Applied and Environmental Microbiology*, **51**, 333–9.

Overton, W. S. (1990). A strategy of model construction. In *Ecosystem Modeling in Theory and Practice: An Introduction With Case Histories*, ed. C. A. S. Hall & J. Day, pp. 49–58. New York: Wiley.

Pitcher, D. G. (1981). *Corynebacterium* and related genera of the normal human skin. *Hautarzt*, **32** (Supplement V), 273–5.

Price, P. B. (1938). The bacteriology of the normal skin: a new quantitative test applied to a study of the bacterial flora and disinfection action of mechanical cleansing. *Journal of Infectious Diseases*, **63**, 301–18.

Retailliau, H. F., Hightower, A. W., Dixon, R. E. & Allen, J. R. (1977). *Acineto-*

bacter calcoaceticus: a nosocomial pathogen with an unusual seasonal pattern. *Journal of Infectious Diseases*, **139**, 371–5.

Ross, R. A., Lee, M.-L. T., Delaney, M. L., Onderdonk, A. B. (1994). Mixed-effect models for predicting interactions in the vaginal ecosystem. *Journal of Clinical Microbiology*, **32**, 871–5.

Sauter, R. L. & Brown, W. J. (1980). Sequential vaginal cultures from normal young women. *Journal of Clinical Microbiology*, **11**, 479–84.

Savage, D. C. (1977). Microbial ecology of the gastrointestinal tract. *Annual Review of Microbiology*, **31**, 107–33.

Savage, D. C., McAllister, J. S. & Davis, C. P. (1971). Anaerobic bacteria on the mucosal epithelium of the murine large bowel. *Infection and Immunity*, **4**, 492–502.

Sengel, P. (1985). Role of extracellular matrix with development of skin and cutaneous appendages. *Progress in Clinical Biological Research*, **171**, 123–35.

Shoemaker, C. A. (1990). Mathematical construction of ecological models. In *Ecosystem Modeling in Theory and Practice: An Introduction With Case Histories*, ed. C. A. S. Hall & J. W. Day, pp. 75–89. New York: Wiley.

Simon, G. L. & Gorbach, S. L. (1986). The human intestinal microflora. *Digestive Diseases and Sciences*, **31**, 147s-162s.

Theilade, E. (1990). Factors controlling the microflora of the healthy mouth. In *Human Microbial Ecology*, ed. M. J. Hill & P. D. Marsh, pp. 1–56. Boca Raton: CRC Press.

Updegraff, D. M. (1964). A cultural method of quantitatively studying microorganisms on skin. *Journal of Investigative Dermatology*, **43**, 129–37.

Wagner, G. & Levin, R. J. (1978). Vaginal fluid. In *Human Reproductive Medicine: The Human Vagina*, vol. 2, ed. E. S. Hafez & T. N. Evans, pp. 121–37. Amsterdam: Elsevier/North Holland Biomedical Press.

Walz, K. A., Metzger, H. & Ludwig, H. (1978). Surface ultrastructure of the vagina. In *Human Reproductive Medicine: The Human Vagina*, vol. 2, ed. E. S. Hafez & T. N. Evans, pp. 55–61. Amsterdam: Elsevier/North Holland Biomedical Press.

Weusthuis, R. A., Pronk, J. T., van der Broek, P. J. A. & van Dijken, J. P. (1994). Chemostat cultivation as a tool for studies on sugar transport in yeasts. *Microbiological Reviews*, **58**, 616–30.

PART 2 PREVENTING DISEASE TRANSMISSION BY WATER AND FOOD

The role of pathogen monitoring in microbial risk assessment

JOAN B. ROSE, JOHN T. LISLE and
CHARLES N. HAAS

Methods for risk assessment

Introduction

There has always been concern with the health risks associated with pathogenic microorganisms in drinking water. Waterborne disease continues to occur in the United States and other developed countries. From 1971 to 1988, 564 outbreaks in the United States involving 138 247 persons were documented (Craun, 1991). It is widely believed that many more diseases and cases result from non-reported infections. The etiological agents most commonly associated with waterborne disease in the United States include, in descending order, undefined cause of gastroenteritis, *Giardia*, *Shigella*, virus causing gastroenteritis and hepatitis A. However, since 1985, seven outbreaks of *Cryptosporidium* have been reported in the USA and this organism has been associated with the largest waterborne outbreak ever documented in the United States, in Milwaukee in 1993 (MacKenzie, 1994). The list of pathogenic microorganisms capable of transmission by the water route is well over 100. There are now over 120 different enteric viruses alone which can be transmitted by fecally contaminated water. In recent years protozoa have become of particular concern, due to the resistant nature of the cyst or oocyst that is produced as a result of the infection. *Giardia* is now the most common cause of waterborne disease outbreaks in the United States when an agent can be identified (Craun, 1991). However, *Cryptosporidium*, although causing fewer outbreaks, has caused several large outbreaks resulting in thousands of cases of disease and deaths in the immunocompromised population from consumption of conventionally treated water (i.e. flocculation, filtration and disinfection) (Smith & Rose, 1990).

Steahr & Roberts (1993) have evaluated data from medical discharge certificates (patients discharged from hospitals) by category of enteric infections of foodborne or waterborne diseases using the codes of the International Classification of Disease, 9th Revision, Clinical Modification (ICD-9-CM). Costs were estimated by the averaged length of stay in the hospital and the 1990 national average cost per day of $687. The numbers of cases per year over a four-year period ranged from 5344 for *Shigella* to 530 689 for unspecified acute gastroenteritis, at costs between $16 million and $2 billion. Various estimates of the percentage of these infections that may be waterborne have been made, ranging from 4% to as much as 60% (Bennett, Homberg & Rogers, 1987). These types of evaluations suggest that contaminated water may be contributing to a significant healthcare burden in the United States that has yet to be sufficiently recognized or assessed.

Risk assessment has become a valuable tool for evaluating a variety of health hazards associated with food and water. Risk estimation can provide a useful tool for decision makers in the development of standards and treatment requirements, for risk management, as well as risk and benefit analysis. However, the risk assessment approach has been used on only a limited scale for judging the risks associated with waterborne pathogenic microorganisms (Haas, 1983; Rose, Haas & Regli, 1991*b*; Regli *et al.*, 1991; Haas *et al.*, 1993). Such a strategy is needed and could be used within the water industry to meet regulatory requirements under the Safe Drinking Water Act, Water Pollution Control Act and other acts of regulatory needs that deal with the treatment of microbial contaminants.

Hazard identification

Although there are many potential waterborne bacteria, *Salmonella typhi* and *Vibrio cholerae* have been the most important two historically. Typhoid fever in the 1920s in the United States was responsible for the most cases and deaths associated with waterborne outbreaks (Craun, 1986). Currently, only a few hundred typhoid cases are reported in the United States each year. While cholera remains a prominent cause of waterborne disease in South America, fewer than 25–50 cases are reported each year in the USA, most of which are imported (Centres for Disease Control, 1990; Herwaldt, Craun & Stokes, 1992; Moore *et al.*, 1993). However, enteric bacteria are still associated with 12% of the waterborne outbreaks (Craun, 1991), and may be responsible for more hospitalizations, severe sequelae and deaths than any other group of waterborne agents (Gerba, Rose & Haas, 1995).

There are at least four genera of bacteria that warrant closer scrutiny, *Campylobacter*, *Escherichia coli*, *Salmonella* non-*typhi* and *Shigella* (Table 3.1). These bacteria cause as many as 4.5 million cases of disease and 5100 deaths each year; it has been estimated that anywhere from 3 to 75% of

Table 3.1. *Waterborne bacterial agents of concern*

Bacteria	Annual reported cases in the USA	Percentage waterborne[a]	Mortality rate (%)[b]
Campylobacter	8 400 000	15	0.1
Pathogenic *E. coli*	2 000 000	75	0.2
Salmonella non-typhi	10 000 000	3	0.1
Shigella	666 667	10	0.2
Yersinia	5025	35	0.05

[a] Based on estimates by the Centers for Disease Control (CDC) (Bennet *et al.*, 1987)
[b] Based on total cases and deaths reported annually to the CDC.

these are waterborne (Bennett *et al.*, 1987). Various studies have reported disease prevalence in United States populations from a low of 0.005% for *Campylobacter* to a high of 32% for *Salmonella*. Multiplying this by the United States population of 242 million suggests that there are large numbers of infected individuals.

Campylobacter, Escherichia coli and *Salmonella* have animal reservoirs, but *Shigella* is associated only with humans. In addition, the bacterium *E. coli*, generally a non-pathogen, includes several strains (enterotoxigenic, hemorrhagic) which are now associated with human disease. There are several other important bacteria, such as enteropathogenic *Aeromonas*, which has been found to cause as much as 10% of the diarrhea in Australia, and *Yersinia*, which causes a few thousand cases each year in the USA.

Morbidity and the severity of disease are perhaps more varied for the bacteria than for any other group of microorganisms. A study of *Salmonella* outbreaks reported that the numbers of infected individuals with resulting illness may range from a low of 6% to a high of 80% (Chalker & Blaser, 1988). The type of illness may be mild diarrhea lasting for a few days or severe gastrointestinal illness.

Hospitalizations during outbreaks have been documented and were used in this context to define the severity rate for the microorganisms as the ratio of hospitalized cases to total number of cases during waterborne outbreaks in the USA. The data are shown in Table 3.2 (Rose, Haas & Gerba, 1996). These levels were 3.4, 4.1, 12.7, 5.9 and 19.4% for *Campylobacter, Salmonella, E. coli, Shigella* and *Yersinia*, respectively. It may be for some of these agents that only the more virulent strains cause an outbreak and therefore are associated with a greater severity rate. *S. typhi*, which causes typhoid fever, for example, had an 84.4% severity rate associated with four recorded outbreaks since 1976 (Rose *et al.*, 1996).

Other issues of interest include the potential for chronic or more serious sequelae, which can occur even without apparent infections and without

Table 3.2. *Severity of disease associated with microorganisms responsible for water-borne outbreaks reported in the United States, 1971–1992*

Etiological agent	Out-breaks[a]	Hospitalizations		Mortality rate[c] (%)
		Ratio[b]	Rate (%)	
Hepatitis A	9	75/265	28.3	0.3
Viral gastroenteritis	4	10/1154	0.9	0.1–0.0001
Salmonella	3	12/293	4.1	0.1
Shigella	24	339/5768	5.9	0.2
Campylobacter jejuni	5	73/2152	3.4	0.1
Yersinia enterocolytica	2	20/103	19.4	0.05
E. coli[d]	2	41/323	12.7	0.2
Typhoid	4	235/277	84.8	6.0
Acute gastrointestinal illness of unknown etiology	50	253/40 039	0.6	

[a]Numbers of outbreaks with hospitalization data.
[b]Numbers of cases hospitalized per number of cases in outbreaks with hospitalization data.
[c]Bennet *et al.* (1987)
[d]Primarily in sensitive populations, very low mortality in non-sensitive populations.

the development of acute disease. It has now been demonstrated that *Campylobacter*, *Shigella*, *Yersinia* and *Salmonella* are associated with reactive arthritides in 2.3% and Reiter's syndrome in 0.23% of cases (Smith *et al.*, 1993). *Campylobacter* has also been associated with Guillain–Barré disease, which causes an acute neuromuscular paralysis; 20–40% of the patients with Guillain–Barré disease had *Campylobacter* infections. In at least one study the morbidity ratio for Guillain–Barré disease was 0.17% associated with the bacterial infection (Mishu & Blaser, 1993).

There are several hundred enteric viruses which are possibly important agents of waterborne disease. However, there is limited information regarding the incidence of virus infections in United States populations and the role of contaminated water in acquiring these. Bennett *et al.* (1987) have reported 20 million cases of enteric viral infections a year and 2010 deaths (Table 3.3). Adenoviruses, which may be transmitted by the respiratory route as well, account for 10 million cases and 1000 deaths, making this the most significant virus affecting United States populations. Rotavirus cases were documented as the second most common virus infection.

In a six-year retrospective study of data from six electron microscopy centers, rotavirus was found in 41–84% (48% for all specimens) of the fecal samples screened, adenoviruses were found in 8–27% (17%) and small round viruses were found in 0–40% (10%) (Lew *et al.*, 1990). More than 95% of the individuals submitting specimens had done so because of

Table 3.3. *Characteristics of enteric viruses*[a,b]

Virus group	Diseases	Incidence[c]	Mortality rate (%)	Morbidity rate (%)
Enterovirus		6 000 000	0.001	
Poliovirus	Paralysis	7	0.90	0.1–1.0
	Aseptic meningitis			
Coxsackievirus A	Aseptic meningitis		0.50	50
	Respiratory illness			
Coxsackievirus B	Paralysis fever		0.59–0.94	
	Pleurodynia			
	Aseptic meningitis			
	Pericarditis			
	Myocarditis			
	Congenital heart anomalies			
Echovirus	Respiratory infection			50
	Aseptic meningitis			
	Diarrhea			
	Pericarditis			
	Myocarditis			
	Fever and rash			
Hepatitis A	Infectious hepatitis	48 000	0.6	75
Adenovirus	Acute conjunctivitis	10 000 000	0.01	
	Diarrhea			
	Respiratory illness			
	Eye infections			
Rotavirus	Infantile gastroenteritis	8 000 000	0.01	56–60
Norwalk agent	Gastroenteritis	6 000 000	0.0001	40–59

[a]Bennet *et al.* (1987)
[b]Gerba & Rose (1990).
[c]Cases reported to the Centers for Disease Control during 1985.

gastrointestinal symptoms. These viruses were found 53% of the time in infants less than one year of age, but only 9% were found in children aged five years and older and adults. However, because young children are more likely to be brought in for diagnosis when they are sick, these data may not reflect the trends in adult infections.

The non-polio enteroviruses are a group of agents that are identified according to serotypes. Surveillance of reported isolates in the USA has found that from March to May (Centers for Disease Control and Prevention, 1993), echovirus 30, coxsackievirus A9 and coxsackievirus B2 accounted for 25, 10 and 10%, respectively and a majority of the identifiable virus isolates (62%). The most common virus among all isolates throughout the year was coxsackievirus B5, accounting for 21% (Centers for Disease Control and Prevention, 1993).

In a nine-year study of fatal viral infections from selected laboratories throughout the world, 78–91% were documented in children (Assaad & Borecka, 1977). Mortality rates were 0.5% for adenoviruses, and for coxsackievirus A and B, echovirus and poliovirus were 0.34, 0.76, 0.28 and 0.9%, respectively. This was prior to the discovery and description of the rotavirus. Currently with widespread vaccination for poliovirus, only a few cases of that disease are reported each year (seven cases and one death in 1985 in the USA). Although this data set is antiquated, it does point to the relative significance of the adenoviruses and non-polio enteroviruses as a cause of mortality. The data here come from every continent; however, between 12 and 27% of the fatalities were reported from the USA and Canada, and between 29 and 61% of the fatalities came from Europe and Israel. Therefore, the contribution to the mortality rates from developing countries in Africa and Asia was minimal.

Hepatitis A virus may be one of the more significant causes of viral mortality according to United States statistics from the Centers for Disease Control, yet better diagnosis of coxsackievirus infections associated with myocarditis is needed (Klingel et al., 1992). In recent studies, enteroviral RNA was detected in endomyocardial biopsies in 32% of the patients with dilated cardiomyopathy and 33% of patients with clinical myocarditis (Kiode et al., 1992). In addition, there is emerging evidence that coxsackievirus B is associated with insulin-dependent diabetes, and this infection may contribute to an increase of 0.0079% in insulin-dependent diabetes (Wagenknecht, Roseman & Herman, 1991).

Bennett et al. (1987) estimated that 5% of all Norwalk virus cases were associated with contaminated water. Severity rates during outbreaks for Norwalk virus were low (0.88%) compared to bacteria and hepatitis A virus (25%) (Table 3.2). It is unclear whether other viruses may be associated with greater severity. Those unknown etiological agents in the acute gastrointestinal illness category had severity rates ranging from 0.02 to 25%. These data along with information on hospitalization and attack rates suggest that the unidentified category is a mixture of viral, protozoan and bacterial agents.

The enteric protozoa have a worldwide distribution. Entamoeba and Giardia reportedly have similar global infection rates of about 10% (Feachem et al., 1983). Entamoeba infections are particularly prevalent in areas without sewerage and with poor hygienic conditions and may be as high as 72%. Although chronic asymptomatic infections may occur, serious diseases such as liver abscess are related to the duration of the Entamoeba infection. Of all the protozoa, Entamoeba has the greatest risk of mortality, which ranges from 0.02 to 6% depending on the virulence of the isolate (Gitler & Mirelman, 1986).

Like Entamoeba, Giardia can produce asymptomatic infections. Approximately 60% of all infections may be without any associated diarrheal illness. Chronic (1–2 month) intermittent diarrhea can develop and must be diagnosed and treated in order to resolve this disease pattern. Unlike *Entamoeba* infection, giardiasis is rarely associated with any mortality.

Cryptosporidium was first diagnosed in humans in 1976. Since that time, it has been well recognized as a cause of diarrhea (Dubey, Speer & Fayer 1990). Reported incidence of *Cryptosporidium* infections in the population ranged from 0.6 to 20%, depending on the geographic locale, with greater prevalence among populations in Asia, Australia, Africa and South America. Serum antibody prevalence has demonstrated much greater exposure with between 17 and 20% of the study populations in the USA and UK positive, while in Australia and South America it was much higher (86 and 64% antibody prevalence, respectively; Fayer & Ungar, 1986; Ungar, 1988). It has now been shown that *Cryptosporidium* is the third most common enteropathogen worldwide, particularly in young children (Current & Garcia, 1991).

Cryptosporidium may cause 7–38% of the diarrhea in immunocompromised patients (Selik, Starcher & Curran, 1987). The disease outcome is much more severe in the immunocompromised, particularly in AIDS patients. Mortality rates in these individuals may be as great as 50% (Navin & Juranek, 1984). Currently there is no treatment for the disease.

Cryptosporidium and *Giardia* are the most significant causes of waterborne disease in the USA today. The Centers for Disease Control (CDC) have estimated that 60% of all giardiasis cases are associated with contaminated water (Bennett *et al.*, 1987). It is also interesting to note that, during surveillances for cryptosporidiosis cases in Oregon and New Mexico, increases in endemic levels were associated with water (Gallaher *et al.*, 1989; Leland *et al.*, 1993). Severity rates during waterborne outbreaks of giardiasis averaged 1.7% except in two situations where there were over 1000 cases reported and the severity rates were 0.17 and 0.2. If these two outbreaks are included in the average the rate drops to 0.45% (Table 3.4). For *Cryptosporidium* in the outbreak in Texas, the severity rate was 0.85% (D'Antonio *et al.*, 1985). In the Carrollton, GA outbreak, although no hospitalizations were documented in the literature, the numbers of visits to the hospital emergency room increased 5–6-fold, giving approximately a 0.80% rate (emergency room visits/case) (Hayes, Matte & O'Brien, 1989). In the Milwaukee outbreak the severity rate was slightly higher at 1.1% (MacKenzie, 1994; Table 3.4).

Dose–response modeling

Haas (1983) was the first to look quantitatively at microbial risks associated with drinking waters. He had examined models that could best estimate

Table 3.4. *Severity rates associated with hospitalizations for illness caused by enteric protozoans during waterborne outbreaks*

Etiological agent	Outbreaks[a]	Hospitalizations		Mortality rate[c] (%)
		Ratio[b]	Rate (%)	
Giardia lamblia	18	60/13 239	0.45	0.0001
Cryptosporidium	3			50.0[d]
In Texas		1/118	0.85	
In Carrollton		104/13 000	0.80[e]	
In Milwaukee		4400/403 000	1.1	

[a]Numbers of outbreaks with hospitalization data.
[b]Numbers of cases hospitalized per number of cases in outbreaks with hospitalization data.
[c]Bennet *et al.* (1987)
[d]Primarily in sensitive populations, very low mortality in non-sensitive populations.
[e]Emergency room visits, no hospitalizations reported.

the probability of infection from the existing database. These data were based on human feeding studies in which the subjects ingested various levels of viruses, protozoan cysts or oocysts, or bacteria. The individuals were then monitored for evidence of infection and in some cases disease. Infection, or the ability of the microorganism to colonize the intestinal tract and be excreted in the feces, was evaluated for most of the virus and protozoa studies. Two models have emerged which have been able to fit the human dose–response data sets. Haas (1983) found that for viruses, the β-Poisson model best described the probability of infection. This model was used to estimate the risk of infection, clinical disease and mortality to hypothetical levels of viruses in drinking water, and to calculate annual and lifetime risks (Haas *et al.*, 1993). Rose *et al.* (1991*b*) used an exponential model to evaluate daily and annual risks of *Giardia* infections after various levels of microbial reduction was achieved through drinking water treatment. This particular study used survey data of surface waters for assessing the needed drinking water treatment for polluted and pristine waters based on *Giardia* cyst occurrence.

Table 3.5 summarizes the dose–response fits for 13 different organisms from various studies.

Exposure and hazard characterization

With the use of the previously defined models for infection and the data from the hazard identification, risks can be estimated on the basis of various exposures. Eight microbial contaminants were compared with regard to

Table 3.5. *Probability of infection models and best fit dose response parameters for various human feeding studies*

Organism	Best model[a]	Model parameters[a]
Rotavirus	β-Poisson	$\alpha = 0.26$ $\beta = 0.042$
Giardia	Exponential	$r = 0.0198$
Salmonella	Exponential	$r = 0.00752$
E. coli	β-Poisson	$\alpha = 0.1705$ $\beta = 1.61 \times 10^6$
Shigella	β-Poisson	$\alpha = 0.248$ $\beta = 3.45$
Cryptosporidium	Exponential	$r = 0.00467$

[a]Models $\qquad P_i = 1 - (1 + N/\beta)^{-\alpha}$ $\qquad$ β-Poisson
$\qquad\qquad\quad P_i = 1 - \exp(-rN)$ $\qquad$ exponential
P_i = Probability of infection (ability of the organism to take hold and reproduce in the intestine)
N = exposure, expressed as numbers (CFU) of microorganisms ingested.
α, β, r = constants for specific organisms that define the dose–response model.

risks of infectivity, severity, mortality in the general population and mortality in sensitive populations (Table 3.6). With an equal exposure to one organism, the risk of infection ranged from a low of 4.1×10^{-7} to a high of 3.1×10^{-1}, depending on the microorganism. *Giardia* infectivity was comparable to those of coxsackie viruses and echoviruses as well as to *Shigella*. *Cryptosporidium* risks were slightly lower, and *Campylobacter* and *Salmonella* were ten-fold lower. However, in the evaluation of endpoints of severity and mortality as opposed to infection, the bacteria and viruses became much more significant than *Giardia*, 1000- to 10 000-fold. Particularly within vulnerable populations risks may become very significant. For example, risk estimates for mortality go from 0.001 to 50% the probability of infection resulting in for *Cryptosporidium* in the general population as compared to the immunocompromised population.

It has been difficult to estimate the level of human exposure to microorganims through drinking water. Table 3.7 shows some of the levels of viruses, protozoa and bacteria that have been reported in North American surface waters. The viruses may be 10–100 time more numerous than protozoan cysts, and levels of *Salmonella* may be 10–10 000 times greater than the viruses. These data are limited, and often extrapolations are needed to go from surface water concentrations to tap water concentrations, taking into account reductions through water treatment (both filtration and disinfection). Some of these data are available for viruses and protozoa (Stetler, Ward & Waltrip, 1984; Payment, Trudel & Plante, 1985; LeChevallier, Norton & Lee, 1991a; Payment & Franco, 1993).

Table 3.6. *Estimated risks of disease and mortality from microbial infections acquired from drinking water*

Etiological agent	Exposure of 2l water/day[a]	Daily risk			
		Infection	Severity[b]	Mortality[bc]	Mortality[bd]
Rotavirus	0.008	4.9×10^{-3}	4.4×10^{-5}	4.0×10^{-7}	4.9×10^{-5}
Giardia	0.132	2.6×10^{-3}	1.8×10^{-5}	2.6×10^{-9}	[e]
Salmonella	0.116	2.7×10^{-4}	1.1×10^{-5}	2.7×10^{-7}	1.0×10^{-5}
Cryptosporidium	0.96	4.5×10^{-3}	4.1×10^{-5}	4.1×10^{-9}	2.1×10^{-5f}

[a]High exposure estimate from Table 3.7., considering 99.9, 99.99 and 99.9999% reductions of *Giardia*, rotavirus and *Salmonella*, respectively.
[b]Taken from Tables 3.2 and 3.4.
[c]General population.
[d]Nursing homes.
[e]No documented increase in mortality in the nursing home population.
[f]Risk in the immunocompromised.

Table 3.7. *Concentrations of enteric viruses and protozoan parasites and* Salmonella *bacteria reported in surface waters in North America*

Micro-organism	Range of concentrations per 100 l	Geometric mean	Arithmetic mean	Reference
Enteroviruses	0–98	14	5.2	Stetler *et al.* (1984)
	0–69	63	0.45	Dahling *et al.* (1979)
	0–14	1.6	0.85	Rose *et al.* (1987)
	<5–4500	Not determined	336	Payment & Trudel (1985)
Rotaviruses	0–215	25	2.0	Rose *et al.* (1987)
	1450–4080		2900	Raphael *et al.* (1985)
Giardia	<1–140	3		Rose *et al.* (1991a)
	4–6600	277		LeChevallier *et al.* (1991a)
Crypto-sporidium	<1–4400	43		Rose *et al.* (1991a)
	7–48 400	270		LeChevallier *et al.* (1991a)
Salmonella	6×10^3–5.8×10^6			Davis & Gloyna (1972)
				Geldreich *et al.*, (1968)
				Oliveri *et al.* (1978)

However, the greatest barrier to the development of adequate risk assessment for microbials remains the lack of monitoring data for evaluating exposure.

Summary

(1) Hazard identification should not rely solely on outbreak or epidemic data. Better health surveillance data are needed on disease (e.g. the association of viruses with heart disease), severity, mortality and chronic outcomes.
(2) More dose–response models are needed. Appropriate animals must be identified to further the studies on virulence mechanisms, and the effects of complex mixtures of microbial organisms.
(3) Occurrence information and exposure assessment are inadequate to develop adequate risk estimates; methods and monitoring for key contaminants are needed.
(4) Quantitative risk estimates will provide an approach for decisions on risk prioritization, risk–risk tradeoffs, cost–benefit analysis, development of appropriate monitoring methods, necessary treatment needs and standards.

Methods for detecting protozoa and viruses in environmental samples and issues of concern for risk assessment

Methods have been developed for the detection of the enteric protozoa, and this has resulted from the increasing recognition of waterborne protozoan disease since its initial identification in 1965 (Rose *et al.*, 1988*b*). Some of the problems associated with this approach have been the variation in the recoveries of the methods as influenced by water quality, the inability to identify the species of origin (bird versus mammalian isolates) and the inability to assess viability of the organisms.

For viruses, the development of detection methods began in 1945 and these methods initially were focused on the enterovirus group. Although methods have developed for other viruses (e.g. rotavirus and hepatitis A virus) there is an insufficient database on many viruses of human health concern, including adenoviruses, Norwalk virus, rotavirus and coxsackieviruses (Hurst, Benton & Stetler, 1989; Gerba & Rose, 1990).

Sample collection, concentration and purification of protozoan cysts and oocysts

In order to detect low concentrations of cysts and oocysts, large volumes of water must be collected through an appropriate filter (10-inch (2.54 cm) cartridge yarn-wound filters with a nominal pore size of 1.0 μM). Unfortunately, unwanted constituents of the sampled water (e.g. particulates, precipitated minerals, etc.) are also concentrated and provide troublesome interference during the ensuing assay procedures (Rose et al., 1989; LeChevallier & Trok, 1990). Both cysts and oocysts, and also debris, are recovered from the filter during the subsequent elution or washing process, with the use of detergents and physical agitation. The eluate containing the cysts and oocysts and debris is concentrated by centrifugation into a single pellet, and an aliquot of this pellet is layered onto a density gradient of Percoll and sucrose (final specific gravity 1.09), where centrifugation separates the cysts and oocysts from much of the debris. This semi-purified sample is then collected from the gradient, and labeled with monoclonal antibodies specific for the cyst and oocyst wall, by means of an indirect fluorescent antibody (IFA) procedure. The sample can then be examined by epifluorescence microscopy for objects that display proper fluorescence (brilliant apple green), shape (ovoid or spherical) and size, or by phase contrast or Nomarski differential interference contrast (DIC) microscopy for internal features (1–4 nuclei, axonemes and median bodies) (LeChevallier, Norton & Lee, 1991b).

The efficiency of recovery of cysts and oocysts during the processing of a sample for *Cryptosporidium* and *Giardia* have been evaluated (Ongerth & Stibbs, 1987; Rose et al., 1988b, 1989; LeChevallier & Trok, 1990; LeChevallier et al., 1991b; Rose, Gerba & Jakubowski, 1991a). The following recovery efficiencies were reported: sample collection, 88–99%; filter elution, 16–78%; concentration and clarification, 66–77%; microscopic detection (IFA only), which also represented the overall recovery efficiency, 9–59%. Current methods for recovering and detecting protozoa underestimate their true concentrations in environmental samples.

In addition to recovery issues, a bigger problem is the sensitivity of the protozoan detection assay. Sensitivity is influenced by the volume of water collected, the volume processed and the volume examined under the microscope. Therefore, if only 10 l is collected, and only 10% of this is processed and ultimately examined, then the level of sensitivity is 1 l. This is not sufficient for examining endemic levels of risk associated with low-level contamination. The 'Surface Water Treatment Rule' initially developed by the Environmental Protection Agency set a goal of microbial safety for drinking water at 10^{-4} infections/consumer on an annual basis. In order to evaluate contamination at that level, 34 000 l of the treated water would need to be collected and examined (Environmental Protection Agency, 1989).

The use of antibodies labeled with fluorescein isothiocyanate (FITC) has greatly enhanced the ability to detect *Cryptosporidium* and *Giardia* microscopically in environmental samples. Fluorescence alone, however, may not be sufficient for their identification. Background fluorescence due to other naturally fluorescing organisms, and nonspecific binding of the antibody may decrease their accuracy for protozoan identification. Although false positives can be problematic, Clancy, Gollnitz & Tabib (1994) found that the bigger problem was false negatives.

A number of IFA systems have been developed for *Cryptosporidium* and *Giardia* (Garcia, Brewer & Bruckner, 1987; Stibbs *et al.*, 1988; Rose *et al.*, 1989). Although some species specificity has been reported, no one antibody system can be used to identify only those protozoan species or isolates that are known to be associated with human infections. In addition, the viability of the cysts and oocysts that are detected in environmental samples and their ability to cause an infection are in question. LeChevallier *et al.* (1991*a,b*) reported that 10–30% of the organisms found were empty cysts and oocysts. It is unclear whether this was related to sample processing.

Collection and concentration of viruses

Viruses may be concentrated from water by the use of either electropositive or electo-negative filters. Both are able to concentrate viruses from large volumes of water, but clogging may occur if the water is high in suspended solids or turbidity (Rose *et al.*, 1989). The adsorbed viruses are eluted off the filter matrix by the passage of a beef extract solution (anywhere from 3 to 8%) through the filter. This appears to be more efficient if done several times, with the use of two or three passages of the eluent through the filter. The beef extract is kept at a high pH of 9.0 to 11.0 during the elution, which facilitates viral desorption. The beef extract eluent is then brought to an acidic pH of 3.5 whereby an organic floc forms. The viruses will adsorb to this floc and can be pelleted with the floc through centrifugation at $500 \times$ g, and resuspended in a smaller volume of a neutral buffer. This final concentrate is used for the viral analysis, by either cell culture or other types of tests including molecular techniques such as the polymerase chain reaction (PCR).

The virus methods have primarily been developed for and tested with the enterovirus group, which consists of poliovirus, echoviruses and coxsackieviruses. They are similar to the protozoan methods in that the sensitivity may be limited by the volume of water collected. However, unlike the protozoa, the analysis of the entire viral concentrate on cell culture is feasible and therefore the analysis step is not the limiting factor. Given this, the analysis of 1000 l of water for presence of viruses is possible.

However, adequate data demonstrating equivalent viral adsorption as larger and larger water volumes are passed through the filters are not available. There is also some information that certain types of enteroviruses, such as poliovirus, may be preferentially adsorbed to filters.

For virus detection, monkey kidney cell lines have generally been used. The viruses are detected by their ability to grow in the cells and cause a destruction of a monolayer of cultured cells, a process known as cytopathic effect (CPE). The various types of viruses are not readily distinguishable unless further tests are run with antibodies, which may be used to identify the different species of enteroviruses. Cell culture also preferentially detects only a small percentage of the viruses which may be found in sewage-polluted waters. The levels and varieties of viruses in polluted water are at present underestimated. Payment, Trudel & Plante (1987) found that when labeled antibodies were used to determine the numbers of infectious viral foci (areas of infection) in cell cultures the levels of viruses were up to 100 times greater than when the older technique of CPE assays was relied on.

Very few studies have identified the specific virus types contained in concentrated water samples. The limited data available show that reoviruses have been found to be at the highest levels, and Hurst, Mc-Clellan & Benton (1988) found that adenoviruses could be detected at levels 94 times higher than enteroviruses when gene probes were used to verify cell cultures. More of this type of data are needed.

Polymerase chain reaction for detection of microbes in environmental samples

PCR, a molecular technique, is now being used to detect a variety of microorganisms in environmental samples (Bej et al., 1990; Mahbubani et al., 1991; Tsai & Olson, 1992a,b; Atlas et al., 1992; Kopecka et al., 1993; Johnson et al., 1995). This method provides rapid results that are highly specific in terms of identification. However, before PCR can be used routinely for environmental monitoring, several areas need to be addressed.

The first of these areas is assay sensitivity, as in most cases very small volumes can be processed through the thermal cyclers (100 μl or less). Therefore, further concentration of the processed water samples is necessary or larger capacity thermal cycling machines must be used. Secondly, it is now clear that PCR detects dead microorganisms. Therefore, its usefulness, without a cultivation procedure being employed in conjunction, may be limited for samples of water that has undergone disinfection. The procedure may be most useful in untreated waters (e.g. source waters, recreational waters, shellfish harvesting waters, ground waters). Thirdly, the problem of interfering compounds in the samples, which lead to false negatives, must be addressed. Antibody capture procedures appear to have great promise in addressing this issue both for protozoa (D. W. Johnson *et al.*, unpublished data) and viruses (Deng, Day & Cliver, 1994). Finally, the test remains only qualitative, with results presented as positive or negative. An accurate and reliable approach for developing a quantitative PCR, whether it be based upon either a most probable number estimation, or same measurable product, is needed.

Summary

(1) Specialized methods have been developed for the detection of specific microbial contaminants, based on concentration of large water volumes, use of antibodies, cell culture and PCR. However, these methods have not been standardized, there are no training courses readily available, and only a few laboratories have the capability of running the tests.
(2) Sensitivity at best is 1 organism/100 l to 1 organism/1000 l, and remains an issue influenced by the volume of water collected and the limitations in the analysis.
(3) Viability assessment and species identification are difficult for the protozoa, but for the viruses, the overall numbers of infectious viruses are measured, but the specific individual viral types go undetermined.
(4) PCR techniques hold great promise; however, the issues of viability, physical or chemical interferences, sensitivity and quantitation need to be addressed.

Monitoring approaches for evaluating risks

Various types of monitoring programs may be used for development of a microbial occurrence database that will be useful in characterizing public health risks and in risk management. Generally, monitoring directed toward source water characterization will prove to be more fruitful, given the sensitivities of the current methods.

Characterization of sources of fecal inputs

The types of microbial contaminants discussed in the previous sections are fecal–oral agents and are excreted in the feces of infected individuals, therefore identification of fecal inputs from wild-life, domestic animals, birds and humans is of interest. Point-source discharges, such as those from sewage treatment plants, are readily accessible and can be monitored.

Types of contaminants, concentrations, variations in the microbial levels and wastewater treatment prior to discharge can be evaluated.

Nonpoint discharges are more difficult to assess than are wastewater discharges. However, key storm water drainage areas, high-density septic tank areas, heavily used recreational sites, watering holes for animals and bird sanctuaries may be identified. In some cases direct monitoring of the animals in a watershed may provide valuable information regarding the potential levels of microbial pathogens that may be washed off land surfaces into the waterways.

Spatial and temporal distributions

The differences in the microbial levels between sites (e.g. the point of waste discharge and a drinking water intake pipe, or waters in California versus those in Florida) are influenced by the hydraulics of the water system, water temperatures, topography, geology and climate in addition to the fecal inputs. The excretion and loading of microbial contaminants is dynamic and therefore changes over time. The changes in microbial levels may be rapid and may occur within hours. The development of monitoring programs to address these spatial and temporal changes and differences may be difficult. Large numbers of samples may need to be collected over different seasons in order to demonstrate significant differences and to establish distributions. The numbers may change from day to day and it is difficult to collect daily samples. In addition to seasonal changes, there may be changes from year to year.

A majority of the samples may be negative or the microbial levels may be below the sensitivity of the methods, and a variety of statistical approaches are needed to handle the 'less-than' data (Helsel, 1990). Average water quality levels of the various microbes, and the likelihood of peak contamination events are key variables to understand, particularly if treatment plants are being designed to handle the average water quality and the potential worst case.

The development of surrogates, and assessment of the potential for changes in the relationship between the levels of the surrogates and the microbial contaminants of concern are needed. Traditionally, coliforms and turbidity have been used as surrogates. However, correlations between these indicators and the microbes of concern have not been strong enough to develop predictive models. The data, to date, do not demonstrate a correlation between the indicator bacteria and enteric protozoa and viruses (Rose et al., 1988a, 1991a; Gerba & Rose, 1990). No indicators have been shown to predict the presence or, more importantly, the absence of viruses or protozoa, nor have they been effective at predicting the treatment requirements. LeChevallier & Norton (1993) found that multiple linear regression models could predict only 57% of the variation of *Giardia* cyst levels on the basis of coliforms and temperature, whereas no model could adequately predict *Cryptosporidium* oocyst levels on the basis of surrogates.

Coliphages are viruses that specifically infect the bacterium *E. coli*, and the various types of coliphage may have some value as indicators, particularly for human enteric viruses (Havelaar, Van Olphen & Drost, 1993; Payment & Franco, 1993). These model viruses behave in the environment and survive many treatment processes (e.g. disinfection) in a manner that is similar to that of the human viruses of health concern. The coliphages are also easily assayed by sampling the water and mixing it with the desired host bacterium. When the mixture is poured into petri dishes, the bacteria produce a layer or 'lawn' of cells, in which the viruses produce holes, termed plaques, in the lawn through lysis of the bacterial cells. The bacterial virus analysis takes less than 24 h, and the quantitative measurement for these viruses is referred to as the plaque-forming unit. Havelaar *et al.* (1993) evaluated wastewater both untreated and treated, river water, treated river water and lake water for the presence of coliphages and human enteric viruses. They found significant correlation between coliphages and enteric viruses which suggests that coliphages could be used to predict the concentration of human viral pathogens. However, no correlation was found in untreated and treated wastewater, and this suggests that there are other unknown factors involved which may complicate the use of this surrogate with recent sewage inputs into a water body.

Fate and transport

Studies on the movement of these microbial contaminants through watersheds have been primarily limited to coliforms in surface waters. In groundwater systems, coliphages or other types of bacterial viruses have been used to model the movement or transport of viruses from septic tanks (Yates & Yates, 1988). Die-off, adsorption, sediment deposition and resuspension (in addition to dilution and dispersion) are all processes that influence the transport and fate of microorganisms. For studying the transport of viruses, a specialized laboratory, bench scale or field sites are needed to undertake these seeded experiments.

Once enteric microorganisms enter the environment, a natural inactivation (or die-off) begins. In water, many factors influence this, including the amounts of solids, oxygen, salinity, ultraviolet light and in particular the temperature. The temperature plays the most significant role in the survival of these microorganisms and most of the data available deal with the effect of varying temperatures.

There are very few data on survival in water for either of the protozoans

Cryptosporidium and *Giardia*. DeRegnier, Cole & Schupp (1989) found that mice could no longer be infected with *Giardia* cysts after the cysts had been exposed to 56 days in river and lake water at 5 °C. However, the level of cysts was below the infectivity level for mice and viability inclusion dyes suggested that an inactivation rate of -0.01 to 0.05 $\log_{10}$ per day occurred at the low temperatures. Therefore, after 60 days, 75–99.9% reduction in cyst viability may be observed. Robertson, Campbell & Smith (1992) have demonstrated that only 55% of the *Cryptosporidium* oocysts were dead in river water after 47 days and 99% were dead after 176 days at temperatures between 5 and 10 °C.

Virus survival studies were evaluated and most of the variation was a result of the water temperature (Kutz & Gerba, 1988). Inactivation rates averaged 0.5 $\log_{10}$ per day. The viruses survived longer in polluted and surface waters than in tap water (Kutz & Gerba, 1988). Therefore, between six and ten days are required for 99.9% viral inactivation at ambient temperatures of between 15 and 25 °C (Kutz & Gerba, 1988). Up to 30 days at 4 °C may be needed before 90% viral inactivations are observed (Kutz & Gerba, 1988).

Treated water monitoring

The monitoring of waters after drinking water treatment may be of limited value. In some cases very large volumes of water may be needed and the level of sensitivity desired may be beyond the limits of the methods. Monitoring programs aimed at treated drinking water should also include an element of public health surveillance, which would be aimed at evaluating the public health impacts. This type of prospective epidemiological surveillance is very costly. For the protozoans, the viability and infectivity are in question; for the viruses, the percentage of individuals who may actually develop acute disease symptoms may be difficult to identify, as the same virus could cause a variety of symptoms with a different symptomatology appearing in different individuals.

Most often treated water monitoring is done at the treatment plant; however, tap-water samples throughout the distribution area may be more appropriate to monitor. The integrity of water distribution systems has rarely been assessed by measurement of the actual microbial contaminants of concern. This may also be a more valid approach for assessing exposure.

Monitoring has been done throughout the drinking water treatment train to examine microbial reductions by the various treatment processes. Surrogates such as particle counts and turbidity, *Clostridium* and coliphage have been used to evelute efficiency with some success. Payment & Franco (1993) were statistically able to relate the removal of coliphages to the removal of enteroviruses through drinking water treatment, specifically the settling stage of the treatment process. The total coliphage removal or inactivation by the complete drinking water process was estimated at 7 $\log_{10}$.

Clostridium is a genus of anaerobic bacteria that produces a resistant spore. *C. perfringens* is found in the feces of animals and humans and has also been suggested as a more conservative indicator of water quality, fecal contamination and water treatment (Payment & Franco, 1993). Membrane filtration methods are available for the enumeration of *C. perfringens* that are fairly rapid and simple to perform. It has recently been found that a significant correlation exists between levels of *C. perfringens* and the enteric viruses and the enteric protozoa for evaluation of the efficiency of drinking water treatment by filtration and disinfection. This association is probably due to the resistant nature of the spores. Therefore, the use of this indicator assures that if it is removed (by 7–8 $\log_{10}$) it is unlikely that enteric protozoa or viruses will be found in very large volumes of water.

Biological tracers such as coliphages may also be useful in that large concentrations of these organisms can be seeded into pilot or full-scale operations and the removals monitored. These types of seeded studies may provide more useful data much more rapidly than the monitoring of indigenous microorganisms can.

Summary

(1) The characterization of sources of microbial contaminants discharged to a given water body may be the easiest type of monitoring program to initiate, particularly with limited financial resources.
(2) Currently, there is a lack of information on the spatial and temporal distributions of microbial contaminants in source waters, and well-designed monitoring studies with long-term goals are needed to address this.
(3) Specialized studies using biological tracers, seeded pilot and full-scale treatment, along with other microbial surrogates provide information on microbial transport and fate, and treatment capabilites for removing microorganisms.
(4) Finished water monitoring should be carried out in conjunction with efforts to improve health surveillance in the exposed population.

Final conclusions

The role of pathogen monitoring in determining public health risks is to provide the necessary data on exposure. Monitoring can provide information on low levels of microbial contamination that may be affecting public health. This type of disease threat has gone unrecognized but may contribute to the endemic level of disease in the population, even though such has not been determined epidemiologically. Currently, more microbiological methods are needed to provide the necessary sensitivity and selectivity, as are more laboratories that can undertake these types of studies. Through research, education and training, better assessment of the microbial disease risks associated with drinking water will be forthcoming.

References

Assaad, F. & Borecka, I. (1977). Nine-year study of WHO virus reports on fatal viral infections. *Bulletin of the World Health Organization*, **55**, 445–53.

Atlas, R. M., Sayler, G., Burlage, R. S. & Bej, A. K. (1992). Molecular approaches for environmental monitoring of microorganisms. *BioTechniques*, **12**, 706–17.

Bej, A. K., Mahbubani, M. H., Miller, R., DiCesare, J. L., Haff, L. & Atlas, R. M. (1990). Multiplex PCR amplification and immobilized capture probes for detection of bacterial pathogens and indicators in water. *Molecular and Cellular Probes*, **4**, 353–65.

Bennett, J. V., Homberg, S. D. & Rogers, M. F. (1987). Infectious and parasitic diseases. *American Journal of Preventive Medicine*, **3**, 102–14.

Centers for Disease Control (1990). Waterborne disease outbreaks. US Department of Health and Human Services, Atlanta, Georgia. *Morbidity and Mortality Weekly Report*, **39**, 1–57.

Centers for Disease Control and Prevention (1993). Aseptic meningitis. *Morbidity and Mortality Monthly Report*, **39**.

Chalker, R. B. & Blaser, M. J. (1988). A review of human salmonellosis: III. Magnitude of *Salmonella* infection in the United States. *Reviews in Infectious Disease*, **10**, 111–24.

Clancy, J. L., Gollnitz, W. D. & Tabib, Z. (1994). Commercial labs: how accurate are they? *Journal, American Water Works Association*, **86**, 89–97.

Craun, G. F. (ed.) (1986). *Waterborne Diseases in the United States*. Boca Raton: CRC Press.

Craun, G. F. (1991). Statistics of waterborne disease in the United States. *Water Science and Technology*, **24**, 10–15.

Current, W. L. & Garcia, L. S. (1991). Cryptosporidiosis. *Clinical Microbiology Reviews*, **4**, 325–58.

Dahling, D. R., Safferman, R. S. (1979). Survival of enteric viruses under natural conditions in a subarctic river. *Applied and Environmental Microbiology*, **30**, 1103–10.

D'Antonio, R. G., Winn, R. E., Taylor, J. P., Gustafson, T. L., Current, W. L., Rhodes, M. M., Gary, G. W. & Zajac, R. A. (1985). A waterborne outbreak of cryptosporidiosis in normal hosts. *Annals of Internal Medicine*, **103**, 886–8.

Davis, E. M. & Gloyna, E. F. (1972). Bacterial die-off in ponds. *Journal of Sanitary Engineering*, **98**, 59–69.

Deng, M. Y, Day, S. P. & Cliver, D. O. (1994). Detection of hepatitis A virus in environmental samples by antigen-capture PCR. *Applied and Environmental Microbiology*, **60**, 1927–33.

DeRegnier, D., Cole, L. & Schupp, D. G. (1989). Viability of *Giardia* cysts suspended in lake, river and tap water. *Applied and Environmental Microbiology*, **55**, 1223–9.

Dubey, J. P., Speer, C. A. & Fayer, R. (eds) (1990). *Cryptosporidiosis of Man and Animals*. Boca Raton: CRC Press.

Environmental Protection Agency (1989). *Surface Water Treatment Rule Guidance Manual for Compliance with the Filtration and Disinfection Requirements for Public Water Systems Using Surface Water Sources*, vol. 54, No. 124, June 29. Washington, DC: Federal Register.

Fayer, R. & Ungar, B. L. P. (1986). *Cryptosporidium* spp and cryptosporidiosis. *Microbiological Reviews*, **50**, 458–83.

Feachem, R. G., Bradley, D. J. & Garelick, H. (eds) (1983). *Entamoeba histolytica* and amebiasis. In *Sanitation and Disease: Health Aspects of Excreta and Wastewater Management*, pp. 337–47. New York: John Wiley and Sons.

Gallaher, M. M., Herndon, J. L., Nims, L. J., Sterling, C. R., Grabowski, D. J. & Hull, H. F. (1989). Cryptosporidiosis and surface water. *American Journal of Public Health*, **79**, 39–42.

Garcia, L. S., Brewer, T. C. & Bruckner, A. (1987). Fluorescence detection of *Cryptosporidium* oocysts in human fecal specimens by using monoclonal antibodies. *Journal of Clinical Microbiology*, **25**, 119–21.

Geldreich, E. E., Best, L. C., Kenner, B. A. & van Donsel, D. J. (1968). The bacteriological aspects of stormwater pollution. *Journal, Water Pollution Control Federation*, **40**, 1861–72.

Gerba, C. P. & Rose, J. B. (1990). Viruses in source and drinking water. In *Drinking Water Microbiology*, ed. G. A. McFeters, pp. 380–96. New York: Springer-Verlag.

Gerba, C. P., Rose, J. B. & Haas, C. N. (1995). Waterborne disease: who is at risk? In *Proceedings of the American Water Works Association's Water Quality Technology Conference*, San Francisco, pp. 57–71. Denver: American Water Works Association.

Gitler, C. & Mirelman, D. (1986). Factors contributing to the pathogenic behavior of *Entamoeba histolytica*. *Annual Reviews in Microbiology*, **40**, 237–61.

Haas, C. N. (1983). Estimation of risk due to low doses of microorganisms: a comparison of alternative methodologies. *American Journal of Epidemiology*, **118**, 573–82.

Haas, C. N., Rose, J. B., Gerba, C. P. & Regli, S. (1993). Risk assessment of virus in drinking water. In *Risk Analysis*, **13**, 545–51.

Havelaar, A. H., Van Olphen, M. & Drost, Y. C. (1993). F-specific RNA bacteriophages are adequate model organisms for enteric viruses in freshwater. *Applied and Environmental Microbiology*, **59**, 2956–62.

Hayes, E. B., Matte, T. D. & O'Brien, T. R. (1989). Contamination of a conventionally treated filtered public water supply by *Cryptosporidium* associated with a large community outbreak of cryptosporidiosis. *New England Journal of Medicine*, **320**, 1372–76.

Helsel, D. R. (1990). Less than obvious: statistical treatment of data below the detection limit. *Environmental Science and Technology*, **24**, 1767–74.

Herwaldt, B. L., Craun, G. F. & Stokes, S. L. (1992). Outbreaks of waterborne disease in the United States: 1989–90. *Journal, American Water Works Association*, **84**, 129–35.

Hurst, C. J., Benton, W. H. & Stetler, R. E. (1989). Detecting viruses in water. *Journal, American Water Works Association*, **9**, 71–80.

Hurst, C. J., McClellan, K. A. & Benton, W. H. (1988). Comparison of cytopathogenicity, immunofluorescence and *in situ* DNA hybridization as methods for the detection of adenoviruses. *Water Research*, **22**, 1547–52.

Johnson, D. W., Pieniazek, N. J., Griffin, D. W., Misener, L. & Rose, J. B. (1995). Development of a PCR protocol for sensitive detection of *Cryptosporidium* in water samples. *Applied and Environmental Microbiology*, **61**, 3849–55.

Kiode, H., Kitaura, Y., Deguchi, H., Ukimura, A., Kawamura, K. & Hirai, K. (1992). Genomic detection of enteroviruses in the myocardium studies on animal hearts with coxsackievirus B3 myocarditis and endomyocardial biopsies from patients with myocarditis and dilated cardiomyopathy. *Japanese Circulation Journal*, **56**, 1081–93.

Klingel, K., Hohenadl, C., Canu, A., Albrecht, M., Seemann, M., Mall, G. & Kandolf, R. (1992). Ongoing enterovirus-induced myocarditis is associated with persistent heart muscle infection: quantitative analysis of virus replication, tissue damage and inflammation. *Proceeding of the National Academy of Science*, **89**, 314–18.

Kopecka, H., Dubrou, S., Prevot, J., Marechal, J. & Lopez-Pila, J. M. (1993). Detection of naturally occurring enteroviruses in waters by reverse transcription, polymerase chain reaction, and hybridization. *Applied and Environmental Microbiology*, **59**, 1213–19.

Kutz, S. M. & Gerba, C. P. (1988). Comparison of virus survival in freshwater sources. *Water Science and Technology*, **20**, 467–71.

LeChevallier, M. W. & Norton, W. D. (1993). Treatment to address source water concerns: protozoa. In *Safety of Water Disinfection: Balancing Chemical and Microbial Risks*, ed. G. F. Craun, pp. 145–64. Washington, DC: ILSI Press.

LeChevallier, M. W., Norton, W. D. & Lee, R. G. (1991a). *Giardia* and *Cryptosporidium* in filtered drinking water supplies. *Applied and Environmental Microbiology*, **57**, 2617–21.

LeChevallier, M. W., Norton, W. D. & Lee, R. G. (1991b). Occurrence of *Cryptosporidium* and *Giardia* spp in surface water supplies. *Applied and Environmental Microbiology*, **57**, 2610–16.

LeChevallier, M. W. & Trok, T. M. (1990). Comparison of the zinc sulfate and immunofluorescence techniques for detecting *Giardia* and *Cryptosporidium*. *Journal, American Water Works Association*, **82**, 75–82.

Leland, D., McAnulty, J., Keene, W. & Sterns, G. (1993). A cryptosporidiosis outbreak in a filtered-water supply. *Journal, American Water Works Association*, **85**, 34–42.

Lew, J. F., Glass, R. I., Petric, M., Lebaron, C. W., Hammond, G. W., Miller, S. E., Robinson, C., Boutilier, J., Riepenhoff-Talty, M., Payne, C. M., Franklin, R., Oshiro, L. S. & Jaqua, M. J. (1990). Six-year retrospective surveillance of gastroenteritis viruses identified at ten electron microscopy centers in the United States and Canada. *Journal of Pediatric Infectious Disease*, **9**, 709–14.

MacKenzie, W. R. (1994). A massive outbreak in Milwaukee of *Cryptosporidium* infection transmitted through the public water supply. *New England Journal of Medicine*, **331**, 161–7.

Mahbubani, M. H., Bej, A. K., Perlin, M., Schaefer, F. W., Jakubowski, W. & Atlas, R. M. (1991). Detection of *Giardia* cysts by using the polymerase chain reaction and distinguishing live from dead cysts. *Applied and Environmental Microbiology*, **57**, 3456–61.

Mishu, B. & Blaser, M. J. (1993). Role of infection due to *Campylobacter jejuni* in the initiation of Guillain-Barré syndrome. *Clinical Infections and Diseases*, **17**, 104–8.

Moore, A. C., Herwaldt, H. G., Craun, G. R., Calderon, R. L., Highsmith, A. K. & Juranek, D. D. (1993). Surveillance for waterborne disease outbreaks – United States, 1991–1992. *Morbidity and Mortality Weekly Report*, **42**, 1–22.

Navin, T. R. & Juranek, D. D. (1984). Cryptosporidiosis: clinical, epidemiological and parasitologic review. *Reviews in Infectious Diseases*, **6**, 313–27.

Oliveri, V. P., Kawata, K. & Kruse, C. W. (1978). Relationship between indicator organisms and selected pathogenic bacteria in urban waterways. *Progress in Water Technology*, **10**, 361–79.

Ongerth, J. E. & Stibbs, H. H. (1987). Identification of *Cryptosporidium* oocysts in river water. *Applied and Environmental Microbiology*, **53**, 672–9.

Payment, P. & Franco, E. (1993). *Clostridium perfringens* and somatic coliphages as indicators of the efficiency of drinking water treatment for viruses and protozoan cysts. *Applied and Environmental Microbiology*, **59**, 2418–24.

Payment, P., Trudel, M. (1985). Detection and health risk associated with low virus concentration in drinking water. *Water Science and Technology*, **17**, 97–103.

Payment, P., Trudel, M. & Plante, R. (1987). Detection and quantitation of human enteric viruses in wastewaters: increased sensitivity using a human immunoserum globulin–immunoperoxidase assay on MA-104 cells. *Canadian Journal of Microbiology*, **33**, 568–70.

Payment, P., Trudel, M. & Plante, R. (1985). Elimination of viruses and indicator bacteria at each step of treatment during preparation of drinking water at seven water treatment plants. *Applied and Environmental Microbiology*, **49**, 1418–28.

Raphael, R. A., Sattar, S. A. & Springthorpe, V. S. (1985). Rotavirus concentration from raw water use in positively charged filters. *Journal of Virological Methods*, **11**, 131–40.

Regli, S., Rose, J. B., Haas, C. N. & Gerba, C. P. (1991). Modeling the risk of *Giardia* and viruses in drinking water. *Journal, American Water Works Association*, **83**, 76–84.

Robertson, L. J., Campbell, A. T. & Smith, H. V. (1992). Survival of *Cryptosporidium parvum* oocysts under various environmental pressures. *Applied and Environmental Microbiology*, **55**, 1519–22.

Rose, J. B., Darbin, H. & Gerba, C. P. (1988a) Correlations of the protozoa, *Cryptosporidium* and *Giardia* with water quality variables in a watershed. *Water Science and Technology*, **20**, 271–6.

Rose, J. B., Gerba, C. P. & Jakubowski, W. (1991a). Survey of potable water supplies for *Cryptosporidium* and *Giardia*. *Environmental Science and Technology*, **25**, 1393–400.

Rose, J. B., Haas, C. N. & Gerba, C. P. (1996). *Microbial Risk Assessment*. Denver: American Water Works Association.

Rose, J. B., Haas, C. N. & Regli, S. (1991*b*). Risk assessment and control of waterborne giardiasis. *American Journal of Public Health*, **81**, 709–13.

Rose, J. B., Kayed, D., Madore, M. S., Gerba, C. P., Arrowood, M. J. & Sterling, C. R. (1988*b*). Methods for the recovery of *Giardia* and *Cryptosporidium* from environmental waters and their comparative occurrence. In *Advances in* Giardia *Research*, ed. P. Wallis & B. Hammond. Calgary: University of Calgary Press.

Rose, J. B., Landeen, L. K., Riley, K. R. & Gerba, C. P. (1989). Evaluation of immunofluorescence techniques for detection of *Cryptosporidium* oocysts and *Giardia* cysts from environmental samples. *Applied and Environmental Microbiology*, **55**, 3189–95.

Rose, J. B., Mullinax, R. L., Singh, S. N., Yates, M. V., Toranzos, G. A. & Gerba, C. P. (1987). Occurrence of rota and enteroviruses in recreational waters of Oak Greek, Arizona. *Water Research*, **21**, 1375–81.

Selik, R. M., Starcher, E. T. & Curran, J. W. (1987). Opportunistic diseases reported in AIDS patients: frequencies, associations and trends. *AIDS*, **1**, 175–82.

Smith, H. V. & Rose, J. B. (1990). Waterborne cryptosporidiosis. *Parasitology Today*, **6**, 8–12.

Smith, J. L., Palumbo, S. A. & Walls, I. (1993). Relationship between foodborne bacterial pathogens and the reactive arthritides. *Journal of Food Safety*, **13**, 209–36.

Steahr, T. E. & Roberts, T. (1993). *Microbial Foodborne Disease: Hospitalizations, Medical Costs and Potential Demand for Safer Food*. Storrs, CT: United States Department of Agriculture, Department of Resource Economics, University of Connecticut.

Stetler, R. E., Ward, R. L. & Waltrip, S. C. (1984). Enteric virus and indicator bacteria levels in a water treatment system modified to reduce trihalomethane production. *Applied and Environmental Microbiology*, **47**, 319–24.

Stibbs, H. H., Riley, E. T., Stockard, J., Riggs, J., Wallis, P. M. & Issac-Renton, J. (1988). Immunofluorescence differentiation between various animal and human source *Giardia* cysts using monoclonal antibodies. In *Advances in* Giardia *Research*, ed. P. Wallis & B. Hammond, pp. 159–63. Calgary: University of Calgary Press.

Tsai, Y. L. & Olson, B. H. (1992*a*). Detection of low numbers of bacterial cells in soils and sediments by polymerase chain reaction. *Applied and Environmental Microbiology*, **58**, 754–7.

Tsai, Y. L. & Olson, B. H. (1992*b*). Rapid method for separation of bacterial DNA from humic substances in sediments for polymerase chain reaction. *Applied and Environmental Microbiology*, **58**, 2292–5.

Ungar, B. L. P. (1988). Seroepidemiology of *Cryptosporidium* infection in two latin American populations. *Journal of Infectious Diseases*, **157**, 551–6.

Wagenknecht, L. E., Roseman, J. M. & Herman, W. H. (1991). Increased incidence of insulin-dependent diabetes mellitus following an epidemic of coxsackievirus B5. *American Journal of Epidemiology*, **133**, 1024–31.

Yates, M. V. & Yates, S. R. (1988). Modeling microbial fate in the subsurface environment. *Critical Reviews in Environmental Control*, **17**, 307–43.

Chapter 4

Estimating the risk of acquiring infectious disease from ingestion of water

CHRISTON J. HURST, ROBERT M. CLARK and
STIG E. REGLI

Introduction to waterborne infectious disease

Humans acquire infections from water through many different routes. These include both recreational or occupational activities performed in contaminated water and the ingestion of contaminated water (Kelly *et al.*, 1982; Koopman *et al.*, 1982; McMillan *et al.*, 1992; Sorvillo *et al.*, 1992). This knowledge has led to the establishment of surveillance programs for assessing both the presence of pathogenic microorganisms in environmental water (LeChevallier, Norton & Lee, 1991*a*; Falcão, Valentini & Leite, 1993) and the occurrence of waterborne diseases in human populations (Tulchinsky *et al.*, 1988; St Louis, 1988; Moore *et al.*, 1993).

Ingestion of contaminated water is perhaps the most notable route by which we acquire waterborne disease, and is the route that will be emphasized in this chapter. This route results in both endemic and epidemic disease. Consumption of microbially contaminated water can lead to bacterial illnesses, such as typhoid (Egoz *et al.*, 1988), cholera (Swerdlow & Ries, 1992) and tularemia (Stewart, 1991); protozoan illnesses, among which are cryptosporidiosis (MacKenzie *et al.*, 1994) and giardiasis (Birkhead & Vogt, 1989; Isaac-Renton *et al.*, 1993); and a variety of viral illnesses (Mosley, 1966), including gastroenteritis (Murphy, Grohmann & Sexton, 1983; Tao *et al.*, 1984) and hepatitis (Bloch *et al.*, 1990; Khuroo, 1991; Ray *et al.*, 1991). Of the four diseases that may cause the greatest risk of suffering and death in underdeveloped parts of the world – cholera, hepatitis, malaria and typhoid – malaria is the only one that is not transmitted by ingestion of water. Infection with the hepatitis E virus is particularly onerous, because it causes a high incidence of death in pregnant women (Ramalingaswami & Purcell, 1988; Khuroo, 1991). A list of some of the microbial illnesses acquired by consumption of contaminated water, along with the names of the causative microorganisms, is presented in Table 4.1.

Table 4.1. *Examples of infectious diseases acquired by ingestion of water*

Disease	Causative microorganism(s)
Bacterial	
Campylosis	*Campylobacter*
Cholera	*Vibrio cholerae*
Paratyphoid	*Salmonella paratyphi*
Shigellosis (bacterial dysentery)	*Shigella*
Tularemia	*Francisella tularensis*
Typhoid	*Salmonella typhi*
Protozoan	
Cryptosporidiosis	*Cryptosporidium parvum*
Entamoebiasis (amoebic dysentery)	*Entamoeba histolytica*
Giardiasis	*Giardia lamblia*
Viral	
Encephalitis	*Enterovirus*
Hepatitis	*Calicivirus, Hepatovirus*
Meningitis	*Enterovirus*
Viral gastroenteritis	*Astrovirus, Calicivirus, Coronavirus, Rotavirus*

Outbreaks of human disease associated with the consumption of untreated drinking water probably have occurred as long as humans have existed. Although it is the outbreaks of epidemic disease that draw the most notoriety, endemic disease is perhaps more insidious. The high death rates associated with cholera (Snow, 1965) and both epidemic as well as endemic typhoid (Johnson, 1916) helped drive the development of municipal and private drinking water treatment utilities that provide a measure of protection for the communities they serve. Unfortunately, the use of community drinking water treatment facilities does not guarantee the absence of waterborne disease outbreaks. There are three major reasons for this situation: first, the types of treatment employed vary from one region to another; second, the many available treatment processes differ with regard to their capacity for physically removing or disinfecting various categories of microorganisms, a fact that has led to the development of treatment facilities that utilize a series of stepwise treatment processes; and third, even when multistep treatment processes are utilized, their operation is not always optimal or reliable.

Inadequate treatment, or outright failures in treatment operation, can lead to microbial contaminants entering the drinking water distribution system. Examples of outbreaks that presumably resulted from inadequate treatment include those of enteritis caused by the protozoans *Cryptosporidium* (D'Antonio et al., 1985; Hayes et al., 1989; Leland et al., 1993) and *Giardia* (McFarlane, 1988). Viruses also can enter the water distribution system in this way and result in human illnesses (Hejkal *et al.*, 1982), among which is gastroenteritis, which has been associated with rotaviruses (Hopkins *et al.*, 1984) and Norwalk virus (Lawson *et al.*, 1991). Microbial contamination can occur following treatment due to cross connections with wastewater plumbing, or seepage into piping and storage tanks. Post-treatment contamination can lead to disease outbreaks (Smith *et al.*, 1989). Contamination can even occur from water taps that are fed by apparently uncontaminated water supplies, possibly representing microbial colonization of the taps (Grundmann *et al.*, 1993). Recently it was estimated that ingesting microbial contaminants in tap water, from a distribution system served by fully operational conventional treatment plants, may have caused 35% of the gastrointestinal illnesses reported among tap water drinkers (Payment *et al.*, 1991).

Not all community distribution water supplies are treated. Often, this represents a matter of cost or an avoidance of disinfected water because of taste preference. At other times, it may result from an assumption that water from streams, wells or springs is intrinsically pure, and so can be consumed without treatment. Even when treated drinking water is available through community distribution systems, people may still prefer to drink water obtained from untreated sources such as springs or wells, thinking that these represent water that somehow is more pure and healthful than the treated water. This assumption often is not valid, as waters from such sources can be contaminated with pathogenic microorganisms (Schwartzbrod *et al.*, 1985) and their consumption can result in outbreaks of waterborne disease (Blake *et al.*, 1977; Bergeisen, Hinds & Skaggs, 1985; Petersen, Cartter & Hadler, 1988). The popular alternative of consuming bottled water may not be much better from a health standpoint, since commercial bottled water can contain pathogens (Blake *et al.*, 1977; Warburton *et al.*, 1992).

In this chapter we present background information on how the causative microorganisms enter source waters, as well as the factors that affect their survival in environmental waters. We then present a modeling exercise for estimating the public health risk from ingesting microbial contaminants in drinking water.

Microbial contaminants in water

The existence of waterborne infectious disease outbreaks provides evidence that the source waters were contaminated. The microorganisms that cause these waterborne infections originate from numerous reservoirs, the list of which includes humans, domesticated animals, wild animals and microbes that naturally inhabit the water.

The origin of microbial contaminants in water

Humans constitute one of the most obvious sources of microbial contaminants in water. We carry microorganisms both on and within our bodies, and we leave a trail of them behind us. The organisms that cause intestinal illnesses are of greatest concern with regard to infections caused by ingestion of water. These organisms are naturally present in feces (Echeverria *et al.*, 1983; Chengqin *et al.*, 1986; Centers for Disease Control, 1993; Mintz *et al.*, 1993), and tend to be transmitted by the fecal–oral route, meaning that infection is acquired either through oral contact with or ingestion of fecally contaminated materials. Humans can become infected by these organisms from sources other than water, such as food (Cameron *et al.*, 1977; Taylor *et al.*, 1993). The excretion of these organisms in human fecal wastes represents a reservoir in the cycle of waterborne disease transmission. These organisms exist as contaminants in wastewater, a factor that holds true for all categories of microorganisms, including bacteria (Weissman *et al.*, 1974; Blake *et al.*, 1980; Pazzaglia *et al.*, 1993) and viruses (Paul, Trask & Gard, 1940; Melnick, 1947; Melnick *et al.*, 1954). Discharge of wastewater into the environment then results in microbial contamination of environmental waters that may subsequently be used as sources of potable water. Drinking water from these sources thereby completes the cycle of infection, from humans to water, and back to humans. Recreational activities in sewage-polluted water is another one of the means by which we become ill from microorganisms in water (McMillan *et al.*, 1992) and has been discussed in Chapter 1. Illnesses are often acquired even from water that has no known sewage input (Sorvillo *et al.*, 1992), suggesting that humans themselves contaminate water through the course of recreational activities.

Other animals besides humans can contribute fecal microbial contaminants to water. Bacterial contaminants can come from warm-blooded animals such as mammals and birds (Geldreich, 1966), as well as from cold-blooded animals such as fish (Geldreich, 1966) and lizards (Pinfold *et al.*, 1993). The fact that some protozoans (Current, 1986; Fayer & Ungar, 1986; Healy, 1990) and viruses (Eiden, Vonderfecht & Yolken, 1985) can infect both humans and animals implies that wild animals living in a watershed area can also contribute these last two categories of contaminating microorganisms to the water (Pacha, Clark & Williams, 1985), leading to human waterborne disease (Isaac-Renton *et al.*, 1993).

The environment itself can serve as a reservoir of bacterial organisms that are pathogenic to humans (Grimes, 1991). Interestingly, both *Legionella pneumophila* and *Vibrio cholerae* seem to exist as natural organisms in water, with *L. pneumophila* existing as an infection within free-living amoebas, and *V. cholerae* existing on the shells of crustacea, where it presumably degrades the chitinous shells (Grimes, 1991). This may be one of the reasons that cholera can exist as an endemic focus even in areas which do not experience outbreaks of the disease (Weissman *et al.*, 1974; Eichold *et al.* 1993).

The fate of microbial contaminants in water

The ability of microbial contaminants to survive after they enter water is a major issue with regard to waterborne infectious disease. Microbial contaminants that are released into water will die-off if they are unable to replicate in that environment (Kehr & Butterfield, 1943; Hurst, 1991*b*; Hurst, Wild & Clark, 1992; Moore, 1993). The results of mathematical analyses have revealed that water temperature is a major factor affecting the survival of microorganisms in water, with survival increasing at lower temperatures (Kehr & Butterfield, 1943; Hurst, 1991*b*; Hurst *et al.*, 1992). Other important factors, at least from the standpoint of viruses, are: the amount of nutrients available in the water that could support the growth of bacterial organisms, generally exhibiting a detrimental effect upon viral survival; the level of turbidity in the water, with higher turbidity generally being beneficial to survival; and water hardness, likewise generally having a beneficial effect upon viral survival (Hurst, 1991*b*). Another major issue is the transport of microbial contaminants, which will occur naturally during the course of water flow. The microorganisms may be conveyed over long distances, far from visible sources of contamination (Dahling & Safferman, 1979). Interestingly, transport also may be aided by human activities, with one possible example being the transport of causative microbial agents of waterborne disease between different parts of the world as contaminants of ship ballast water (McCarthy & Khambaty, 1994).

Estimating the risk of microbial disease

Two items of information are helpful in order for us to estimate the risk of infectious disease caused by ingestion of microbially contaminated water. The first item is the levels of pathogenic microorganisms present in source waters that are likely to be consumed by individuals. The second item is a means of calculating the probability of acquiring infection from drinking untreated water, and the accompanying probabilities that infection will lead to illness, and illness ultimately progress to death. Having used this information to calculate the risk of acquiring infections from ingesting untreated water, we can then determine the extent to which these risks could be reduced by using treatment processes to remove microorganisms before the water is consumed.

What are the levels of microbial contaminants in water?

The first item of information needed for the risk modeling exercise presented in this chapter is an estimation of the level of pathogenic microbial contamination in source waters. In a previous statistical review of published literature (Hurst, 1991*a*), it was found that the levels of human enteric viruses reported for untreated freshwater sources varied widely. The range of average virus levels reported in the different studies was from 9.34 to 0.002 infectious units of virus/l of water examined, with the highest reported value being 315/l of water examined. An overall estimate was made for the level of human enteric viruses by adopting the median statistic of the average values from the different studies. That estimate was 1.4 infectious units of virus/l.

For the risk modeling exercise contained in this chapter, we similarly reviewed published literature reporting the levels of human pathogenic bacteria and protozoa in freshwater sources. Table 4.2 presents the findings regarding the levels of pathogenic bacteria in source waters. The findings for human pathogenic protozoan cysts and oocysts are presented in Table 4.3. It can be seen that, as is the case with viruses, the levels of bacteria and protozoans in environmental waters varies widely. The average level of pathogenic human enteric bacteria in source waters for individual reported studies ranged from 0.13 to 6.92/l. Two of the cited reports represented water from estuarine sites (Colwell, Kaper & Joseph, 1977; Kaper *et al.*, 1979), and the other four represented only freshwater sites. With the use of the median statistic for the average bacterial levels from the different studies, it is estimated that the overall level of human pathogenic bacteria in these waters was 2.16 culturable organisms/l. Deleting the values from the studies by Colwell *et al.* (1977) and Kaper *et al.* (1979) would not change the median statistic. This median level, which rounds to 2.2 culturable organisms/l, is surprisingly close to the estimate of 2.0 culturable organisms/l used by Kehr & Butterfield (1943) when they attempted to assess the risk of disease from bacterial contaminants in water 50 years ago. The average levels of pathogenic human enteric protozoans reported in freshwater sources varied from 1975.93 to 0.01 microscopically countable organisms/l of water examined. With the use of the median statistic for the average protozoan levels from the different studies, it can be estimated that the overall level of pathogenic protozoans in these waters was 0.17/l.

It should be noted that published data that were indicated by authors of the individual studies to represent sewage contaminated waters were not included in either the previously published viral table (Hurst, 1991*a*) or the bacterial and protozoan data summary tables presented in this chapter. It is not always possible to avoid using sewage-contaminated waters as a source of drinking water. This is one of two reasons why water that might be used for human ingestion may contain levels of microbial contami-

Table 4.2. *Levels of pathogenic human enteric bacteria in source water*

Bacteria investigated	Number of samples examined	Percentage of samples positive for bacteria	Average level per liter (includes all negative samples)	Reference
Salmonella typhi (surface water)	65	12	0.13[a]	Stewart & Ghosal, 1938
Vibrio cholerae (surface water, type not specified)	262	26	3.07[a]	Spira *et al.*, 1980
Vibrio cholerae (surface water, type non-01)	14	50	0.87	Kaper *et al.*, 1979
Vibrio cholerae (surface water, type non-01)	—[b]	—[b]	3.30	Colwell *et al.*, 1977
Vibrio cholerae (surface water, type non-01)	92	12	6.92[a]	Islam *et al.*, 1994
Vibrio cholerae (surface and well water, type 01)	16	6	1.25	Brayton *et al.*, 1987
			Median = 2.16 [2.2] culturable organisms per liter	

Note: This table does not include data representing water samples from sources that were stated to be sewage contaminated. Data from the studies by Kaper *et al.* (1979) and Colwell *et al.* (1977) included some samples of estuarine water. Deleting the data from these two studies would not change the median value of 2.16. The data cited for all other studies represent water from only fresh (non-saline) sources.
[a]The titer value presented here was calculated using the Thomas formula for most probable number analysis (Greenberg, Clesceri & Eaton, 1992) based upon presence versus absence values reported in the referenced study.
[b]Values were not given in the reference.

nants substantially higher than the estimates listed here, as has been found by LeChevallier *et al.* (1991*a*) for protozoans, and by Payment, Trudel & Plante (1985) for viruses. The second major reason is that bacterial, protozoan and viral recovery methods can not be assumed to be completely efficient for determining the level of microorganisms present in water samples. Because of inherent inaccuracies in techniques used for assessing the efficiency of microbial detection, the microbial levels listed in the viral (Hurst, 1991*a*), bacterial and protozoan data tables do not include any microbial titer values that were reported to have been adjusted to 'correct' for efficiency of microbial recovery methods.

Table 4.3. *Levels of pathogenic human enteric protozoans in source water*

Protozoas investigated	Number of samples examined	Percentage of samples positive for protozoa	Average level per liter (includes all negative samples)	Reference
Cryptosporidium	3	100	1975.93	Madore *et al.*, (1987)
Cryptosporidium	11	100	1.37	Ongerth & Stibbs (1987)
Cryptosporidium	3	—[a]	0.08	Rose *et al.* (1988)
Cryptosporidium	7	71	0.05	Fogel *et al.* (1993)
Cryptosporidium	90	76	0.93	Rose (1988)
Giardia	3	100	0.17	Ongerth (1989)
Giardia	3	—[a]	0.01	Rose *et al.* (1988)
Giardia	7	43	0.07	Fogel *et al.* (1993)
Giardia lamblia	—[a]	—[a]	3.36	Payment & Franco (1993)
			Median = 0.17 [1.7 × 10^{-1}] microscopically counted organisms per liter	

Note: Data listed in this table represent concentrations of cysts or oocysts in water samples that had been collected from fresh (non-saline) sources. This table does not include data representing water samples from sources that were stated to be sewage contaminated.
[a]Values were not given in the reference.

How can we estimate the risk from ingesting microbially contaminated water?

Several previous efforts have been made at estimating the disease risk to humans from consuming drinking water that contained microbial contaminants. These efforts have a long history, dating back at least to 1943 (Kehr & Butterfield, 1943). More recent efforts, covering various aspects of human health hazards associated with consuming microbially contaminated drinking water, include those by Payment *et al.* (1991); Haas *et al.* (1993); and Gerba & Haas (1988). An excellent literature survey regarding this topic has been published by Teunis, Havelaar & Medema (1994).

Equations 4.1–4.4 present the format that we have chosen to use for this chapter to estimate the individual human health risk of infection, illness, and death on an annual basis from drinking untreated source waters. These risks are calculated separately for bacterial, protozoan and viral contaminants. Later in this chapter, we will use the results from these equations to estimate how using treatment processes to achieve various levels of microbial removal from water prior to consumption can reduce those annual risks.

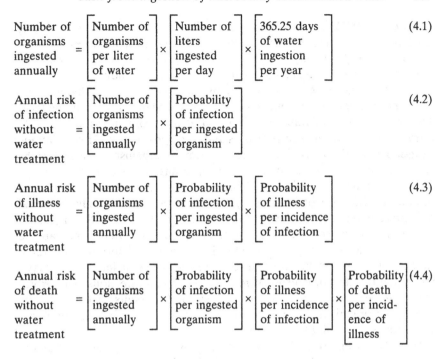

$$
\begin{bmatrix} \text{Number of} \\ \text{organisms} \\ \text{ingested} \\ \text{annually} \end{bmatrix} = \begin{bmatrix} \text{Number of} \\ \text{organisms} \\ \text{per liter} \\ \text{of water} \end{bmatrix} \times \begin{bmatrix} \text{Number of} \\ \text{liters} \\ \text{ingested} \\ \text{per day} \end{bmatrix} \times \begin{bmatrix} \text{365.25 days} \\ \text{of water} \\ \text{ingestion} \\ \text{per year} \end{bmatrix} \quad (4.1)
$$

$$
\begin{bmatrix} \text{Annual risk} \\ \text{of infection} \\ \text{without} \\ \text{water} \\ \text{treatment} \end{bmatrix} = \begin{bmatrix} \text{Number of} \\ \text{organisms} \\ \text{ingested} \\ \text{annually} \end{bmatrix} \times \begin{bmatrix} \text{Probability} \\ \text{of infection} \\ \text{per ingested} \\ \text{organism} \end{bmatrix} \quad (4.2)
$$

$$
\begin{bmatrix} \text{Annual risk} \\ \text{of illness} \\ \text{without} \\ \text{water} \\ \text{treatment} \end{bmatrix} = \begin{bmatrix} \text{Number of} \\ \text{organisms} \\ \text{ingested} \\ \text{annually} \end{bmatrix} \times \begin{bmatrix} \text{Probability} \\ \text{of infection} \\ \text{per ingested} \\ \text{organism} \end{bmatrix} \times \begin{bmatrix} \text{Probability} \\ \text{of illness} \\ \text{per incidence} \\ \text{of infection} \end{bmatrix} \quad (4.3)
$$

$$
\begin{bmatrix} \text{Annual risk} \\ \text{of death} \\ \text{without} \\ \text{water} \\ \text{treatment} \end{bmatrix} = \begin{bmatrix} \text{Number of} \\ \text{organisms} \\ \text{ingested} \\ \text{annually} \end{bmatrix} \times \begin{bmatrix} \text{Probability} \\ \text{of infection} \\ \text{per ingested} \\ \text{organism} \end{bmatrix} \times \begin{bmatrix} \text{Probability} \\ \text{of illness} \\ \text{per incidence} \\ \text{of infection} \end{bmatrix} \times \begin{bmatrix} \text{Probability} \\ \text{of death} \\ \text{per incid-} \\ \text{ence of} \\ \text{illness} \end{bmatrix} \quad (4.4)
$$

The next item of information needed for the exercise presented in this chapter is the amount of water ingested daily per individual. The value used in Equation 4.1 as an estimate for the volume of water ingested on a daily basis per individual consumer is 2 l/day. This volume is the estimate recommended by the US Environmental Protection Agency (1989) for use in calculating the risks from ingestion of water.

Probability of infection and the blindfolded bowler analogy

Once we know the concentration of microorganisms in source water, and the amount of water ingested per day, the next factor to consider is the likelihood that ingesting individual pathogenic microorganisms in water will lead to infection. It is important to note that infection and illness are not synonymous terms. While one cannot become ill without first becoming infected, one can become infected without that infection progressing to illness. Studies have been carried out in which microorganisms were fed to otherwise healthy individuals, and those individuals were then examined for evidence of infection. Two criteria can be used for determining the existence of an intestinal infection. The first, and most important, criterion is the development of a response in the individual's antibody titer against the ingested organism (Cash *et al.*, 1974; Bernstein, Ziegler & Ward, 1986;

Clark *et al.*, 1986). Development of an antibody titer when none previously existed, or an increase in an existing antibody titer, indicates that the individual has been infected. The second criterion is the determination that the number of indicated organisms shed in the individual's feces is greater than could be accounted for by the dose that the individual was fed. Finding appreciable numbers of the same type of organism in the feces suggests that the individual was infected, if those organisms are obligate intracellular parasites such as viruses and some protozoans. This technique cannot accurately be used to assess infection by organisms that live in the lumen of the intestine, such as other protozoans and bacteria.

Table 4.4 lists the minimum number of microorganisms required to initiate an infection of humans by the route of ingestion. The bacterial data presented in Table 4.4 were determined by cultivation of the organisms in the laboratory, so that each ingested organism was viable and capable of growth in a bacterial medium. The viral data were determined by quantitation of the organisms in cultured cells by means of infectivity assays, indicating that the organisms were both viable and infectious. The protozoan data presented in Table 4.4 represent the results of counting the number of organisms under a microscope, which by itself does not give a value for either viability or infectivity.

It can be seen from the information presented in Table 4.4 that the minimum infectious dose is generally higher for bacterial organisms than for protozoans and viruses. Also, it can be seen that the minimum infectious dose varies at both the genus and species levels for all three categories of microorganisms examined – bacteria, protozoans and viruses. The viral data indicate that there are additional strain-dependent differences regarding the minimum infectious dose for microorganisms belonging to the same species. It was gratifying to note that when values were available from two different research studies for the same strain within a single species of microorganism – the SM strain of human poliovirus 1 – both studies reported the same minimum infectious dose level. The median value for the probability of infection by viruses would not be changed by deleting either of these two values for the SM strain.

In a consideration of the lowest number of viruses required to cause infection, it can be seen that in two instances – the human poliovirus 3 results published by Katz & Plotkin (1967) and the *Rotavirus* results published by Ward *et al.* (1986) – the infectious dose was less than one infectious unit as measured by cultured cells. This may reflect the fact that these viruses evolved to replicate within cells of the intestinal wall as part of a living organ, preceded by exposure to the chemical environments within the stomach and intestines. Consequently, these viruses are not as well suited to replicating under the conditions extant in the laboratory, where cultured cells are used to grow and quantitate them.

For many of the microorganisms listed in Table 4.4, the minimum infec-

Table 4.4. *Minimum infectious dose of enteric microorganisms for humans as determined by ingestion*

Microorganism examined	Lowest number of microorganisms required to cause infection	Probability of infection per individual microorganism ingested	Reference
Bacteria[a]			
Salmonella typhi			
(Quailes strain)	100 000	0.00001	Hornick *et al.* (1970a)
Vibrio cholerae	1000	0.001	Levine, Black &
(type 01)		Median = 0.0001	Clements (1984)
		$[1.0 \times 10^{-4}]$	
Protozoans[b]			
Cryptosporidium parvum	30[c]	0.033	DuPont *et al.* (1995)
Giardia lamblia	10	0.1	Rendtorff (1954)
		Median = 0.067	
		$[6.7 \times 10^{-2}]$	
Viruses[d]			
Human echovirus 12	17	0.059	Schiff *et al.* (1984)
Human poliovirus 1			
(SM strain)	2	0.5	Koprowski (1955)
Human poliovirus 1			
(SM strain)	2	0.5	Katz & Plotkin (1967)
Human poliovirus 1			
(strain not identified)	20	0.05	Minor *et al.* (1981)
Human poliovirus 3 (Fox			
strain)	0.5[c]	2.0	Katz & Plotkin (1967)
Rotavirus	0.9	1.1	Ward *et al.* (1986)
		Median = 0.5	
		$[5.0 \times 10^{-1}]$	

Note: The median statistic for bacterial infectious dose was calculated after performing a $\log_{10}$ transformation on the values for probability of infection per individual ingested microorganism. This method provides a conservative estimate for median value, and was chosen for use with the bacterial infectious dose data, because the two numbers being used to determine the midpoint differed by a factor of 100. The median statistics for protozoan and virus dosage data were calculated in the standard manner. The two probability of infection values used to determine the midpoint for the protozoans differed by a factor of only 3, and the two values used to determine the midpoint for the viruses were identical.
Infection was determined serologically.
[a]Number of organisms was determined by culturing.
[b]Number of organisms was determined by microscopy.
[c]Lower dosages were not tested.
[d]Number of organisms was determined by infectivity in cultured cells.
[e]Actual dose was 1 $TCID_{50}$, divided by 2 to yield estimated ID_{100} of 0.5 as reported in this table, using the technique described by Hurst, McClellan & Benton (1988).

tious dose is much greater than one. This is particularly true in the case of bacteria, where the minimum infectious doses reported in Table 4.4 range from 1000 to 100000 organisms. Higher minimum infectious dose values can be found in the literature for many of the bacteria and protozoans listed in Table 4.4. However, in those studies, it is generally the case that lower doses of microorganisms were not examined. If a unit of 100 microorganisms is sufficient to cause an infection, then so would 1000 or even 1000 000. For this reason, we tried to cite studies that not only determined a minimum infectious dose, but also tested lower dosage levels of the same organism and found the lower doses insufficient to cause an infection. One of us previously used a different approach that yielded far lower estimates for the likelihood of infection from ingesting microorganisms (Regli et al., 1991).

A cursory examination of the published scientific literature suggests that there exists an apparent independence between the ingested dose of *Vibrio cholerae* and the likelihood of resulting infection. This issue is examined in Table 4.5, where the lowest ingested doses are reported for two different studies, and which seem to indicate that the proportion of individuals infected by a unit 'minimum infectious dose' may consistently be from 50 to 100%, whereas the ingested doses vary from 1000 to 10000 000 microorganisms. Much of this apparent independence between ingested dose and the chance of becoming infected can be explained by strain-dependent differences between microorganisms belonging to the same phylogenetic species. This is particularly evidenced in the data for three different strains referenced from the study by Morris et al. (1990). Table 4.5 also presents a comparison between the infectivity of types 01 and non-01 strains of *V. cholerae*. The single type 01 strain tested evidenced a lower infectious dose than did any of the three non-01 strains. This suggests a possible reason why type 01 strains have caused the majority of cholera pandemics.

In order for a microorganism to infect an individual such as a human, that microorganism must first survive the chemical environment within the

Table 4.5. *Apparent independence of ingested dosage of* Vibrio cholerae *in humans and likelihood of infection*

Strain of microorganism	Lowest dosage of microorganisms ingested	Proportion of individuals infected by that dosage	Reference
Strain A (type non-01)	1.0×10^6	2/4 (50%)	Morris et al. (1990)
Strain B (type non-01)	1.0×10^7	2/2 (100%)	Morris et al. (1990)
Strain C (type non-01)	1.0×10^5	1/2 (50%)	Morris et al. (1990)
Strain N1691 (type 01)	1.0×10^3	6/6 (100%)	Levine et al. (1984)

body long enough that by random probability it will encounter the proper biochemical attachment site on a susceptible cell. These factors involve probability distributions. Haas (1983) has concluded from statistical probability models that '... it is impossible to rule out the hypothesis that a single microorganism when ingested has the potential of inducing infection or disease'. For the purpose of the modeling exercise presented in this chapter, we have assumed that any single microorganism is capable of causing infection, and that the differences in minimum infectious dose between the various microorganisms simply represent the random statistical probabilities that the necessary combination of events leading to an infection by those microorganisms would occur.

One way to explain this assumption regarding the probability of infection is to introduce an analogy that we will term 'The Blindfolded Bowler'. For this analogy, it is assumed that a bowler is placed into a bowling alley and blindfolded. He is then pointed in the direction of a bowling lane and successively given objects that he is asked to roll down the lane. If the bowler is given ten balls all of the same size, and not told whether any of them resulted in a strike until he has rolled all ten, he may find that, on the whole, ten balls is not sufficient to result in a strike. Thus, for that set of conditions, the minimum 'unit' number of balls required to generate a strike is greater than ten. If the bowler is allowed to use a larger number of balls, perhaps 20, 1000 or even 100 000, he will eventually find the minimum 'unit' number of balls required to generate a strike. If, for example, under a defined set of conditions, there was a reliable chance that a unit of 100 balls could consistently yield a strike, and an assumption could be made that each individual ball was physically capable of producing the strike, then the probability of any individual ball producing a strike would be 0.01, the numerical inverse of the unit number of balls. It may be expected that using balls of larger diameter would decrease the minimum unit number of balls required to cause a strike, and that using balls of smaller diameter might increase the minimum unit number required to cause a strike. If the bowler instead were given cylinders to roll down the lane, and the length of those cylinders was greater than the diameter of the bowling balls, it might be the case that a unit of only 20 cylinders would be required to produce a strike. In that case, the probability of a strike would be 0.05 per object rolled down the lane. Rolling cylinders whose length was equal to the width of the bowling lane might reduce the minimum unit number to 1, giving a probability of 1.0 per cylinder.

This analogy is not used to suggest that size, or overall shape, determines the minimum infectious dose for microorganisms. Instead, factors that are likely to influence the probability of infection per ingested organism include the number of susceptible cells in the intestinal wall, the relative number and accessibility of receptor sites on those cells, plus random genetic and phenotypic variation on the part of both the microorganism and the host.

We can, of course, remove the bowler's blindfold, which enables him to understand the situation. We cannot, however, remove our own blindfold with respect to the variation in minimum unit number of viruses, protozoans or bacteria required to cause infection if ingested by a human. For the purposes of this risk modeling effort, we have applied our Blindfolded Bowler Analogy, and assumed that the probability of infection per individual microorganism ingested is the numerical inverse of the minimum infectious dose.

If we were to place a barrier across the lane used by our blindfolded bowler, we would severely decrease the probability of a rolled object producing a strike. The acidity of stomach secretions likewise represents a barrier to microorganisms that infect cells lining the intestines. Some organisms, such as enteroviruses, have evolved a natural resistance to low pH exposures, which enables them to get past the barrier posed by the stomach. Protozoan cysts and oocysts also seem to have evolved some resistance to degradation by acids. Ingesting food along with pathogenic microorganisms seems to help the microorganisms survive their trip through the stomach (Blake *et al.*, 1980). This finding is important, because humans often ingest food along with water and other beverages. Individuals who consume antacids eliminate this barrier from their stomach, a particular problem if they consume microbially contaminated water along with those antacids. Individuals who have the disease hypochlorhydria, and therefore produce an abnormally low amount of stomach acid, are correspondingly at greater risk of enteric infectious diseases such as cholera (Nalin *et al.*, 1978), as are those individuals who take medications that block production of stomach acid.

The organism *Vibrio cholerae* has an interesting means of natural protection from stomach acids. In addition to causing massive outbreaks of disease associated with the consumption of contaminated water, this microorganism is known to cause illness associated with the consumption of macrocrustaceans, such as crabs (Ragazzoni *et al.*, 1991; Roman *et al.*, 1991). This bacterial organism's natural habitat is on the chitinous shells of crustaceans (Grimes, 1991). Copepods and other taxa of microcrustaceans can reside as natural microfauna within community drinking water distribution systems (van Lieverloo, van der Kooij & Veenendaal, 1995). The association of *Vibrio* organisms with chitin may facilitate their survival during passage through the stomach, and could be an important factor in waterborne disease, if *V. cholerae* in source waters were attached to microcrustacean copepods and these microcrustaceans were inadvertently ingested along with the water (Nalin, 1976). This knowledge suggests that a potentially supportive environment may exist for *Vibrio* contaminants that gain entrance to water distribution systems.

Probability of infection progressing to illness and death

As mentioned above, infection and illness are not synonymous. The likelihood that an infection will result in the development of symptoms associated with illness varies greatly between the many pathogenic microorganisms. From the median values derived from the data listed in Table 4.6, the overall likelihood of infection leading to illness is approximately 0.50–0.60, or 50–60%, for all three categories of microorganisms. This probability value varies from one category of microorganisms to another, between different species within a genus, and even between different strains of the same species.

Illnesses that do not lead to death cover a broad range in terms of the severity of symptoms. The risk modeling exercise presented in this chapter does not address severity of illness, except in distinguishing the likelihood of infection leading to illness (Table 4.6) from the likelihood that illness

Table 4.6. *Likelihood of infection resulting in illness for enteric microbial diseases*

Causative microorganism	Ratio of illnesses to total number of infections	Probability of illness	Reference
Bacteria			
Salmonella typhi	34/69	0.4928	Hornick *et al.* (1970*a*)
Vibrio cholerae (type 01)	15/55	0.2727	Levine *et al.* (1984)
Vibrio cholerae (type non-01)	8/15	0.5333	Morris *et al.* (1990)
		Median = 0.4928 $[4.9 \times 10^{-1}]$	
Protozoans			
Cryptosporidium	36/65	0.5538	Bongard *et al.* (1994)
Cyclospora	11/50	0.2200	Ortega *et al.* (1993)
Giardia lamblia	11/34	0.3235	López *et al.* (1980)
Giardia lamblia	3/5	0.6000	Nash *et al.* (1987)
Giardia[a]	26/133	0.1955	Islam (1990)
Giardia	977/1211	0.8068	Birkhead & Vogt
		Median = 0.4386 $[4.4 \times 10^{-1}]$	(1989)
Viruses			
Human coxsackie-virus A21	24/26	0.9231	Couch *et al.* (1965)
Human echovirus 12	3/59	0.0508	Schiff *et al.* (1984)
Rotavirus	4/5	0.8000	Kapikian *et al.* (1983)
Rotavirus	17/30	0.5667	Ward *et al.* (1986)
Rotavirus	15/26	0.5769	Ward *et al.* (1989)
		Median = 0.5769 $[5.8 \times 10^{-1}]$	

Note: Infection was defined serologically.
[a] Data cited here were those listed for India.

Table 4.7. *Likelihood of illness resulting in death for enteric microbial diseases*

Causative microorganism	Ratio of deaths from illness to total cases of illness	Probability of death	Reference
Bacteria			
Campylobacter	2/6441	0.0003	Tauxe *et al.* (1988)
Escherichia coli (type 0111)	1/9	0.1111	Caprioli *et al.* (1994)
Escherichia coli (type 0157-H7)	1/10	0.1000	Turney *et al.* (1994)
Escherichia coli (type 0157-H7)	4/243	0.0165	Geldreich *et al.* (1992)
Salmonella typhi	20 000/300 000[a]	0.0667	Johnson (1916)
Salmonella typhi (Zermatt, phage type E_1)	3/437	0.0069	Bernard (1965)
Shigella flexneri	1/610	0.0016	Centers for Disease Control and Prevention (1994)
Vibrio cholerae (Burundi, Zimbabwe type 01)	292/1177	0.2481	Ndayimirje *et al.* (1993)
Vibrio cholerae (West Africa type 01)	20 000/150 000	0.1333	Swerdlow & Ries (1993)
Vibrio cholerae (Bangladesh type non-01)	1473/297	0.0137	Cholera Working Group (1993)
Vibrio cholerae (Portugal type 01)	48/2467	0.0195	Blake *et al.* (1977)
Vibrio cholerae (Americas type 01)	6323/731 312	0.0086	Centers for Disease Control and Prevention (1993)
		Median = 0.0180 [1.8×10^{-2}]	
Protozoans			
Cryptosporidium parvum	100/403 000	0.0002 Only value = 0.0002 [2.0×10^{-4}]	Marchione (1994); MacKenzie *et al.* (1994)
Viruses			
Human coxsackievirus A2	2/403	0.0050	Assaad & Borecka (1977)
Human coxsackievirus A3	2/27	0.0741	Assaad & Borecka (1977)

Human coxsackievirus A4	3/580	0.0052	Assaad & Borecka (1977)
Human coxsackievirus A6	2/204	0.0098	Assaad & Borecka (1977)
Human coxsackievirus A9	8/3039	0.0026	Assaad & Borecka (1977)
Human coxsackievirus A10	6/351	0.0171	Assaad & Borecka (1977)
Human coxsackievirus A16	2/1654	0.0012	Assaad & Borecka (1977)
Human coxsackievirus A19	3/18	0.1667	Assaad & Borecka (1977)
Human echovirus 6	14/4774	0.0029	Assaad & Borecka (1977)
Human echovirus 9	14/5237	0.0027	Assaad & Borecka (1977)
Human poliovirus 1	72/8074	0.0089	Assaad & Borecka (1977)
Human poliovirus 2	43/2360	0.0182	Assaad & Borecka (1977)
Human poliovirus 3	27/2427	0.0111	Assaad & Borecka (1977)
Human polioviruses 1, 2 and 3 (data had been combined)	11/204	0.0539	Langmuir, Nathanson & Hall (1956)
Human poliovirus 1	95/1031	0.0921	Kim-Farley *et al.* (1984)
Human poliovirus 3	1/54	0.0185	van Wijngaarden *et al.* (1992)
Hepatitis A virus	—[b]	0.0060[a]	Centers for Disease Control (1990)
Hepatitis E virus (Kashmir)	—[b]	0.0280[c]	Ramalingaswami & Purcell (1988)
Hepatitis E virus (Somalia)	87/2000	0.0435	Gove *et al.* (1987)
Hepatitis E virus (Delhi)	90/97 600	0.0009	Viswanathan (1957)
Rotavirus	2/222	0.0090	Fang *et al.* (1989)
Rotavirus	7/3439	0.0020	Foster *et al.* (1980)
		Median = 0.0094 [9.4 × 10⁻³]	

[a]Relationship between number of deaths from illness to total cases of illnesses was estimated by the cited author.
[b]Values were not given in the reference.
[c]Value reported for adult males.

subsequently results in death (Table 4.7). By looking at the information presented in Table 4.7, it can be seen that the likelihood of illness leading to death varies by category of microorganisms, species and strain.

Many types of bacteria produce toxins that contribute to their pathogenicity. Even different strains within the same bacterial species can vary with respect to the type and level of toxins produced (Venkateswaran et al., 1989), and this may play a role in the observed intraspecies differences for probability that infection by *Vibrio cholerae* will lead to illness. Aside from factors such as microbial toxins, the severity of illness largely depends upon the ability of the infected individual to mount an effective immunological defense, resulting in successfully destroying either the pathogens themselves, or those cells of the individual's body that are infected by the pathogens. Ability to mount a protective immune response naturally varies in relationship to normal factors such as increasing age or pregnancy (Gove et al., 1987). Other, unnatural conditions which can reduce the body's ability to generate an adequate immune response, resulting in a more severe disease outcome, are underlying malnutrition (Crawford & Vermund, 1988) and genetic immunosuppressive disorders (Saulsbury, Winkelstein & Yolken (1980), as well as immunosuppressive viral diseases including acquired immune deficiency syndrome (Clifford et al., 1990) and measles (Crawford & Vermund, 1988).

From an examination of the calculated median values listed in Table 4.7, it can be estimated that the overall probability of infection leading to death for illnesses caused by either enteric bacteria or viruses is approximately 0.01–0.02, or 1–2%. However, the probability of illness from hepatitis E virus leading to death can be 0.17 for women in their second or third trimester of pregnancy (Gove et al., 1987). Some waterborne hepatitis E virus disease outbreaks have resulted in even higher probabilities of death for pregnant women (Khuroo, 1991). Only one value could be found, that represented a general population, for use in estimating the probability that illness from an enteric protozoan infection would lead to death. This was from a waterborne outbreak of cryptosporidiosis, for which the estimated probability of death is 0.0002, or 0.02%. We have used this value to represent probability of death from protozoan illness for the purpose of our modeling exercise. However, use of this value for protozoan illness may represent a severe understatement when it comes to estimating the risk for specific subpopulations of individuals. In particular, numbers from a publication by Crawford & Vermund (1988) indicate that the probability of cryptosporidial illness leading to death can be 0.14 (2/14) in the case of underlying malnutrition, and 0.20 in the case of underlying measles.

Multiple concurrent infections

For any given population, there will be individuals who are already infected with one pathogenic organism, at the time when they become infected by yet another organism. Thus, individuals may experience episodes of infectious disease caused by more than a single type of microorganism. Concurrent infections can deleteriously affect the outcome from a person's encounter with the second, or even third, organism (Mata *et al.*, 1983; Crawford & Vermund, 1988; Grohmann *et al.*, 1993). It is important to understand this concept when examining the results of our risk assessment modeling. The probabilities of infection per ingested organism, and of infection leading to illness, which respectively are listed in Tables 4.4 and 4.6, were derived from experimentally infected individuals who neither had obvious underlying illness, nor were known to have immunosuppressive health conditions.

There are reports of concurrent enteric infections in individuals caused by multiple genera of protozoans (Cristino, Carvalho & Salgado, 1988) or bacteria (Mata *et al.*, 1983); and of simultaneous infections with protozoa and virus (Taylor *et al.*, 1985; Crawford & Vermund, 1988; Grohmann *et al.*, 1993), as well as with bacteria and virus (Mata *et al.*, 1983). It is even possible to become concurrently infected with two different enteric viruses of the same genus, as has been noted for infections by rotaviruses (Rodriguez *et al.*, 1983) and enteroviruses (Koprowski, 1955; Tambini *et al.*, 1993). In fact, the capability of simultaneously being infected with multiple species of the genus *Enterovirus* is what allows the administration of oral poliovirus vaccine in a trivalent form, containing infectious viruses of all three human poliovirus species (Abraham *et al.*, 1993).

The possibility of reinfection

In order to understand the risk values from this modeling effort, it is also important to realize that infection with a given species of microorganism does not offer assurance against subsequent reinfection by that same species at some later time. Reinfection can occur even if the individual has developed a measurable immune response to that organism as determined by antibody titers (Cash *et al.*, 1974). Studies have shown that it is possible for individuals to be reinfected with the same subgroup (Simhon *et al.*, 1981), or even the same strain of enteric virus (Koprowski, 1956), with the same strain of enteric protozoa (Nash *et al.*, 1987), or the same phage type of enteric bacteria (Hornick *et al.*, 1970b). It is possible for reinfection to occur as soon as 12 weeks after the initial infection (Nash *et al.*, 1987). No information was found that indicated an upper limit to the number of times an individual could be reinfected by the same species, serotype or strain of microorganism. When just viruses alone are considered, the possibility of reinfection, compounded by the existence of 140 different human enteric viruses (Hurst, Benton & Stetler, 1989), presents an enormous potential health risk.

Results of the risk modeling exercise

As stated above, the risk modeling exercise developed for this chapter was based upon Equations 4.1–4.4. The value used to represent the daily amount of water ingested per individual was 2 l, as recommended by the US Environmental Protection Agency (1989). The other values substituted into the equations are listed in Table 4.8. The values used to represent the estimated number of pathogenic microorganisms per liter of untreated source water for the categories of bacteria and protozoans are the median statistics derived from Tables 4.2 and 4.3, respectively. The value used to represent the estimated number of pathogenic viruses per liter is an analogous median statistic from the review by Hurst (1991a). The values used to represent the estimated probabilities of infection per ingested organism, illness per incidence of infection, and death per incidence of illness, are median statistics respectively taken from Tables 4.4, 4.6 and 4.7. The results derived from substituting these values into Equations 4.1–4.4, respectively are presented in Tables 4.9–4.11 for the three categories of microorganisms: bacteria, protozoans and viruses.

Table 4.8. *Values used for estimation of human health risk from ingesting pathogenic microorganisms*[a]

	Bacteria	Protozoans	Viruses
Estimated number of pathogenic organisms per liter of untreated source water	2.2	1.7×10^{-1}	1.4^{b}
Probability of infection per ingested organism	1.0×10^{-4}	6.7×10^{-2}	5.0×10^{-1}
Probability of illness per incidence of infection	4.9×10^{-1}	4.4×10^{-1}	5.8×10^{-1}
Probability of death per incidence of illness	1.8×10^{-2}	2.0×10^{-4}	9.4×10^{-3}

[a]Values represent medians derived in Tables 4.2–4.7 with the exception of the value indicated by footnote b.
[b]Median statistic calculated for level of human enteric viruses in naturally occurring freshwaters that were not indicated to be heavily sewage impacted, including all virus-negative samples (Hurst, 1991a).

Table 4.9. *Estimated individual annual risk of infection, illness and death from ingestion of bacterial pathogens in drinking water*

Level of reduction in viable microorganisms	Risk of infection	Risk of illness	Risk of death
No water treatment	0.1607	0.0787	0.0014
90%	0.0161	0.0079	0.0001
99%	0.0016	0.0008	<0.0001
99.9%	0.0002	<0.0001	
99.99%	<0.0001		

Note: These calculations were made by substituting into Equations 4.1–4.4 the values listed for bacteria in Table 4.8.

Table 4.10. *Estimated individual annual risk of infection, illness and death from ingestion of protozoan pathogens in drinking water*

Level of reduction in viable microorganisms	Risk of infection	Risk of illness	Risk of death
No water treatment	1.0000[a]	1.0000[b]	0.0007
90%	0.8320	0.3661	<0.0001
99%	0.0832	0.0366	
99.9%	0.0083	0.0037	
99.99%	0.0008	0.0004	
99.999%	<0.0001	<0.0001	

Note: These calculations were made by substituting into Equations 4.1–4.4 the values listed for protozoans in Table 4.8.
[a]The actual calculation yielded a value of 8.3201, but the number listed in this table for annual risk is not allowed to exceed 1.0.
[b]The actual calculation yielded a value of 3.6608, but the number listed in this table for annual risk is not allowed to exceed 1.0.

Sample calculations

We have chosen to use the protozoan data when presenting sample risk calculations for this section of the chapter. The actual calculated value derived from employing Equations 4.1 and 4.2 for estimating an individual's annual risk of infection from protozoan pathogens by ingesting untreated drinking water would be 8.3201. This suggests the possibility that, during the course of a year, an individual could suffer more than eight individual incidences of infection, or 'infection events', caused by protozoa if they drank untreated water that contained the indicated median level of 1.7×10^{-1} pathogenic protozoan organisms per liter. The derivation of this value was as follows: $[(1.7 \times 10^{-1}$ organisms/l) × (2.0 l of water ingested/day) ×

Table 4.11. *Estimated individual annual risk of infection, illness and death from ingestion of viral pathogens in drinking water*

Level of reduction in viable microorganisms	Risk of infection	Risk of illness	Risk of death
No water treatment	1.0000[a]	1.0000[b]	1.0000[c]
90%	1.0000	1.0000	0.2788
99%	1.0000	1.0000	0.0279
99.9%	0.5113	0.2966	0.0028
99.99%	0.0511	0.0297	0.0003
99.999%	0.0051	0.0030	<0.0001
99.9999%	0.0005	0.0003	
99.99999%	<0.0001	<0.0001	

Note: These calculations were made by substituting into Equations 4.1–4.4 the values listed for viruses in Table 4.8.
[a]The actual calculation yielded a value of 511.3500, but the number listed in this table for annual risk is not allowed to exceed 1.0.
[b]The actual calculation yielded a value of 296.5830, but the number listed in this table for annual risk is not allowed to exceed 1.0.
[c]The actual calculation yielded a value of 2.7879, but the number listed in this table for annual risk is not allowed to exceed 1.0.

(365.25 days of water ingestion/year) × (6.7 × 10⁻² probability of infection per ingested organism)]. According to statistical rules, the value listed in Table 4.10 for risk of protozoan infection with no water treatment is limited at the upper end to 1.000. For this exercise, the risks are being estimated on an annual basis. Thus, this listed value of 1.000 indicates that every individual is likely to suffer at least one protozoan infection event during any given year.

In order to determine the effect of using drinking water treatment processes to achieve different levels of microbial removal from the source waters, we multiplied the annual values for estimated risk of infection (Equation 4.2), risk of illness (Equation 4.3) and risk of death (Equation 4.4) by a mathematical value representing the extent of microbial removal (1 minus the level of microbial removal, expressed as percentage in a decimal format). Thus, estimation of the risk of protozoan infection from ingesting this same water, if treatment was 90% effective at achieving removal of pathogenic protozoa, would be as follows: [8.320 × (1 − 0.90)]. The resulting calculated value is 0.8320, suggesting that each individual would have an 83% chance of suffering a protozoan infection event during any given year. This calculated annual risk value of 0.8320 can be listed in Table 4.10 directly, since the value does not exceed 1.0. Our lower limit cutoff level for listing annual values in Tables 4.9–4.11 was arbitrarily set at 0.0001, representing a risk level of 1 per 10000 consumers per year.

Calculations for bacterial disease

The estimated annual risk of bacterial infection from drinking untreated water that contains 2.2 pathogenic enteric bacteria per liter is less than 1.000 (Table 4.9). The estimated annual risk of death from bacterial disease drops below the lower cutoff level if water treatment is effective at achieving 99% (two $\log_{10}$ units) of bacterial reduction through physical removal or disinfection. With four $\log_{10}$ units of treatment effectiveness, the estimated annual risk of bacterial infection drops below the lower cutoff. The fact that these risk levels are reduced below the lower cutoff level so readily for bacterial disease may explain why the practice of water filtration was adequate at reducing the incidence of typhoid death in the United States at the beginning of the 20th century (Johnson, 1916).

Calculations for protozoan disease

The results of calculations estimating the annual risks for infection and illness from protozoans, if individuals ingested untreated water containing 0.17 pathogenic enteric protozoan cysts or oocysts per liter, are both listed in Table 4.10 as 1.000. This suggests that all consumers who drink untreated water containing that level of pathogenic protozoan cysts or oocysts per liter would be likely to suffer at least one infection event per year caused by the protozoan contaminants in their water and, in addition, all consumers would probably have at least one of those annual protozoan-related infection events lead to illness. The estimated annual risk of death from protozoan disease drops below the lower cutoff level if water treatment is effective at achieving one $\log_{10}$ unit of protozoan removal. The estimated annual risk of protozoan infection drops below the lower cutoff level by achieving five $\log_{10}$ units of treatment effectiveness.

Calculations for viral disease

The results of calculations estimating the annual risks of viral infection, illness and death from ingesting untreated water containing 1.4 infectious units of virus per liter are all listed as 1.000 (Table 4.11). This suggests that all consumers who drink untreated water containing that level of pathogenic enteric viruses would be likely to suffer one infection event per year caused by the viral contaminants in their water and, in addition, all would be likely to suffer a viral infection event that results in death. The estimated annual risk of death from viral disease drops below the lower cutoff level if water treatment is effective at achieving five $\log_{10}$ units of viral removal. The estimated annual risk of viral infection drops below the lower cutoff level with seven $\log_{10}$ units of treatment effectiveness.

Discussion of the model parameters

For the purpose of this risk modeling exercise, all enteric bacterial pathogens have been grouped into a single category. Similarly, there is a single category apiece for enteric protozoan pathogens and enteric viral pathogens. For this modeling exercise, the probability of each individual occurrence of infection, termed here an 'infection event', is assumed to be independent of any other infection event caused by either the same or another microorganism. The probability of an infection event leading to illness is conditionally dependent upon an infection having occurred. The probability of illness progressing to death is in turn dependent upon an infection event having led to illness. It is possible that the probability of individual infection events leading to illness, and the accompanying severity of illness including progression to death, may decrease with subsequent reinfections by the same species, serotype or strain of microorganism (Nash et al., 1987). This represents the basis for the phenomenon of naturally acquired immunity, which assumes that after a sufficient number of antigenic exposures to, or infections by, the same serotype of microorganism, an individual will develop a level of immunity that affords protection against the likelihood that subsequent reinfection events caused by that same microorganism will progress to severe illness. Naturally acquired immunity is presumably not protective against infection, only against the severity of the resulting illness. Naturally acquired immunity has not been taken into account by the modeling exercise presented here, for three reasons. First, different strains of even the same species of microorganism vary in their immunogenicity, which is defined as the ability to illicit an immune response. Second, the duration in terms of years for the protective effect attributed to naturally acquired immunity remains uncertain and is probably variable. Third, it has been proven that reinfection of an individual by even the same strain of microorganism can result in sequential episodes of illness (Hornick et al., 1970b; Simhon et al., 1981). Considering the implied severity of risk, particularly from viruses, why do we not all die during infancy? Humans have evolved three mechanisms to protect us during infancy while we begin developing our naturally acquired immunity. First, by initially drinking breast milk we avoid ingesting contaminated water. Second, from maternal antibodies acquired transplacentally plus antibodies received in colostrum and breast milk we acquire some temporary immunological protection against illness. Third, our immune systems function better during infancy than at any subsequent time in our lives.

If a community accurately knew the level of pathogenic bacteria, protozoans or viruses in their source water, they could substitute those known levels into Equation 4.1 in place of our estimated pathogen levels of 2.2 bacteria, 0.17 protozoan cysts and oocysts and 1.4 viruses per liter. Also, if an individual or community wanted to estimate the infectious disease risks associated with one particular microorganism, then it would be more accurate for them to model those risks by substituting values for the known probabilities of infection, illness and death for that specific pathogen into Equations 4.2–4.4, instead of using the median values that we have derived by grouping the pathogens according to whether they were bacteria, protozoans or viruses.

We considered it important to use only data developed in humans for this modeling exercise. Our reason for this decision is that the minimum number of microorganisms required to cause an infection can differ between humans and other animals as much as several hundred-fold (Koprowski, 1955).

It is important to understand that not all of the individuals who are thought to receive the minimum infectious dose of microorganisms during an infection study will necessarily develop an infection. For several of the feeding studies cited in this chapter, 100% of the individuals were infected by the stated minimum infectious dose. However, the percentage of individuals infected in general averaged between 50 and 67. There was variability even between studies that fed to volunteers the same dosage of the same strain of microorganism. This variability was evidenced by comparing the results from two feeding studies that used the SM strain of human poliovirus 1 (Koprowski, 1955; Katz & Plotkin, 1967). Both studies determined that the minimum infectious dose for humans was equivalent to two infectious units for cell culture. One of these studies reported that this dose infected 67% of volunteers (Katz & Plotkin, 1967), but the other study reported this dose to have infected 100% of volunteers (Koprowski, 1955). Three main sources of error can affect volunteer feeding studies. The first source of error is random distribution, which occurs when a suspension of microorganisms (or any other type of particles) is subdivided into a set of samples of smaller volume; not all of those samples will contain the same number of organisms. For example, if a volume of liquid containing 20 organisms were divided into ten smaller samples, some of those samples could contain three or possibly even more organisms and, correspondingly, some of the samples could contain only one or even no organisms. If the minimum infectious dose were two organisms, then infections might not develop in those individuals who by random chance were fed samples containing either one or no organisms. The second source of error lies in the fact that there are both genotypic and phenotypic differences between individuals, and these differences may play a role in susceptibility to infection by a given microorganism. For example, although most individuals might become infected as a result of ingesting two organisms, other individuals might require a dose of three or more organisms before they became infected, and still other individuals might be susceptible to infection by only a single organism. As a third factor, it is possible that an individual's susceptibility to infection will vary with age or health status.

Accurate mathematical compensation is difficult to achieve for all three sources of error. Therefore, we have adopted the reported minimum infectious dose levels without attempting to modify or mathematically correct those values.

The number used in Equation 4.1 as an estimate for the daily volume of water ingested per individual consumer is 2 l. This is the volume that the United States Environmental Protection Agency has recommended for use when calculating the risks from ingestion of water (US Environmental Protection Agency, 1989). Daily water consumption may, however, vary greatly, due to factors such as environmental temperature, differences in individual body mass and the amount of physical exercise performed. Other estimates for water ingestion are available, including one by Payment et al. (1991) that consumers drink the equivalent of 46.5–47.3 glasses of water per week. Making an assumption that the average glass of water represents 250 ml, the value published by Payment et al. (1991) would represent approximately 1.7 l of water ingested per day. It is possible, and perhaps likely, that not all of this ingested water will be consumed plain. Instead, some water will be used in soups, sauces and as the basis for other beverages (Birkhead et al., 1993).

There is a great deal of confidence in the numerical values that we have used for estimating the probability of infection per ingested organism (Table 4.4) and the probability of an infection event resulting in illness (Table 4.6). The reason for this confidence is that the values listed in these tables were derived from studies performed in a controlled, prospective manner. There is somewhat less confidence in the numerical values available for calculating the probability of illness progressing to death (Table 4.7), because these values represent statistics from epidemiological surveys that generally were compiled retrospectively. There must also be some uncertainty as to whether the cited epidemiological studies considered the total number of illness cases attributed to a particular pathogen, or only those cases for which severe symptomatology was demonstrated. Given a fixed number of deaths for a particular disease, varying the criteria used in establishing the number of cases of illness will, in turn, change the calculated probability of illness progressing to death. The 1955–1956 waterborne disease outbreak in Delhi, India, caused by hepatitis E virus represents an example of this variation in reporting the number of cases of illness. Viswanathan (1957) reported that 90 deaths were attributable to hepatitis contracted during this outbreak. We have chosen to cite his estimate of 97 600 cases of illness for this outbreak, since he was an epidemiologist who studied that epidemic at the time when it occurred (Viswanathan, 1957). Some more recent publications (Ramalingaswami & Purcell, 1988; Khuroo, 1991) cite 29 300 cases of illness for this same outbreak, a number that represents only those individuals whose illness was sufficiently severe that it resulted in jaundice. If we instead had used the lower number of cases of illness in our risk modeling exercise, then the probability of illness progressing to death for that outbreak would have risen from 0.0009 to 0.0031. Our choice of using overall median values to represent the probability of illness progressing to death for the categories of bacterial, protozoan and viral diseases, rather than using corresponding mean values, helps to lessen the impact upon our modeling exercise of any cited study that may represent a particularly high or low probability value.

Validation of the risk modeling approach

Validation of the risk modeling approach described in this chapter can be attempted by using data which Payment and colleagues have published for a section of Quebec Province, Canada. Results of an epidemiological study (Payment *et al.*, 1991) indicated that the annual risk of consumers acquiring gastroenteritis from ingesting microorganisms contained in the community distributed tap water was 0.26. In an earlier study, it was determined that the average level of enteric viruses in the raw water entering that area's drinking water treatment plants was 3.3 cell culture infectious units per liter (Payment *et al.*, 1985). In that same 1985 publication, it was indicated that the average cumulative viral reduction achieved by those community drinking water treatment plants was 99.97%. Substituting into Equations 4.1–4.3 this reported value for the level of viruses in the raw water, and taking into account the reduction in viral level achieved by community drinking water treatment, results in a predicted individual annual risk of viral illness equal to 0.2097. The calculations used for this prediction are as follows: [(3.3 viral organisms per liter) × (2.0 liters of water ingested per day) × (365.25 days of water ingestion per year) × (5.0×10^{-1} probability of infection per ingested virus) × (5.8×10^{-1} probability of illness per viral infection) × (1.0000 – 0.9997 as the estimate of residual viral disease risk following community drinking water treatment)].

An estimate for the amount of disease risk represented by pathogenic protozoan cysts and oocysts in the community treated drinking water for that same geographical area can be made using results published later by Payment & Franco (1993). In this publication, the authors presented information regarding the levels of protozoans belonging to two genera, *Giardia* and *Cryptosporidium*, in the finished (treated) water produced by three water treatment plants. The levels of these protozoans in finished water from two of the treatment plants was below the researchers' limit of detection. However, the level of protozoans in finished water from the third treatment plant was 0.02 *Giardia* cysts and 0.02 *Cryptosporidium* oocysts/100 l. Added together, the levels of these two organisms gives an estimate of 0.04 protozoan cysts or oocysts/100 l, equivalent to 0.0004 organisms/l. Substituting this level of protozoans into Equations 4.1–4.3

yields an estimated annual protozoan disease risk of 0.0086. The calculations for this prediction are as follows: [(0.0004 protozoan organisms/l) × (2.0 l of water ingested/day) × (365.25 days of water ingestion/year) × (6.7 × 10^{-2} probability of infection per ingested protozoan cyst or oocyst) × (4.4 × 10^{-1} probability of illness per protozoan infection).

Adding together the estimated 0.2097 annual risk of viral illness per individual consumer, and the estimated 0.0086 annual risk from protozoan illness per consumer, gives an annual risk estimate of 0.2183 for illnesses attributed to these two categories of microorganisms. This indicates that, by using the risk modeling approach presented in this chapter, we are able to account for 84% of the actual epidemiologically determined risk. The calculation for this is as follows: (0.2183 / 0.26 × 100).

If we are able to make the assumption that the remaining 0.0417 level of annual illness risk is due to pathogenic bacterial contaminants in the distribution system water, then we can perform a backwards calculation to estimate the associated level of pathogenic bacteria in the distribution system water. The process for this back calculation is as follows: [(0.0417 level of risk presumed due to bacterial pathogens) / (4.9 × 10^{-1} probability of bacterial infection leading to illness) / (1.0 × 10^{-4} probability of bacterial infection per ingested organism) / (365.25 days of water ingestion per year) / (2 l of water ingested/day)]. According to this modeling approach, the residual individual annual risk of 0.0417 would represent a level of 1.16 pathogenic bacteria/l in the community distributed tap water. As mentioned earlier in this chapter, some of these bacteria could represent organisms that have persisted through the processing regimen at the community drinking water treatment plants. A second source of bacteria in water distribution systems is accidental contamination, which can result from breaks in water distribution lines and also from inadvertant cross connections with pipes that carry sewage. A third source of bacterial contaminants is the biofilm that grows on the inside walls of water distribution pipes and water storage tanks. It is important to note that published data on levels of largely non-pathogenic 'indicator' bacteria, such as coliforms, in water cannot directly be used to estimate disease risk.

Methods available for treating drinking water to reduce the risk of infectious disease

The numbers presented in Tables 4.9–4.11 for estimated annual risk of infection, illness and death seem dire, but drinking water treatment can be achieved by many different processes. Some of these, such as full conventional treatment, which is defined as coagulation, sedimentation, rapid filtration and chemical disinfection, performed in series, are only practical when conducted on a community basis. This conventional treatment process clearly is effective at achieving microbial removal (Payment *et al.*, 1985; Hurst, 1991*a*; Stetler, Waltrip & Hurst, 1992; Payment & Franco; 1993). However, this is not the only available means of treating drinking water. Microbial removal also can be achieved by less complicated community-based operations such as slow sand filtration (Fogel *et al.*, 1993).

Drinking water treatment can be performed with varying degrees of effectiveness on a household basis. Household treatments include heating water to a temperature that is below the boiling point but high enough to destroy the microbial contaminants (Rice & Johnson, 1991; Fayer, 1994), heating water still higher to achieve the boiling point (Akhter *et al.*, 1994), and perhaps even using distillation (Van Steenderen, 1977). Unintentional water treatment is a side benefit from heating or boiling water for the purpose of preparing foods and beverages. Household cartridge filtration systems can effectively remove microbial pathogens from water (Payment *et al.*, 1991). Laboratory evaluation has indicated that the practice of using clay as a coagulant for treating water on a household basis may have some success (Lund & Nissen, 1986). Chemical coagulants such as alum can also have some degree of effectiveness in the treatment of drinking water (Ahmad, Jahan & Huq, 1984). Chemical coagulants can be used in conjunction with chlorine compounds, the later type of mixture being available as tablets (Powers *et al.*, 1994). Powdered plant materials also have been evaluated as coagulants that could be used on a household basis both alone and in conjunction with other chemicals (Al Azharia Jahn & Dirar, 1979). Chemical disinfectants that can be used effectively on a small scale include chlorine and iodine (Ellis, Cotton & Khowaja, 1993; Powers *et al.*, 1994).

Summary

Pathogenic microorganisms can frequently be found in source waters, and the ingestion of microbially contaminated water clearly has the potential for causing infectious disease. Microorganisms can escape even well-operated, community based drinking water treatment processes, resulting in the presence in treated drinking water of pathogenic organisms belonging to three categories: bacteria (Payment, Coffin & Paquette, 1994), protozoans (Hibler, 1988; LeChevallier, Norton & Lee, 1991*b*) and viruses (Hurst, 1991*a*). These residual microbial contaminants can be sufficient to cause appreciable levels of disease among consumers. Payment and co-workers (Payment *et al.*, 1991) have calculated that individuals ingesting conventionally treated, community distributed tap water had a 0.76 annual risk for incidence of gastrointestinal illness. This contrasted with an annual risk of 0.50 among a control group of consumers who drank water from the same community distribution system, but whose water had gone through an in-home filtration system prior to ingestion. The difference in annual risk attributed to the tap water was 0.26, representing 35% of gastrointestinal illness. This represents endemic disease, because there were no epidemics of gastroenteritis reported in the community during that time.

Our goal in this chapter was to present a modeling excercise that could be used to estimate the risks of infection, illness and death, on an annual basis, from ingesting pathogenic bacteria, protozoans and viruses contained in drinking water. The models presented in this chapter represent an example of how this risk can be estimated, and also the extent to which the risk can be lessened by reducing the level of microorganisms in water prior to its ingestion. In order to perform this exercise, we examined published studies regarding the level of pathogenic microorganisms in environmental waters that were likely to be used as sources of potable water. Based upon this occurrence data, we estimated an annual ingested dosage of these organisms per individual consumer, and multiplied that dose by an estimated probability of infection per ingested organism. Those probabilities of infection were derived from studies in which human volunteers had been fed microorganisms belonging to the three different categories. The resulting estimates for risk of infection from consuming untreated water were then multiplied by literature-derived estimates for probability of illness per incidence of infection, and probability of death per incidence of illness. We also determined how the resulting calculated estimates for annual risk of infection, illness and death would be reduced by achieving various levels of microbial removal from source waters prior to ingestion of the water.

The calculations presented in this chapter suggest that appreciable human health risks may exist from ingesting microbial contaminants in water, and that extensive treatment of drinking water may be required in order to reduce these disease hazards to acceptable levels. The potential reality of these risks is borne out by the fact that epidemics of waterborne disease do occur in communities served by drinking water treatment facilities, and by published calculations for endemic waterborne disease (Payment et al., 1991). The approach presented in this chapter for modeling the risks of acquiring infectious disease from contaminants in drinking water can be used by communities to estimate the danger to their population. Communities will be able to improve their estimates of disease risk if they know the levels of pathogens in their source water, and the efficiency of microbial removal by their treatment processes. Alternatively, by knowing the levels of pathogenic microorganisms in their source water, and by determining a target value for the maximum desired disease risk to their community, this modeling approach could be used to estimate the level of water treatment needed to decrease the disease risk to that target level.

References

Abraham, R., Minor, P., Dunn, G., Modlin, J. F. & Ogra, P. L. (1993). Shedding of virulent poliovirus revertants during immunization with oral poliovirus vaccine after prior immunization with inactivated polio vaccine. *Journal of Infectious Diseases*, **168**, 1105–9.

Ahmad, K., Jahan, K. & Huq, I. (1984). Decontamination of drinking water by alum for the preparation of oral rehydration solution. *Food and Nutrition Bulletin*, **6** (2), 54–7.

Akhter, M. N., Levy, M. E., Mitchell, C., Boddie, R., Donegan, N., Griffith, B., Jones, M. & Stair, T. O. (1994). Assessment of inadequately filtered public drinking water – Washington, DC, December, 1993. *Morbidity and Mortality Weekly Report*, **43**, 661–9.

Al Azharia Jahn, S. & Dirar, H. (1979). Studies on natural water coagulants in the Sudan, with special reference to Moringa oleifera seeds. *Water SA* (South Africa), **5** (2), 90–7.

Assaad, F. & Borecka, I. (1977). Nine-year study of WHO virus reports on fatal virus infections. *Bulletin of the World Health Organization*, **55**, 445–53.

Bergeisen, G. H., Hinds, M. W. & Skaggs, J. W. (1985). A waterborne outbreak of hepatitis A in Meade County, Kentucky. *American Journal of Public Health*, **75** (2), 161–4.

Bernard, R. P. (1965). The Zermatt typhoid outbreak in 1963. *Journal of Hygiene, Cambridge*, **63**, 537–63.

Bernstein, D. I., Ziegler, J. M. & Ward, R. L. (1986). Rotavirus fecal IgA antibody response in adults challenged with human rotavirus. *Journal of Medical Virology*, **20**, 297–304.

Birkhead, G. & Vogt, R. L. (1989). Epidemiologic surveillance for endemic *Giardia lamblia* infection in Vermont: the roles of waterborne and person-to-person transmission. *American Journal of Epidemiology*, **129**, 762–8.

Birkhead, G. S., Morse, D. L., Levine, W. C., Fudala, J. K., Kondracki, S. F., Chang, H.-G., Shayegani, M., Novick, L. & Blake, P. A. (1993). Typhoid fever at a resort hotel in New York: a large outbreak with an unusual vehicle. *Journal of Infectious Diseases*, **167**, 1228–32.

Blake, P. A., Allegra, D. T., Snyder, J. D., Barrett, T. J., McFarland, L., Caraway, C. T., Feeley, J. C., Craig, J. P., Lee, J. V., Puhr, N. D. & Feldman, R. A. (1980). Cholera – a possible endemic focus in the United States. *New England Journal of Medicine*, **302**, 305–9.

Blake, P. A., Rosenberg, M. L., Costa, J. B., Ferreira, P. S., Guimaraes, C. L. & Gangarosa, E. J. (1977). Cholera in Portugal, 1974. I. Modes of transmission. *American Journal of Epidemiology*, **105**, 337–43.

Bloch, A. B., Stramer, S. L., Smith, J. D., Margolis, H. S., Fields, H. A., McKinley, T. W., Gerba, C. P., Maynard, J. E. & Sikes, R. K. (1990). Recovery of hepatitis A virus from a water supply responsible for a common source outbreak of hepatitis A. *American Journal of Public Health*, **80** (4), 428–30.

Bongard, J., Savage, R., Dern, R., Bostrum, H., Kazmierczak, J., Keifer, S., Anderson, H. & Davis, J. P. (1994). *Cryptosporidium* infections associated with swimming pools – Dane County, Wisconsin, 1993. *Morbidity and Mortality Weekly Report*, **43**, 561–72.

Brayton, P. R., Tamplin, M. L., Huq, A. & Colwell, R. R. (1987). Enumeration of *Vibrio cholerae* 01 in Bangladesh waters by fluorescent-antibody direct viable count. *Applied and Environmental Microbiology*, **53**, 2862–5.

Cameron, J. M., Hester, K., Smith, W. L., Caviness, E., Hosty, T. & Wolf, F. S. (1977). *Vibrio cholerae* – Alabama. *Morbidity and Mortality Weekly Report*, **26**, 159–60.

Caprioli, A., Luzzi, I., Rosmini, F., Resti, C., Edefonti, A., Perfumo, F., Farina, C., Goglio, A., Gianviti, A. & Rizzoni, G. (1994). Communitywide outbreak of hemolytic–uremic syndrome associated with non-0157 verocytotoxin-producing *Escherichia coli*. *Journal of Infectious Diseases*, **169**, 208–11.

Cash, R. A., Music, S. I., Libonati, J. P., Snyder, M. J., Wenzel, R. P. & Hornick, R. B. (1974). Response of man to infection with *Vibrio cholerae*. I. Clinical, serologic, and bacteriologic responses to a known inoculum. *Journal of Infectious Diseases*, **129**, 45–52.

Centers for Disease Control (1990). Protection against viral hepatitis: recommendations of the immunization practices advisory committee (ACIP). *Morbidity and Mortality Weekly Report*, **39** (RR-2), 1–26.

Centers for Disease Control (1993). Isolation of wild poliovirus type 3 among members of a religious community objecting to vaccination – Alberta, Canada, 1993. *Morbidity and Mortality Weekly Report*, **42**, 337–9.

Centers for Disease Control and Prevention (1993). Update: cholera – Western hemisphere, 1992. *Morbidity and Mortality Weekly Report*, **42**, 89–91.

Centers for Disease Control and Prevention (1994). Outbreak of *Shigella flexneri* 2a infections on a cruise ship. *Morbidity and Mortality Weekly Report*, **43**, 657.

Chengqin, S., Yilun, W., Hongkai, S., Daibao, W., Yunhua, C., Dogmin, W., Lina, H. & Zaolun, Y. (1986). An outbreak of epidemic diarrhoea in adults caused by a new rotavirus in Anhui province of China in the summer of 1983. *Journal of Medical Virology*, **19**, 167–73.

Cholera Working Group, International Centre for Diarrhoeal Diseases Research, Bangladesh (1993). Large epidemic of cholera-like disease in Bangladesh caused by *Vibrio cholerae* 0139 synonym Bengal. *Lancet*, **342**, 387–90.

Clark, H. F., Offit, P. A., Dolan, K. T., Tezza, A., Gogalin, K., Twist, E. M. & Plotkin, S. A. (1986). Response of adult human volunteers to oral administration of bovine and bovine/human reassortant rotaviruses. *Vaccine*, **4** (March), 25–31.

Clifford, C. P., Crook, D. W. M., Conlon, C. P., Fraise, A. P., Day, D. G. & Peto, T. E. A. (1990). Impact of waterborne outbreak of cryptosporidiosis on AIDS and renal transplant patients. *Lancet*, **335**, 1455–6.

Colwell, R. R., Kaper, J. & Joseph, S. W. (1977). *Vibrio cholerae*, *Vibrio parahaemolyticus*, and other vibrios: occurrence and distribution in Chesapeake Bay. *Science*, **198**, 394–6.

Couch, R. B., Cate, T. R., Gerome, P. J., Fleet, W. F., Lang, D. J., Griffith, W. R. & Knight, V. (1965). Production of illness with a small-particle aerosol of Coxsackie A21. *Journal of Clinical Investigation*, **44**, 535–42.

Crawford, F. G. & Vermund, S. H. (1988). Human cryptosporidiosis. *CRC Critical Reviews in Microbiology*, **16**, 113–59.

Cristino, J. A. G. M., Carvalho, M. I. P. & Salgado, M. J. (1988). An outbreak of cryptosporidiosis in a hospital day-care centre. *Epidemiology and Infection*, **101**, 355–9.

Current, W. L. (1986). *Cryptosporidium*: its biology and potential for environmental transmission. *CRC Critical Reviews in Environmental Control*, **17**, 21–51.

Dahling, D. R. & Safferman, R. S. (1979). Survival of enteric viruses under natural conditions in a subarctic river. *Applied and Environmental Microbiology*, **38**, 1103–10.

D'Antonio, R. G., Winn, R. E., Taylor, J. P., Gustafson, T. L., Current, W. L., Rhodes, M. M., Gary, G. W. Jr & Zajac, R. A. (1985). A waterborne outbreak of cryptosporidiosis in normal hosts. *Annals of Internal Medicine*, **103** (6 pt 1), 886–8.

DuPont, H. L., Chappell, C. L., Sterling, C. R., Okhuysen, P. C., Rose, J. B. & Jakubowski, W. (1995). The infectivity of *Cryptosporidium parvum* in healthy volunteers. *New England Journal of Medicine*, **332**, 855–9.

Echeverria, P., Blacklow, N. R., Cukor, G. G., Vibulbandhitkit, S., Changchawalit, S. & Boonthai, P. (1983). Rotavirus as a cause of severe gastroenteritis in adults. *Journal of Clinical Microbiology*, **18**, 663–7.

Egoz, N., Shihab, S., Leitner, L. & Lucian, M. (1988). An outbreak of typhoid fever due to contamination of the municipal water supply in Northern Israel. *Israeli Journal of Medical Sciences*, **24**, 640–3.

Eichold, B. H., Williamson, J. R., Woernle, C. H. & McPhearson, R. M. (1993). Isolation of *Vibrio cholerae* 01 from oysters – Mobile Bay, 1991–1992. *Morbidity and Mortality Weekly Report*, **42**, 91–3.

Eiden, J., Vonderfecht, S. & Yolken, R. H. (1985). Evidence that a novel rotavirus-like agent of rats can cause gastroenteritis in man. *Lancet*, **ii**, 8–10.

Ellis, K. V., Cotton, A. P. & Khowaja, M. A. (1993). Iodine disinfection of poor quality waters. *Water Research*, **27**, 369–75.

Falcão, D. P., Valentini, S. R. & Leite, C. Q. F. (1993). Pathogenic or potentially pathogenic bacteria as contaminants of freshwater from different sources in Araraquara, Brazil. *Water Research*, **27**, 1737–41.

Fang, Z. Y., Ye, Q., Ho, M.-S., Dong, H., Qing, S., Penaranda, M. E., Hung, T., Wen, L. & Glass, R. I. (1989). Investigation of an outbreak of adult diarrhea rotavirus in China. *Journal of Infectious Diseases*, **160**, 948–53.

Fayer, R. (1994). Effect of high temperature on infectivity of *Cryptosporidium parvum* oocysts in water. *Applied and Environmental Microbiology*, **60**, 2732–5.

Fayer, R. & Ungar, B. L. P. (1986). *Cryptosporidium* spp. and cryptosporidiosis. *Microbiological Reviews*, **50**, 458–83.

Fogel., D., Isaac-Renton, J., Guasparini, R., Moorehead, W. & Ongerth, J. (1993). Removing giardia and cryptosporidium by slow sand filtration. *Journal, American Water Works Association*, **85** (11), 77–84.

Foster, S. O., Palmer, E. L., Gary, G. W. Jr, Martin, M. L., Herrmann, K. L., Beasley, P. & Sampson, J. (1980). Gastroenteritis due to rotavirus in an isolated Pacific island group: an epidemic of 3,439 cases. *Journal of Infectious Diseases*, **141**, 32–9.

Geldreich, E. E. (1966). *Sanitary Significance of Fecal Coliforms in the Environment*. Cincinnati, Ohio: US Department of the Interior Federal Water Pollution Control Administration.

Geldreich, E. E., Fox, K. R., Goodrich, J. A., Rice, E. W., Clark, R. M. & Swerdlow, D. L. (1992). Searching for a water supply connection in the Cabool, Missouri disease outbreak of *Escherichia coli* 0157:H7. *Water Research*, **26**, 1127–37.

Gerba, C. P. & Haas, C. N. (1988). Assessment of risks associated with enteric viruses in contaminated drinking water. In *Chemical and Biological Characterization of Sludges, Sediments, Dredge-spoils, and Drilling Muds, ASTM STP 976*, ed. J. J. Lichtenberg, J. A. Winter, C. I. Weber & L. Fradkin, pp. 489–94. Philadelphia: American Society for Testing and Materials.

Gove, S., Ali-Salad, A., Farah, M. A., Delaney, D., Roble, M. J., Walter, J. & Aziz, N. (1987). Enterically transmitted non-A, non-B hepatitis – East Africa. *Morbidity and Mortality Weekly Report*, **36**, 241–4.

Greenberg, A. E., Clesceri, L. S. & Eaton, A. D. (eds.) (1992). *Standard Methods for the Examination of Water and Wastewater*, 18th edn, Method 9510G, pp. 9–113 through 9–116. Washington: American Public Health Association, American Water Works Association and Water Environment Federation.

Grimes, D. J. (1991). Ecology of estuarine bacteria capable of causing human disease: a review. *Estuaries*, **14**, 345–60.

Grohmann, G. S., Glass, R. I., Pereira, H. G., Monroe, S. S., Hightower, A. W., Weber, R. & Bryan, R. T. (1993). Enteric viruses and diarrhea in HIV-infected patients. *New England Journal of Medicine*, **329**, 14–20.

Grundmann, H., Kropec, A., Hartung, D., Berner, R. & Daschner, F. (1993). *Pseudomonas aeruginosa* in a neonatal intensive care unit: reservoirs and ecology of the nosocomial pathogen. *Journal of Infectious Diseases*, **168**, 943–7.

Haas, C. N. (1983). Estimation of risk due to low doses of microorganisms: a comparison of alternative methodologies. *American Journal of Epidemiology*, **118**, 573–82.

Haas, C. N., Rose, J. B., Gerba, C. & Regli, S. (1993). Risk assessment of virus in drinking water. *Risk Analysis*, **13**, 545–52.

Hayes, E. B., Matte, T. D., O'Brien, T. R., McKinley, T. W., Logsdon, G. S., Rose, J. B., Ungar, B. L. P., Word, D. M., Pinsky, P. F., Cummings, M. L., Wilson, M. A., Long, E. G., Hurwitz, E. S. & Juranek, D. D. (1989). Large community outbreak of cryptosporidiosis due to contamination of a filtered public water supply. *New England Journal of Medicine*, **320**, 1372–6.

Healy, G. R. (1990). Giardiasis in perspective: the evidence of animals as a source of human *Giardia* infections. In *Giardiasis*, ed. E. A. Meyer, pp. 305–13. Amsterdam: Elsevier Science Publishers (Biomedical Division).

Hejkal, T. W., Keswick, B., LaBelle, R. L., Gerba, C. P., Sanchez, Y., Dreesman, G., Hafkin, B. & Melnick, J. L. (1982). Viruses in a community water supply associated with an outbreak of gastroenteritis and infectious hepatitis. *Journal, American Water Works Association*, **74**, 318–21.

Hibler, C. P. (1988). Analysis of municipal water samples for cysts of *Giardia*. In *Advances in Giardia Research*, ed. P. M. Wallis & B. R. Hammond, pp. 237–45. Calgary: University of Calgary Press.

Hopkins, R. S., Gaspard, G. B., Williams, F. P. Jr, Karlin, R. J., Cukor, G. & Blacklow, N. R. (1984). A community waterborne gastroenteritis outbreak: evidence for rotavirus as the agent. *American Journal of Public Health*, **74**, 263–5.

Hornick, R. B., Greisman, S. E., Woodward, T. E., DuPont, H. L., Dawkins, A. T. & Snyder, M. J. (1970a). Typhoid fever: pathogenesis and immunologic control, first of two parts. *New England Journal of Medicine*, **283**, 686–91.

Hornick, R. B., Greisman, S. E., Woodward, T. E., DuPont, H. L., Dawkins, A. T.

& Snyder, M. J. (1970*b*). Typhoid fever: pathogenesis and immunologic control, second of two parts. *New England Journal of Medicine*, **283**, 739–46.

Hurst, C. J. (1991*a*). Presence of enteric viruses in freshwater and their removal by the conventional drinking water treatment process. *Bulletin of the World Health Organization*, **69**, 113–19.

Hurst, C. J. (1991*b*). Using linear and polynomial models to examine the environmental stability of viruses. In *Modeling the Environmental Fate of Microorganisms*, ed. C. J. Hurst, pp. 137–59. Washington: American Society for Microbiology.

Hurst, C. J., Benton, W. H. & Stetler, R. E. (1989). Detecting viruses in water. *Journal, American Water Works Association*, **81** (9), 71–80.

Hurst, C. J., McClellan, K. A. & Benton, W. H. (1988). Comparison of cytopathogenicity, immunofluorescence and *in situ* DNA hybridization as methods for the detection of adenoviruses. *Water Research*, **22**, 1547–52.

Hurst, C. J., Wild, D. K. & Clark, R. M. (1992). Comparing the accuracy of equation formats for modeling microbial population decay rates. In *Modeling the Metabolic and Physiologic Activities of Microorganisms*, ed. C. J. Hurst, pp. 149–75. New York: John Wiley & Sons.

Isaac-Renton, J. L., Cordeiro, C., Sarafis, K. & Shahriari, H. (1993). Characterization of *Giardia duodenalis* isolates from a waterborne outbreak. *Journal of Infectious Diseases*, **167**, 431–40.

Islam, A. (1990). Giardiasis in developing countries. In *Giardiasis*, ed. E. A. Meyer, pp. 235–66. Amsterdam: Elsevier Science Publishers (Biomedical Division).

Islam, M. S., Hasan, M. K., Miah, M. A., Yunus, M., Zaman, K. & Albert, M. J. (1994). Isolation of *Vibrio cholerae* 0139 synonym Bengal from the aquatic environment in Bangladesh: implications for disease transmission. *Applied and Environmental Microbiology*, **60**, 1684–6.

Johnson, G. A. (1916). The typhoid toll. *Journal, American Water Works Association*, **3** (2), 249–326.

Kaper, J., Lockman, H., Colwell, R. R. & Joseph, S. W. (1979). Ecology, serology, and enterotoxin production of *Vibrio cholerae* in Chesapeake Bay. *Applied and Environmental Microbiology*, **37**, 91–103.

Kapikian, A. Z., Wyatt, R. G., Levine, M. M., Yolken, R. H., VanKirk, D. H., Dolin, R., Greenberg, H. B. & Chanock, R. M. (1983). Oral administration of human rotavirus to volunteers: induction of illness and correlates of resistance. *Journal of Infectious Diseases*, **147**, 95–106.

Katz, M. & Plotkin, S. A. (1967). Minimal infective dose of attenuated poliovirus for man. *American Journal of Public Health and the Nation's Health*, **57**, 1837–40.

Kehr, R. W. & Butterfield, C. T. (1943). Notes on the relation between coliforms and enteric pathogens. *Public Health Reports*, **58**, 589–607.

Kelly, M. T., Peterson, J. W., Sarles, H. E. Jr, Romanko, M., Martin, D. & Hafkin, B. (1982). Cholera on the Texas gulf coast. *Journal of the American Medical Association*, **247**, 1598–9.

Khuroo, M. S. (1991). Hepatitis E: the enterically transmitted non-A, non-B hepatitis. *Indian Journal of Gastroenterology*, **10** (3), 96–100.

Kim-Farley, R. J., Rutherford, G., Lichfield, P., Hsu, S.-T., Orenstein, W. A., Schonberger, L. B., Bart, K. J., Lui, K.-J. & Lin, C.-C. (1984). Outbreak of paralytic poliomyelitis, Taiwan. *Lancet*, **ii**, 1322–4.

Koopman, J. S., Eckert, E. A., Greenberg, H. B., Strohm, B. C., Isaacson, R. E. & Monto, A. S. (1982). Norwalk virus enteric illness acquired by swimming exposure. *American Journal of Epidemiology*, **115**, 173–7.

Koprowski, H. (1955). Living attenuated poliomyelitis virus as an immunizing agent of man. *South African Medical Journal*, **29**, 1134–42.

Koprowski, H. (1956). Immunization against poliomyelitis with living attenuated virus. *American Journal of Tropical Medicine and Hygiene*, **5**, 440–52.

Langmuir, A. D., Nathanson, N. & Hall, W. J. (1956). Surveillance of poliomyelitis in the United States in 1955. *American Journal of Public Health and the Nation's Health*, **46**, 75–88.

Lawson, H. W., Braun, M. M., Glass, R. I. M., Stine, S. E., Monroe, S. S., Atrash, H. K., Lee, L. E. & Englender, S. J. (1991). Waterborne outbreak of Norwalk virus gastroenteritis at a southwest US resort: role of geological formations in contamination of well water. *Lancet*, **337**, 1200–4.

LeChevallier, M. W., Norton, W. D. & Lee, R. G. (1991a). Occurrence of *Giardia* and *Cryptosporidium* spp. in surface water supplies. *Applied and Environmental Microbiology*, **57**, 2610–6.

LeChevallier, M. W., Norton, W. D. & Lee, R. G. (1991b). *Giardia* and *Cryptosporidium* spp. in filtered drinking water supplies. *Applied and Environmental Microbiology*, **57**, 2617–21.

Leland, D., McAnulty, J., Keene, W. & Stevens, G. (1993). A cryptosporidiosis outbreak in a filtered-water supply. *Journal, American Water Works Association*, **85** (6), 34–42.

Levine, M. M., Black, R. & Clements, M. L. (1984). Pathogenesis of enteric infections caused by *Vibrio*. In *Vibrios in the Environment*, ed. R. Colwell, pp. 109–22. New York: John Wiley & Sons.

López, C. E., Dykes, A. C., Juranek, D. D., Sinclair, S. P., Conn, J. M., Christie, R. W., Lippy, E. C., Schultz, M. G. & Mires, M. H. (1980). Waterborne giardiasis: a communitywide outbreak of disease and a high rate of asymptomatic infection. *American Journal of Epidemiology*, **112**, 495–507.

Lund, E. & Nissen, B. (1986). Low technology water purification by bentonite clay flocculation as performed in Sudanese villages virological examinations. *Water Research*, **20**, 37–43.

MacKenzie, W. R., Hoxie, N. J., Proctor, M. E., Gradus, M. S., Blair, K. A., Peterson, D. E., Kazmierczak, J. J., Addiss, D. G., Fox, K. R., Rose, J. B. & Davis, J. P. (1994). A massive outbreak in Milwaukee of cryptosporidium infection transmitted through the public water supply. *New England Journal of Medicine*, **331**, 161–7.

Madore, M. S., Rose, J. B., Gerba, C. P., Arrowood, M. J. & Sterling, C. R. (1987). Occurrence of *Cryptosporidium* oocysts in sewage effluents and selected surface waters. *Journal of Parasitology*, **73**, 702–5.

Marchione, M. (1994). Crypto report cites faulty equipment. *Milwaukee Journal*, **July 21**, Al-9.

Mata, L., Simhon, A., Padilla, R., Gamboa, M. D., Vargas, G., Hernández, F., Mohs, E. & Lizano, C. (1983). Diarrhea associated with rotaviruses, enterotoxigenic *Escherichia coli*, *Campylobacter*, and other agents in Costa Rican children, 1976–1981. *American Journal of Tropical Medicine and Hygiene*, **32**, 146–53.

McCarthy, S. A. & Khambaty, F. M. (1994). International dissemination of epidemic *Vibrio cholerae* by cargo ship ballast and other nonpotable waters. *Applied and Environmental Microbiology*, **60**, 2597–601.

McFarlane, S. A. M. (1988). Routine monitoring of watersheds for *Giardia* cysts in Northeastern Pennsylvania. In *Advances in* Giardia *Research*, ed. P. M. Wallis & B. R. Hammond, pp. 223–5. Calgary: University of Calgary Press.

McMillan, N. S., Martin, S. A., Sobsey, M. D., Wait, D. A., Meriwether, R. A. & MacCormack, J. N. (1992). Outbreak of pharyngoconjunctival fever at a summer camp – North Carolina, 1991. *Morbidity and Mortality Weekly Report*, **41**, 342–4.

Melnick, J. L. (1947). Poliomyelitis virus in urban sewage in epidemic and in nonepidemic times. *American Journal of Hygiene*, **45**, 240–53.

Melnick, J. L., Emmons, J., Opton, E. M. & Coffey, J. H. (1954). Coxsackie viruses from sewage methodology including an evaluation of the grab sample and gauze pad collection procedures. *American Journal of Hygiene*, **59**, 185–95.

Minor, T. E., Allen, C. I., Tsiatis, A. A., Nelson, D. B. & D'Alessio, D. J. (1981). Human infective dose determinations for oral poliovirus type 1 vaccine in infants. *Journal of Clinical Microbiology*, **13**, 388–9.

Mintz, E. D., Hudson-Wragg, M., Mshar, P., Cartter, M. L. & Hadler, J. L. (1993). Foodborne giardiasis in a corporate office setting. *Journal of Infectious Diseases*, **167**, 250–3.

Moore, A. C., Herwaldt, B. L., Craun, G. F., Calderon, R. L., Highsmith, A. K. & Juranek, D. D. (1993). Surveillance for waterborne disease outbreaks – United States, 1991–1992. *Morbidity and Mortality Weekly Report*, **42** (SS-2), 1–22.

Moore, B. E. (1993). Survival of human immunodeficiency virus (HIV), HIV-infected lymphocytes, and poliovirus in water. *Applied and Environmental Microbiology*, **59**, 1437–43.

Morris, J. G. Jr, Takeda, T., Tall, B. D., Losonsky, G. A., Bhattacharya, S. K., Forrest, B. D., Kay, B. A. & Nishibuchi, M. (1990). Experimental non-O group 1 *Vibrio cholerae* gastroenteritis in humans. *Journal of Clinical Investigation*, **85**, 697–705.

Mosley, J. W. (1966). Transmission of viral diseases by drinking water. In *Transmission of Viruses by the Water Route*, ed. G. Berg, pp. 5–23. New York: Interscience Publishers, John Wiley & Sons.

Murphy, A. M., Grohmann, G. S. & Sexton, M. F. H. (1983). Infectious gastroenteritis in Norfolk Island and recovery of viruses from drinking water. *Journal of Hygiene, Cambridge*, **91**, 139–46.

Nalin, D. R. (1976). Cholera, copepods, and chitinase. *Lancet*, **ii**, 958.

Nalin, D. R., Levine, R. J., Levine, M. M., Hoover, D., Bergquist, E., McLaughlin, J., Libonati, J., Alam, J. & Hornick, R. B. (1978). Cholera, non-vibrio cholera, and stomach acid. *Lancet*, **ii**, 856–9.

Nash, T. E., Herrington, D. A., Losonsky, G. A. & Levine, M. M. (1987). Experimental human infections with *Giardia lamblia*. *Journal of Infectious Diseases*, **156**, 974–84.

Ndayimirije, N., Maregeya, M., Nshimirimana, D., Nkurikiye, S., Dzuda, C., Chigumira, S., Matanhire, D., Mwenye, K., Mashayamombe, S., Todd, C., Bassett, M., Munochiveyi, R. & Grady, S. (1993). Epidemic cholera – Burundi and Zimbabwe, 1992–1993. *Morbidity and Mortality Weekly Report*, **42**, 407–16.

136   C. J. HURST, R. M. CLARK AND S. E. REGLI

Ongerth, J. E. (1989). Giardia cyst concentrations in river water. *Journal, American Water Works Association*, **81** (9), 81–6.

Ongerth, J. E. & Stibbs, H. H. (1987). Identification of *Cryptosporidium* oocysts in river water. *Applied and Environmental Microbiology*, **53**, 672–6.

Ortega, Y. R., Sterling, C. R., Gilman, R. H., Cama, V. A. & Díaz, F. (1993). Cyclospora species – a new protozoan pathogen of humans. *New England Journal of Medicine*, **328**, 1308–12.

Pacha, R. E., Clark, G. W. & Williams, E. A. (1985). Occurrence of *Campylobacter jejuni* and *Giardia* species in muskrat (*Ondatra zibethica*). *Applied and Environmental Microbiology*, **50**, 177–8.

Paul, J. R., Trask, J. D. & Gard, S. (1940). II. Poliomyelitic virus in urban sewage. *Journal of Experimental Medicine*, **71**, 765–77.

Payment, P., Coffin, E. & Paquette, G. (1994). Blood agar to detect virulence factors in tap water heterotrophic bacteria. *Applied and Environmental Microbiology*, **60**, 1179–83.

Payment, P. & Franco, E. (1993). *Clostridium perfringens* and somatic coliphages as indicators of the efficiency of drinking water treatment for viruses and protozoan cysts. *Applied and Environmental Microbiology*, **59**, 2418–24.

Payment, P., Richardson, L., Siemiatycki, J., Dewar, R., Edwardes, M. & Franco, E. (1991). A randomized trial to evaluate the risk of gastrointestinal disease due to consumption of drinking water meeting current microbiological standards. *American Journal of Public Health*, **81**, 703–8.

Payment, P., Trudel, M. & Plante, R. (1985). Elimination of viruses and indicator bacteria at each step of treatment during preparation of drinking water at seven water treatment plants. *Applied and Environmental Microbiology*, **49**, 1418–28.

Pazzaglia, G., Lesmana, M., Tjaniadi, P., Subekti, D. & Kay, B. (1993). Use of vaginal tampons in sewer surveys for non-01 *Vibrio cholerae*. *Applied and Environmental Microbiology*, **59**, 2740–2.

Petersen, L. R., Cartter, M. L. & Hadler, J. L. (1988). A food-borne outbreak of *Giardia lamblia*. *Journal of Infectious Diseases*, **157**, 846–8.

Pinfold, J. V., Horan, N. J., Wirojanagud, W. & Mara, D. (1993). The bacteriological quality of rainjar water in rural northeast Thailand. *Water Research*, **27**, 297–302.

Powers, E. M., Hernandez, C., Boutros, S. N. & Harper, B. G. (1994). Biocidal efficacy of a flocculating emergency water purification tablet. *Applied and Environmental Microbiology*, **60**, 2316–23.

Ragazzoni, H., Mertz, K., Finelli, L., Genese, G., Dunston, F. J., Russell, B., Riley, W., Feller, E., Ares, M. B., Fernandez, M., Sfakianaki, E., Simmons, J. A. & Hopkins, R. S. (1991). Cholera – New Jersey and Florida. *Morbidity and Mortality Weekly Report*, **40**, 287–9.

Ramalingaswami, V. & Purcell, R. H. (1988). Waterborne non-A, non-B hepatitis. *Lancet*, **i**, 571–3.

Ray, R., Aggarwal, R., Salunke, P. N., Mehrotra, N. N., Talwar, G. P. & Naik, S. R. (1991). Hepatitis E virus genome in stools of hepatitis patients during large epidemic in north India. *Lancet*, **338**, 783–4.

Regli, S., Rose, J. B., Haas, C. N. & Gerba, C. P. (1991). Modeling the risk from giardia and viruses in drinking water. *Journal, American Water Works Association*, **83** (11), 76–84.

Rendtorff, R. C. (1954). The experimental transmission of human intestinal protozoan parasites. II. *Giardia lamblia* cysts given in capsules. *American Journal of Hygiene*, **59**, 209–20.

Rice, E. W. & Johnson, C. H. (1991). Cholera in Peru. *Lancet*, **338**, 455.

Rodriguez, W. J., Kim, H. W., Brandt, C. D., Gardner, M. K. & Parrott, R. H. (1983). Use of electrophoresis of RNA from human rotavirus to establish the identity of strains involved in outbreaks in a tertiary care nursery. *Journal of Infectious Diseases*, **148**, 34–40.

Roman, R., Middleton, M., Cato, S., Bell, E., Ong, K. R., Gruenewald, R., Ferguson, A. & Ramon, A. (1991). Cholera – New York, 1991. *Morbidity and Mortality Weekly Report*, **40**, 516–18.

Rose, J. B. (1988). Occurrence and significance of cryptosporidium in water. *Journal, American Water Works Association*, **80** (2), 53–8.

Rose, J. B., Kayed, D., Madore, M. S., Gerba, C. P., Arrowood, M. J., Sterling, C. R. & Riggs, J. L. (1988). Methods for the recovery of *Giardia and Cryptosporidium* from environmental waters and their comparative occurrence. In *Advances in* Giardia *Research*, ed. P. M. Wallis & B. R. Hammond, pp. 205–9. Calgary: University of Calgary Press.

Saulsbury, F. T., Winkelstein, J. A. & Yolken, R. H. (1980). Chronic rotavirus infection in immunodeficiency. *Journal of Pediatrics*, **97**, 61–5.

Schiff, G. M., Stefanović, G. M., Young, E. C., Sander, D. S., Pennekamp, J. K. & Ward, R. L. (1984). Studies of echovirus-12 in volunteers: determination of minimal infectious dose and the effect of previous infection on infectious dose. *Journal of Infectious Diseases*, **150**, 858–66.

Schwartzbrod, L., Finance, C., Aymard, M., Brigaud, M. & Lucena, F. (1985). Recovery of reoviruses from tap water. *Zentralblatt für Bakteriologie und Hygiene, 1. Abteilung Originale Reihe B*, **181**, 383–9.

Simhon, A., Chrystie, I. L., Totterdell, B. M., Banatvala, J. E., Rice, S. J. & Walker-Smith, J. A. (1981). Sequential rotavirus diarrhoea caused by virus of same subgroup. *Lancet*, **ii**, 1174.

Smith, H. V., Patterson, W. J., Hardie, R., Greene, L. A., Benton, C., Tulloch, W., Gilmour, R. A., Girdwood, R. W. A., Sharp, J. C. M. & Forbes, G. I. (1989). An outbreak of waterborne cryptosporidiosis caused by post-treatment contamination. *Epidemiology and Infection*, **103**, 703–15.

Snow, J. (1965). *Snow on Cholera*. New York: Hafner Publishing.

Sorvillo, F. J., Fujioka, K., Nahlen, B., Tormey, M. P., Kebabjian, R. & Mascola, L. (1992). Swimming-associated cryptosporidiosis. *American Journal of Public Health*, **82**, 742–4.

Spira, W. M., Khan, M. U., Saeed, Y. A. & Sattar, M. A. (1980). Microbiological surveillance of intra-neighbourhood El Tor cholera transmission in rural Bangladesh. *Bulletin of the World Health Organization*, **58**, 731–40.

Stetler, R. E., Waltrip, S. C. & Hurst, C. J. (1992). Virus removal and recovery in the drinking water treatment train. *Water Research*, **26**, 727–31.

Stewart, A. D. & Ghosal, S. C. (1938). On the value of Wilson and Blair's bismuth sulphite medium in the isolation of *Bact. typhosum* from river water. *Indian Journal of Medical Research*, **25**, 591–3.

138   C. J. HURST, R. M. CLARK AND S. E. REGLI

Stewart, S. J. (1991). *Francisella.* In *Manual of Clinical Microbiology*, 5th edn, ed. A. Balows, W. J. Hausler Jr, K. L. Herrmann, H. D. Isenberg & H. J. Shadomy, pp. 454–6. Washington: The American Society for Microbiology.

St Louis, M. E. (1988). Water-related disease outbreaks, 1985. *Morbidity and Mortality Weekly Report*, **37** (SS-2), 15–24.

Swerdlow, D. L. & Ries, A. A. (1992). Cholera in the Americas. Guidelines for the clinician. *Journal, American Medical Association*, **267**, 1495–9.

Swerdlow, D. L. & Ries, A. A. (1993). *Vibrio cholerae* non-01 – the eighth pandemic? *Lancet*, **342**, 382–33.

Tambini, G., Andrus, J. K., Marques, E., Boshell, J., Pallansch, M., de Quadros, C. A. & Kew, O. (1993). Direct detection of wild poliovirus circulation by stool surveys of healthy children and analysis of community wastewater. *Journal of Infectious Diseases*, **168**, 1510–4.

Tao, H., Guangmu, C., Changan, W., Henli, Y., Zhaoying, F., Tungxin, C., Zinyi, C., Weiwe, Y., Xuejian, C., Shuasen, D., Xiaoquang, L. & Weicheng, C. (1984). Waterborne outbreak of rotavirus diarrhoea in adults in China caused by a novel rotavirus. *Lancet*, **i**, 1139–42.

Tauxe, R. V., Hargrett-Bean, N., Patton, C. M. & Wachsmuth, I. K. (1988). *Campylobacter* isolates in the United States, 1982–1986. *Morbidity and Mortality Weekly Report*, **37** (SS-2), 1–13.

Taylor, J. L., Tuttle, J., Pramukul, T., O'Brien, K., Barrett, T. J., Jolbitado, B., Lim, Y. L., Vugia, D., Morris, J. G. Jr, Tauxe, R. V. & Dwyer, D. M. (1993). An outbreak of cholera in Maryland associated with imported commercial frozen fresh coconut milk. *Journal of Infectious Diseases*, **167**, 1330–5.

Taylor, J. P., Perdue, J. N., Dingley, D., Gustafson, T. L., Patterson, M. & Reed, L. A. (1985). Cryptosporidiosis outbreak in a day-care center. *American Journal of Diseases of Childhood*, **139**, 1023–5.

Teunis, P. F. M., Havelaar, A. H. & Medema, G. J. (1994). *A Literature Survey on the Assessment of Microbiological Risk for Drinking Water*, Report number 734301006. Bilthoven: Rijksinstituut voor Volksgezondheid en Milieuhygiene [National Institute of Public Health and Environmental Protection].

Tulchinsky, T. H., Levine, I., Abrookin, R. & Halperin, R. (1988). Waterborne enteric disease outbreaks in Israel, 1976–1985. *Israeli Journal of Medical Sciences*, **24**, 644–51.

Turney, C., Green-Smith, M., Shipp, M., Mordhorst, C., Whittingslow, C., Brawley, L., Koppel, D., Bridges, E., Davis, G., Voss, J., Lee, R., Jay, M., Abbott, S., Bryant, R., Reilly, K., Werner, S. B., Barrett, L., Jackson, R. J., Rutherford, G. W. III. & Lior, H. (1994). *Escherichia coli* 0157:H7 outbreak linked to home-cooked hamburger – California, July 1993. *Morbidity and Mortality Weekly Report*, **43**, 213–16.

US Environmental Protection Agency (1989) *Exposure Factors Handbook*, pp. 2–1 through 2–10. Washington: United States Environmental Protection Agency Office of Health and Environmental Assessment.

van Lieverloo, J. H. M., van der Kooij, D. & Veenendaal, G. (1995). National survey of invertebrates in drinking water distribution systems in The Netherlands. In *Proceedings of the 1994 Water Quality Technology Conference*, pp. 2065–81. Denver: American Water Works Association.

Van Steenderen, R. A. (1977). Studies on the efficiency of a solar distillation still for supplementing drinking water supplies in South West Africa. *Water SA* (South Africa), **3**, 1–4.

van Wijngaarden, J. K., van Loon, A. M., Oostvogel, P., Mulders, M. N., Buitenwerf, J. & Engelhard, C. F. (1992). Update: poliomyelitis outbreak – Netherlands, 1992. *Morbidity and Mortality Weekly Report*, **41**, 917–19.

Venkateswaran, K., Kiiyukia, C., Takaki, M., Nakano, H., Matsuda, H., Kawakami, H. & Hashimoto, H. (1989). Characterization of toxigenic vibrios isolated from the freshwater environment of Hiroshima, Japan. *Applied and Environmental Microbiology*, **55**, 2613–18.

Viswanathan, R. (1957). Epidemiology. *Indian Journal of Medical Research*, **45**, Supplementary numbers 1–29.

Warburton, D. W., Dodds, K. L., Burke, R., Johnston, M. A. & Laffey, P. J. (1992). A review of the microbiological quality of bottled water sold in Canada between 1981 and 1989. *Canadian Journal of Microbiology*, **38**, 12–19.

Ward, R. L., Bernstein, D. I., Shukla, R., Young, E. C., Sherwood, J. R., McNeal, M. M., Walker, M. C. & Schiff, G. M. (1989). Effects of antibody to rotavirus on protection of adults challenged with a human rotavirus. *Journal of Infectious Diseases*, **159**, 79–88.

Ward, R. L., Bernstein, D. I., Young, E. C., Sherwood, J. R., Knowlton, D. R. & Schiff, G. M. (1986). Human rotavirus studies in volunteers: determination of infectious dose and serological response to infection. *Journal of Infectious Diseases*, **154**, 871–80.

Weissman, J. B., DeWitt, W. E., Thompson, J., Muchnick, C. N., Portnoy, B. L., Feeley, J. C. & Gangarosa, E. J. (1974). A case of cholera in Texas, 1973. *American Journal of Epidemiology*, **100**, 487–98.

Chapter 5

Bacterial resistance to potable water disinfectants

MIC H. STEWART and
BETTY H. OLSON

Introduction

Historically, the primary goal of potable water treatment was to remove microorganisms responsible for waterborne disease, and this remains as a primary goal today. Several treatment processes, including coagulation, flocculation, sedimentation, filtration and disinfection, are commonly used to remove pathogens from source waters. The addition of disinfectant agents such as chlorine, chloramines, chlorine dioxide, or ozone to potable water is one of the most important steps in pathogen removal and may, in many cases, be the only treatment employed. Disinfectant agents have been used since the early 1900s and have significantly reduced waterborne disease (National Research Council, 1977, 1980). Indeed, the disinfectant residual is generally considered to be the primary variable controlling microbial growth in the distribution system and has appropriately been termed the 'last line of defense' for potable water (White, 1975).

In spite of the demonstrated effectiveness of disinfectant agents under controlled laboratory conditions, numerous field investigations have indicated that microorganisms, including indicator bacteria and human pathogens, may persist in water systems, maintaining a disinfectant residual considered adequate for the control of microorganisms (Boring, Martin & Elliot, 1971; Geldreich et al., 1972; LeChevallier, Seidler & Evans, 1980; National Research Council, 1980; Means et al., 1981; Goshko et al., 1983a; Goshko, Pipes & Christian, 1983b; Olson & Nagy, 1984; Wierenga, 1985; Maki et al., 1986). Whereas the presence of pathogenic organisms is of significant health concern, high levels of heterotrophic bacteria in disinfected potable water can also be problematic. Some of these organisms can interfere with the detection of traditional indicators of microbial pollution (Lamka, LeChevallier & Seidler, 1980) or act as opportunistic pathogens.

It is essential to understand how disinfectants inactivate microorganisms and what factors may promote microbial resistance to these agents in the potable water environment in order to devise effective control strategies for the elimination or reduction of organisms in disinfected potable water systems. The rate of microbial inactivation by disinfectant agents is controlled, in part, by such variables as the concentration and chemical form of the disinfectant, the temperature, the type and physiological composition of the microorganisms, and environmental conditions that modify the disinfectant–microbial interaction (White, 1972; National Research Council, 1980). Historically, investigators have described microbial inactivation by disinfectants as a simple log–linear function (Chick, 1908; Watson, 1908; Morris, 1966), whereas nonlinear inactivation kinetics have been reported more recently (Prokop & Humphrey, 1970; Scarpino *et al.*, 1977; Wolfe, Ward & Olson, 1985). These studies provide important information concerning mechanisms of microbial inactivation and, indirectly, mechanisms of microbial resistance. The differences observed in disinfectant kinetic studies emphasize the need for further understanding of the conditions that control the disinfectant–microorganism interaction. These conditions may significantly alter the ability of the disinfectant to inactivate the organism and thus enhance microbial resistance.

Microbial resistance to disinfectants in the potable water environment may be a function of the presence of physical or chemical agents that impair the ability of the disinfectant to inactivate the organism. Physicochemical factors known to afford microbial protection from disinfectants include (1) the presence of H_2S, Fe^{2+}, or other compounds that chemically reduce the disinfectant (National Research Council, 1980); (2) tubercule formation and growth within other microenvironments promoted by corrosion processes (Tuovinen *et al.*, 1980; Martin *et al.*, 1982; Tuovinen & Hsu, 1982; Victoreen, 1984); (3) the presence of particulate matter such as detritus or clay particles that can provide physical protection from disinfection (Neefe *et al.*, 1947; Stagg, Wallis & Ward, 1977; Hoff, 1978; National Research Council, 1980; LeChevallier, Evans & Seidler, 1981; Ridgway & Olson, 1982; Herson, Marshall & Victoreen, 1984); and (4) the association with nematode or other vectors that may provide a physical barrier from the disinfectant (Hoerhammer, 1911; Chang *et al.*, 1960*a*; Tracy, Camarena & Wing, 1966; Levy *et al.*, 1984; Fields *et al.*, 1986; King *et al.*, 1988; Kilvington & Price, 1990).

Microbial resistance to disinfectants may also be due to factors that can be classified as cell-mediated processes. This form of resistance is physiological in nature and is based on the organism's ability to develop adaptive phenotypes to survive under adverse environmental conditions. Examples of cell-mediated processes include (1) polymer or capsule production, which may act to limit the penetration of disinfectant agents into the cell (Earnhardt, 1980; Ridgway & Olson, 1982; Reilly & Kippin, 1983; Olivieri *et al.*, 1985*b*; LeChevallier, Cawthon & Lee, 1988*a*); (2) cellular aggregation, which provides physical protection to internal organisms

(Galasso & Sharp, 1962; Poduska & Hershey, 1972; Katzenelson, Kletter & Shuval, 1974; Wei & Chang, 1975; Young & Sharp, 1977; Scarpino, 1984; Stewart & Olson, 1986, 1992*b*); and (3) alteration of cell envelope properties, restricting entrance of disinfectants into the cell (Farkas-Himsley, 1964; Hugo & Stretton, 1966; Gilbert & Brown, 1978, 1980).

This category of cell-mediated events suggests the importance of examining the response of microorganisms to disinfectants under environmental conditions. The conditions experienced by bacteria growing in natural water environments are appreciably different from those experienced by organisms grown in a nutrient-rich laboratory medium. Numerous studies have indicated that growth conditions can significantly alter the physiology of the cell (Robinson & Tempest, 1973; Berg, Matin & Roberts, 1982; Kjelleberg, Humphrey & Marshall, 1983; Sterkenburg, Vlegels & Wouters, 1984; Brown & Williams, 1985). In general, these studies indicated that the cell envelope was the site most frequently altered. This finding is highly significant, considering that components located in the cellular envelope, such as enzymes, are frequently the primary target sites for disinfectant agents.

Importantly, many researchers have observed that growth under reduced-nutrient conditions can enhance an organism's resistance to disinfectants (Favero & Drake, 1966; Carson *et al.*, 1972, 1978; Favero *et al.*, 1971; Berg *et al.*, 1982; Berg, Matin & Roberts, 1983; Wolfe & Olson, 1985; LeChevallier *et al.*, 1988*a*). For example, Wolfe & Olson (1985) observed that a *Flavobacterium* isolated from a finished drinking-water reservoir and exposed to 0.75 mg/l free chlorine was very resistant to disinfection (e.g. less than 0.5 log inactivation within 60 min). However, after this strain was subcultured once in R2A medium, it was readily inactivated (greater than 4 logs within 5 min) by 0.4 mg/l free chlorine. Interestingly, R2A is considered a low-nutrient medium formulated for the recovery of environmental isolates (Reasoner & Geldreich, 1985). Under the low-nutrient environment of the distribution system, organisms may be under selective pressure to modify physiological attributes to maximize nutrient acquisition. Inadvertently, these cellular modifications may provide protection against stressor compounds such as disinfectants.

Microbial inactivation by disinfectant agents

Kinetics of inactivation

Initial investigations to describe the kinetics of microbial inactivation considered the importance of fundamental variables such as (1) chemical species and concentration of disinfectant, (2) contact time with disinfectant,

(3) temperature, (4) type and concentration of microorganism, and (5) pH (Butterfield *et al.*, 1943; Butterfield & Wattie, 1946; Butterfield, 1948*a,b*; White, 1972; National Research Council, 1980), but did not fully address the complex effects of other modifying environmental factors on the organism. Chick (1908) formulated the following model to describe microbial inactivation by first-order kinetics:

$$\ln N/N_0 = -kt \tag{5.1}$$

where N is the number of organisms present at time t; N_0 is the number of organisms present at time 0; k is a rate constant characteristic of the type of disinfectant, microorganism and water quality; and t is the contact time.

Rapidly recognizing that the concentration and type of disinfectant were critical factors in predicting microbial inactivation, Watson (1908) modified Chick's original equation to include a lethality coefficient (i.e. the potential inactivating ability of the test disinfectant), as follows:

$$\ln (N/N_0) = -\Lambda C^n t \tag{5.2}$$

where C is the concentration of disinfectant, Λ is the coefficient of specific lethality, n is the coefficient of dilution, and N, N_0, and t are as described above.

Because the lethality coefficient could be used as a reference for a particular disinfectant's properties, Morris (1966) later defined the lethality coefficient to reflect 99% inactivation after 10 min contact time. Calculation of these lethality coefficients allows a comparison of the relative effectiveness of different disinfectants against various microorganisms. Table 5.1 illustrates an order of disinfection efficiency, from most to least potent, of $0_3 > \text{HOCl} > \text{ClO}_2 > \text{OCl}^- > \text{NHCl}_2 > \text{NH}_2\text{Cl}$ of microbial activity. In general, vegetative bacteria are the most susceptible, followed by viruses, with bacterial spores and protozoan cysts being the most resistant. It should be noted, however, that this order of disinfection efficiency is only an approximation, as lethality coefficients do not account for the many complex variables occurring under environmental conditions.

Comparison of disinfectant effectiveness has more recently been described by multiplying the concentration of the disinfectant with the contact time. This product (referred to as Ct) is generally based on a predetermined level of inactivation. Typically, a level of 99% inactivation is used when comparing Ct values. In general, the lower the Ct value, the more effective the disinfectant. The Ct method allows (as do lethality coefficients) a general comparison of the effectiveness of various disinfectant agents on different microbial agents. Table 5.1 lists Ct values for various microorganisms.

Specific lethality coefficients and graphs constructed to determine 99% inactivation rely on a first-order kinetic relationship. Numerous studies,

Table 5.1. *Specific lethality and Ct values for potable water disinfectants*[a]

Disinfectant	Organism	pH	Temperature (°C)	Concentration (mg/l)	Contact time[b] (min)	Ct[c]	Λ[d]
O_3	E. coli	7.0	12	0.0125	0.33	0.004	1153.0
	Poliovirus 1	7.0	25	0.042	10.0	0.42	11.0
	E. histolytica	6.5–8.0	10–27	0.8	5.0	4.5	1.0
ClO_2	E. coli	7.0	25	0.5	0.55	0.28	16.0
	Poliovirus 1	7.0	25	0.8	1.5	1.2	3.8
HOCl	E. coli	6.0	5	0.1	0.4	0.04	115.0
	Poliovirus 1	6.0	15	1.0	1.0	1.0	4.6
	E. histolytica	6.0	30	1.0	4.0	4.0	0.25
OCl⁻	E. coli	10.0	5	1.0	0.9	0.9	5.0
	Poliovirus 1	10.0	15	1.0	3.5	3.5	1.3
	E. histolytica	8.0	30	1.0	15.0	15.0	0.31
$NHCl_2$	E. coli	4.5	15	1.0	5.5	5.5	1.2
	Poliovirus 1	4.5	15	100.0	50.0	5000.0	0.001
NH_2Cl	E. coli	9.0	25	1.2	33.5	40.2	0.11
	Poliovirus 1	9.0	25	10.0	32.0	320.0	0.01

[a]Adapted from National Research Council (1980).
[b]Contact time required for 99% inactivation.
[c]Product of disinfectant concentration and time required for 99% inactivation.
[d]Lethality coefficient; calculated as $\ln (N/N_0 = -\Lambda C^n t$, where $\ln (N/N_0)$ is the natural logarithm of the number of organisms present at time t divided by the number of organisms present initially [at 99% inactivation $\ln (N/N_0 = -4.61]$; Λ is the lethality coefficient; C is the concentration of disinfectant; n is the coefficient of dilution [assume $n = 1$]; and t is the contact time.

however, have indicated that microbial inactivation does not always conform to log–linear or first-order kinetics (Prokop & Humphrey, 1970; Scarpino et al., 1977; Wolfe et al., 1985). Figure 5.1 represents possible nonlinear inactivation curves. Curve A depicts a convex survival curve, which may be a function of clumped organisms or the presence of several sites within the cell that require inactivation before cellular death occurs. Curve B represents a linear inactivation relationship. Curve D represents a concave survival curve, which may result with mixed populations. Finally, curve C represents a complex combination of the events represented in curves A and D. Nonlinear inactivation kinetics have been attributed to several factors, including (1) the time required for a disinfectant agent to reach critical target areas within the cell (Collins & Selleck, 1972), (2) the time necessary for irreversible damage to a number of cellular sites (Kimball, 1953), (3) heterogeneous distribution of organisms with differing sensitivity (Hess, Diachisin & DeFalco, 1953; Wolfe et al., 1985), (4) aggregation of microorganisms in suspension, preventing simultaneous contact with the disinfectant agent (Poduska & Hershey, 1972; Katzenelson *et al.*, 1974; Wei & Chang, 1975; Stewart & Olson, 1986), and (5) chemical alteration of disinfectant species over a period of time, resulting in reduced inactivation ability (Gard, 1957).

Several models have been proposed to address nonlinear inactivation kinetics (Fair *et al.*, 1948; Kimball, 1953; Gard, 1957; Collins & Selleck, 1972; Hom, 1972; Haas & Karra, 1984*a,b*). These models were formulated into equations to account for the effect of a particular disinfectant species over time (Kimball, 1953); the requirement of inactivation of several sites to effect inactivation (referred to as the multi-hit or multi-target model) (Atwood & Norman, 1949; Kimball, 1953; Moats, 1971; Wei & Chang, 1975; Severin, Suidan & Engelbrecht, 1983); and the inactivation of enough critical cellular sites to produce a threshold effect for death (known as the series-event model) (Severin *et al.*, 1983). Other models, however, have also been proposed to analyze nonlinear inactivation data based on a best-fit mathematical model that does not use the actual mechanistic events leading to cellular inactivation (Collins & Selleck, 1972; Selleck, Saunier & Collins, 1978; Haas & Karra, 1984*a,b*).

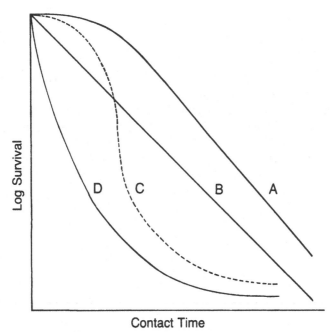

Figure 5.1 Theoretical inactivation curves. Curve A depicts a convex survival curve, which may be a function of clumped organisms or the presence of several sites within the cell that require inactivation before cellular death occurs. Curve B represents a linear inactivation relationship. Curve D represents a concave survival curve, which may result from mixed populations. Curve C represents a complex combination of the events represented in curves A and D. Adapted from Moats (1971).

The model developed by Collins & Selleck (1972) and Selleck *et al.* (1978) addresses nonlinear kinetics caused by an initial lag and a nonlinear, declining rate of inactivation. The data in this model are constructed by plotting the negative logarithm of survival, N/N_0, against the log of the product of concentration and time, Ct. The Collins–Selleck model has proved useful in describing and comparing nonlinear disinfection data of different disinfectants under varying conditions of pH, contact time and other physical/chemical factors that occur under field conditions (Montgomery, 1985).

The observation and mathematical modeling of microbial inactivation provides indirect information concerning the physiological mechanism of inactivation and, conversely, mechanisms of resistance. For example, organisms that are resistant by virtue of their ability to form large cellular aggregates demonstrate nonlinear inactivation kinetics (Wei & Chang, 1975). Presumably, the nonlinear portion of the curve represents internally located cells that remain viable for longer periods of contact time because of their reduced exposure to the disinfectant. The following section provides further details on mechanisms of microbial inactivation by disinfectant agents.

Early theories of inactivation mechanisms

The first recorded study of the germicidal activity of disinfectants was performed by Koch in 1881 (Chang, 1944). Preliminary work to describe the mechanisms of microbial inactivation started in the mid-1900s (White, 1972). During this period, three inactivation theories were advanced by investigators in the field: (1) the nascent oxygen theory, (2) the nascent oxygen and direct chlorination theory, and (3) the toxic substance theory (Chang, 1944). In the nascent oxygen theory, germicidal action was caused by the formation of an oxygen radical as hypochlorous acid dissociated in water, as illustrated below:

$$Cl_2 + H_2O = HOCl + HCl \qquad (5.3)$$

or

$$OCl^- + H^+ = HOCl \qquad (5.4)$$

where $HOCl$ is defined as $HCl + O\cdot$.

This theory, based on microbial inactivation of cellular constituents by short-lived oxygen radicals, was disproved in the mid-1940s by Chang (1944), who observed that inactivation levels of *Entamoeba histolytica* cysts were not inactivated following exposure to high concentrations (e.g. 3000 mg/l) of H_2O_2, which contained 1500 times more nascent oxygen than did similar levels of $HOCl$.

During this period, another theory, termed the nascent oxygen and direct chlorination theory, was also advanced. This theory held that microorganisms were inactivated by nascent oxygen and by direct chlorination of the protoplasm within the organism. This theory was based on observations by Holwerda (1928), who noted that chloramines could act as effective microbicides. Holwerda (1928) postulated that chloramines could hydrolyze and form hypochlorous acid, even though he was unable to detect HOCl in his studies. This modified nascent oxygen theory, however, was not based on experimentally derived data but rather on theoretical considerations of known chemical phenomena; as such, its scientific acceptance was short-lived.

Baker (1926) observed the formation of chloramines in sewage by the reaction of chlorine with organic compounds. This led him to formulate the toxic substance theory, which, supported by the fact that chloramines exert a measurable germicidal effect, suggested that chlorine compounds inactivate bacteria by combining with the lipoprotein substance in the bacterial cell wall, forming toxic chloro-compounds. These compounds, in turn, were hypothesized to inactivate organisms by interfering with cell division and other cell-wall-associated processes (Chang, 1944). Although this theory gained acceptance by researchers in the field, it also lacked the essential experimental evidence required for scientific verification.

It was not until the early 1940s that researchers began producing experimental data elucidating the mechanisms of microbial inactivation for disinfectant agents such as chlorine and chloramines. Since the 1940s, investigations with chlorine indicated that these compounds could inactivate microorganisms by reacting with various cellular sites.

The inactivating action of these disinfectants is based on their ability to react with sensitive target sites, generally located at the cell envelope or within the cytoplasmic region of the cell. Therefore, the magnitude or rate of penetration by a disinfectant is considered to be a critical characteristic by many investigators (Fair, Morris & Chang, 1947; Friberg, 1956, 1957; Bekhtereva & Krainova, 1975; Olivieri *et al.*, 1975; Kaminski *et al.*, 1976; Dennis, 1977).

Mechanism of microbial inactivation by chlorine

Microbial inactivation by chlorine is a function of (1) altered permeability of the outer cellular envelope, resulting in leakage of critical cell components (Friberg, 1957; Venkobachar, Iyengar & Prabhakara Rao, 1977; Kulikovsky, Pankratz & Sadoff, 1975; Camper & McFeters, 1979; Haas & Engelbrecht, 1980*a*,*b*; McFeters & Camper, 1983); (2) interference of cell-associated membrane functions (e.g. phosphorylation of high-energy compounds) (Venkobachar, Iyengar & Prabhakara Rao, 1975); (3) impairment of enzyme and protein function as a result of irreversible binding of the sulfhydryl groups (Green & Stumpf, 1946; Knox *et al.*, 1948; Bekhtereva & Krainova, 1975; Venkobachar *et al.*, 1975, 1977); and (4) nucleic acid denaturation (Bocharov, 1970; Bocharov & Kulikovsky, 1971; Patton *et al.*, 1972; Hoyano *et al.*, 1973; Rosenkranz, 1973; Olivieri, 1974; Olivieri *et al.*, 1975; Dennis, Olivieri & Krusé, 1979*a*,*b*; Haas & Engelbrecht, 1980*a*; Gould, Richards & Miles, 1984).

The actual mechanism of chlorine inactivation may result from the concerted action of all of these events or merely from the effect of chlorine on a few critical cellular sites. The series of steps leading to complete cessation of cell function is a complex process and may vary appreciably from one microorganism to another. Evidence to date suggests that bacterial inactivation by chlorine is primarily caused by the impairment of physiological functions associated with the bacterial cell membrane. Chlorine can alter the permeability state of the cell membrane, which may lead to the loss of critical cellular components or modification of the ionic environment, resulting in enzyme dysfunction. Chlorine may act directly and inhibit enzyme function by oxidation of the sulfhydryl group. This step may be reversible if oxidation does not proceed beyond the disulfide state. Chlorine may also react with nucleic acids and thus impair replication processes or induce lethal mutation points. Whereas reaction with nucleic acids may lead to cell inactivation, this is probably secondary to the action of chlorine on cell-membrane-associated functions.

Viral inactivation by chlorine is believed to result from its reaction with either the protein coat or the nucleic acid. The relative number of sensitive sites and associated steric positioning may well account for some of the variation observed in the literature on viral inactivation by chlorine.

Mechanism of inactivation by chloramines

Although numerous studies have been conducted to determine the mode of microbial inactivation by free chlorine, there have been fewer studies concerning chloramine inactivation mechanisms. It should be noted, however, that because of the poorly controlled experimental conditions employed by early investigators, many of the postulated chlorine inactivation mechanisms may have been caused by the action of chloramines rather than free chlorine. Research to date indicates that chloramines primarily inactivate microorganisms by irreversible denaturation of proteins (Boyle, 1963; Fujioka, Tenno & Loh, 1983; Jacangelo & Olivieri, 1985; Jacangelo, Olivieri & Kawata, 1987, 1991) and to a lesser extent by denaturation of the nucleic acid (Ingols, 1958; Fetner, 1962; Boyle, 1963; Shih & Lederberg, 1976, 1979; Olivieri *et al.*, 1980; Jacangelo & Olivieri, 1985; Jacangelo *et al.*, 1991). Chloramine inactivation of bacteria is caused primarily by the oxidation of sulfhydryl-containing enzymes and, to a lesser extent, a reaction with the nucleic acid. By contrast with the available data on chlorine, there are no existing data to suggest that chloramines can modify the permeability state of the cell. Viral inactivation by chloramines is similar to the mechanism of inactivation by chlorine, in which primary targets consist of both capsid proteins and nucleic acid.

Mechanism of inactivation by chlorine dioxide

As is the case with chlorine, chlorine dioxide inactivates microorganisms by denaturation of the sulfhydryl groups contained in proteins (Ingols & Ridenour, 1948; Ingols *et al.*, 1953; Noss, Dennis & Olivieri, 1983; Olivieri *et al.*, 1985*a*), inhibition of protein synthesis (Bernarde, Snow & Olivieri, 1967; Roller, Olivieri & Kawata, 1980), denaturation of nucleic acid (Roller *et al.*, 1980; Alvarez & O'Brien, 1982) and impairment of permeability control (Berg, Roberts & Matin, 1986). Although the specific mechanism of bacterial inactivation has not been resolved, it has been demonstrated that chlorine dioxide disrupts the permeability gradient of the membrane and is also reactive with sulfhydryl groups. Further research may indicate that chlorine dioxide is indeed similar to chlorine and chloramines in its disruption of protein function. Viral studies with chlorine dioxide suggest that protein components of the capsid are irreversibly denatured.

Mechanism of inactivation by ozone

Numerous reports have discussed the considerable biocidal effectiveness of ozone (Venosa, 1972; National Research Council, 1980, 1987; Hoff & Geldreich, 1981; Wolfe *et al.*, 1989). However, only limited information is available on the mechanism of inactivation by this disinfectant. Research has indicated that, like chlorine-based disinfectants, ozone inactivates microorganisms by disruption of membrane permeability (Scott & Lesher, 1963; Domingue *et al.*, 1988); impairment of enzyme function and/or protein integrity by oxidation of sulfhydryl groups of proteins (Kim, Gentile & Sproul, 1980; Sproul, Pfister & Kim, 1982; Domingue *et al.*, 1988; Shinriki *et al.*, 1988); and nucleic acid denaturation (Prat & Cier, 1968; Hamelin & Chung, 1978; Roy *et al.*, 1981; Sproul *et al.*, 1982). The inactivation of bacteria by ozone appears to be broad-based in nature. Disruption of membrane permeability, oxidation of sulfhydryl groups contained in membrane proteins and damage to nucleic acid have been reported by various investigators as the mechanism of bacterial inactivation by ozone. Viral ozone inactivation results from the action of this disinfectant on the protein coat and the nucleic acid. Recent evidence also indicates that the nucleic acid may be cross-linked to the protein, thus preventing uncoating and subsequent infection of host microorganisms.

Microbial disinfection resistance mechanisms

Microbial survival in treated water supplies has been a topic of interest since the early use of these systems to provide water for consumers. Reasons for this interest centered on the deterioration of water quality and increased maintenance problems associated with the presence of certain microorganisms. A number of investigators have documented the occurrence of microorganisms in the presence of disinfectants used for drinking water supplies (Geldreich et al., 1972; LeChevallier et al., 1980; Means et al., 1981; Ridgway & Olson, 1982; Goshko et al., 1983a,b; Hudson, Hankins & Battaglia, 1983; Reilly & Kippin, 1983; Olson & Nagy, 1984; Wierenga, 1985; Maki et al., 1986).

Microorganisms have also been recovered from chlorinated and brominated swimming pools (Favero & Drake, 1966) and in disinfectant solutions used in healthcare settings (Adair, Geftic & Gelzer, 1969; Anderson et al., 1984). Importantly, the occurrence of these organisms in the presence of disinfectants would not be expected on the basis of traditional laboratory disinfection studies.

Microbial survival in the presence of disinfectant agents may be mediated by physicochemical factors, cell-mediated factors, or a combination of both. Common examples of physicochemical factors would include (1) corrosion processes, which may produce disinfectant demand compounds, protective microenvironments on the pipe wall, or extensive tuberculation; (2) attachment of the microorganisms onto particles; and (3) association with nematodes, protozoa, or invertebrates. In general, physicochemical events enhance an organism's survival by providing a protective environment that limits the contact or reduces the concentration of disinfectant reaching the organism. The following section summarizes physicochemical events that are important in aiding the survival of microorganisms in the presence of disinfectants.

Physically mediated protection

Corrosion

Microorganisms capable of reducing sulfate and nitrate and producing methane have been recognized as promoting corrosion in water distribution systems (O'Connor, Hash & Edwards, 1975; Allen, Taylor & Geldreich, 1980; Tuovinen et al., 1980; Victoreen, 1984; Montgomery, 1985). Corrosion contributes to the degeneration of pipes in distribution systems and can also produce undesirable color and taste properties in the water. Importantly, the areas affected by corrosion may develop a unique microbial ecosystem, often promoting the growth of allochthonous bacterial species (Victoreen, 1969; Ridgway & Olson, 1981a). Corrosion sites may produce protective microhabitats that stimulate the growth of organisms via the presence of compounds with a high chlorine demand resulting from corrosion processes.

Iron bacteria, such as *Gallionella* and *Sphaerotilus*, oxidize the ferrous ions on iron pipes to ferric ions and other hydroxides (Mackenthun & Keup, 1970; Ridgway & Olson, 1981b; Montgomery, 1985). This process can lead to the precipitation of ferric salts, which is known as tuberculation. The formation of these tubercules is greatly affected by flow rates, disinfectant residuals, nutrient availability and pH (Allen *et al.*, 1980). Under favorable growth conditions, *Gallionella* can rapidly multiply and produce nutrients that may promote the growth of various heterotrophic bacteria in treated water systems (Ridgway & Olson, 1981b; Martin *et al.*, 1982).

Tuovinen & Hsu (1982) indicated that tubercle formation, promoted by corrosion processes, provided protection for bacteria against 0.2–1.0 mg/l free chlorine in a water distribution system. Tubercle formation was observed to aid the survival of the coliform *Klebsiella pneumoniae* (Martin *et al.*, 1982). Ground tubercle material, added to distilled water, was determined to provide growth nutrients for this coliform and exert a chlorine demand of approximately 3 mg/l.

Particle association

Particulate materials consisting of clays, silt, organic compounds, metal oxides, plankton and other microscopic organisms are commonly found in water supplies and, collectively, determine the degree of turbidity (National Research Council, 1977; American Public Health Association, 1989). Microbial attachment to particulate material has been considered to aid in survival against disinfectant agents (Neefe *et al.*, 1947; Tracy *et al.*, 1966; Stagg *et al.*, 1977; Hejkal *et al.*, 1979; Foster *et al.*, 1980; National Research Council, 1980; LeChevallier *et al.*, 1981; Ridgway & Olson, 1982; Scarpino, 1984; Berman, Rice & Hoff, 1988). Under optimal chemical and physical conditions, maximum disinfection efficiency is reached when the disinfectant agent has unhindered access to the target organism. Particulate matter may interfere with this process either by acting chemically to create a chlorine demand, thus neutralizing the action of the disinfectant, or by physically shielding the organism from the disinfectant (US Environmental Protection Agency, 1983). The particulate–microbial complex may be theorized as (1) adsorption of microorganisms onto larger particles, (2) adsorption of small particles onto the surface of the microorganism, and

(3) encasement of the microorganism by one or more large particles or many associated small particles (Hoff, 1978). The relative size and stability of this particulate–microbial complex under disinfectant exposure will govern the level of protection against disinfectant agents. Disinfectant protection is enhanced with decreasing organism size and increasing particle availability (Hoff, 1978). Therefore, small organisms such as viruses may be afforded much greater disinfectant protection than larger organisms at lower turbidity conditions. Recognition of this phenomenon has been responsible for the turbidity standard mandated by the National Primary Drinking Water Regulation (US Environmental Protection Agency, 1975).

Particle association by bacteria

Bacterial attachment to particulate material or inert surfaces in water systems has been documented by several investigators (Marshall, 1976; Hoff, 1978; LeChevallier et al., 1981; Ridgway & Olson, 1982; Herson et al., 1984). Hoff (1978) indicated that naturally occurring coliforms associated with solids from an unchlorinated primary sewage effluent demonstrated significant resistance to chlorine. Coliforms associated with the solids were still viable after 60-min exposure to 0.5 mg/l chlorine. By comparison, a laboratory strain of Escherichia coli, not associated with solids, was completely inactivated. However, these results are somewhat inconclusive, as other researchers (Ward et al., 1984) have demonstrated that E. coli strains isolated from different environments can demonstrate considerably different sensitivities to disinfectants.

Foster et al. (1980) conducted experiments using fecal-associated E. coli and coliforms to determine relative protection from ozone disinfection. At suspensions of 1 nephelometric turbidity unit (NTU), there was no protection from 0.011–0.067 mg/l residual ozone after 30 s contact time. However, when a 5-NTU suspension was tested, there was a moderate level of protection after 30 s contact time with 0.014 mg/l residual ozone.

In a study similar to that of Foster et al. (1980), Berman et al. (1988) examined the survival of naturally occurring coliforms associated with sewage particles undergoing chlorine and chloramine disinfection. Coliforms associated with particles greater than 7 μm were three times more resistant to chlorine (0.5 mg/l, 5 °C, pH 7) compared to coliforms associated with particles smaller than 7 μm. Particle size had little protective effect when coliforms were exposed to the same concentration of chloramines. However, 0.5 mg/l chlorine was more effective in inactivating particle-associated coliforms than was 1.0 mg/l chloramines. For example, at pH 7 and 5 °C, the Ct_{99} values for chlorine- and chloramine-disinfected, particle-associated coliforms were 2.7 mg min/l and 87 mg min/l, respectively.

LeChevallier et al. (1981) evaluated the relationship between increased turbidity and bacterial survival in the presence of chlorine. Bacteria isolated from surface water were mixed with various amounts of surface-water sediment to produce turbidities ranging from 0.2 to 15 NTU. Following exposure to chlorine, it was observed that protection occurred only at the higher turbidities (e.g. 13 NTU) and not at lower turbidities (e.g. 6 NTU or less). Chlorine demand and physical protection were concluded to be the significant variables affecting bacterial protection at high turbidities. The chlorine demands ranged from 0.8 mg/l at the 1.5 NTU condition to 1.9 mg/l at the 13 NTU condition. It is important to note that the relationship between turbidity and chlorine demand may vary appreciably, depending on the relative amounts of inorganic and organic material present. The results of this study indicate that particle presence can provide significant protection from disinfectant agents, but only when present at levels exceeding the turbidity standards of the US Environmental Protection Agency.

In a study performed by Ridgway & Olson (1982), the chlorine resistance patterns of bacteria isolated from a chlorinated and an unchlorinated water distribution system were compared. It was observed that bacterial species recovered from the chlorinated system were more resistant to both combined and free chlorine than were those recovered from the unchlorinated system. This led the authors to postulate that chlorinated water tends to select for chlorine-tolerant bacteria. In addition, differential filtration experiments revealed that bacteria retained on the surface of 2.0-μm filters were significantly more resistant to chlorine than was the total microbial population retained on 0.2-μm filters. These investigators theorized that the observed resistance was a result of bacterial aggregation or attachment to particulate material. Scanning microscopy analyses confirmed the presence of bacterial aggregates and bacterially colonized particles.

In a similar study conducted by Herson *et al.* (1984), attachment of bacteria to particulate matter was shown to aid bacterial survival in the presence of chlorine. Using a simulated distribution system and a free-chlorine residual of 2.5 mg/l, these authors observed higher bacterial levels and lower chlorine residuals when the turbidity was increased. Scanning electron micrographs revealed the presence of colonized particles. Furthermore, upon being autoclaved, these particles were shown to release nutrients capable of supporting the growth of *Enterobacter cloacae*.

Bacteria associated with granular activated carbon (GAC) have also been observed to be protected from disinfectants (LeChevallier *et al.*, 1984; Stewart, Wolfe & Means, 1990). LeChevallier *et al.* indicated that GAC-attached organisms were approximately five times more resistant to 1.6 mg/l free chlorine than were unattached bacteria. Importantly, only a half-log reduction of pathogens such as *Salmonella typhimurium*, *Yersinia enterocolitica* and *Shigella sonnei* was achieved after 60 min of exposure to 1.6 mg/l free chlorine when these organisms were attached to GAC. In a similar study by Stewart *et al.* (1990), it was observed that less than 0.5 log of bacteria associated with GAC was inactivated by either chlorine (1.5 mg/l, pH 7.0) or chloramines (1.5 mg/l, pH 7.0) after 40 min contact time.

Particle association by viruses

Neefe *et al.* (1947) gave early evidence of particles' protective effect on viruses against chlorination. Feces from patients with infectious hepatitis, suspended in water and exposed to a chlorine residual of 1.1 mg/l for 30 min, was capable of infecting two of five volunteers. After treatment by coagulation and filtration, followed by chlorination (1.1 mg/l residual), this same water was incapable of producing hepatitis in any of five volunteers, indicating that with a reduction of particles there was a concomitant reduction in viral number.

Hoff & Geldreich (1978) reported a study by Boardman (1976), who investigated the protective effects of inorganic particles (alum, calcium carbonate, kaolin) on bacteriophage T7 and poliovirus. In this study, the viruses were mixed with the inorganic particles and exposed to chlorine at 0.2 mg/l. The addition of inorganic particles to the suspension of T7 did not protect this virus from chlorine. Poliovirus, however, was shown to be partially protected by the presence of inorganic particles. When mixed with alum, kaolin and calcium carbonate, 78%, 97%, and 85% of the virus, respectively, was inactivated. The differences in protection among the various inorganic particles, or between bacteriophage T7 and poliovirus, may have resulted from the particular surface charge characteristics of these two viruses.

Hoff (1978) examined the relationship between turbidity and viral inactivation by adsorbing poliovirus onto bentonite clay and aluminum phosphate to produce turbidities of approximately 5 NTU. The level of virus attachment was determined to be between 65 and 80% for bentonite and greater than 90% for aluminum phosphate. Even with the high attachment level, there was virtually no protection from exposure to chlorine at 0.5 mg/l. Studies with Hep-2 cell-associated virus, however, indicated significant protection from chlorine. Hoff concluded that the degree of protection afforded by particulate matter is determined largely by the type and not the amount of available turbidity.

Investigations by Stagg *et al.* (1977) indicated that when the bacteriophage MS-2 was associated with bentonite clay, it could be protected from chlorine. In these studies, the virus was mixed with bentonite clay to produce a turbidity level of approximately 3 NTU. It was determined that more than 80% of the viruses were attached to those particles that were less than 2 μm in diameter. Exposure to chlorine at 0.02–1.0 mg/l indicated that attached viruses were two times more resistant to chlorine than were freely suspended viruses.

In further work with poliovirus inactivation by chlorine, Hejkal *et al.* (1979) determined that poliovirus contained in fecal material was four times more resistant to inactivation than freely suspended poliovirus exposed under the same conditions. They also concluded that viruses associated with organic particles did not encounter greater protection from chlorine than did freely suspended viruses. In a similar study by Foster *et al.* (1980), poliovirus associated with fecal material at a turbidity level of 5 NTU was found to be protected from ozone disinfection. In this study, less than 1 log of virus was inactivated with ozone at 0.013 mg/l after 30 s contact time.

The above studies suggest that although the association of virus with cells may aid in protection from disinfection, the presence of inorganic or even organic particles may not necessarily enhance protection from chlorine. These results tend to suggest that viruses encased within cells will not be exposed to disinfectant agents. However, the relationship between particulate matter and viruses seems highly dependent on the specific surface characteristics of the viruses and the chemical nature of the particulate material. This relationship, in turn, may be critical in determining whether or not the virus will be inactivated by the disinfectant.

Vector-harbored organisms

The introduction of bacteria into potable water supplies via nematodes or crustaceans has been suggested by various authors (Hoerhammer, 1911; Chang *et al.*, 1960*a*; Chang, Woodward & Kabler, 1960*b*; Tracy *et al.*, 1966; Levy *et al.*, 1984; Olson & Nagy, 1984; King *et al.*, 1988; Shotts & Wooley, 1990). Macroinvertebrates, including the groups of Crustacea, Nematoda, Platyhelminthes and Insecta, have been considered by many to be capable of living and growing in distribution systems (Chang *et al.*, 1960*b*; Small & Greaves, 1968; Levy *et al.*, 1984). Because bacteria and viruses may be contained within invertebrates, it was reasonable to assume that these organisms could act as vectors for the transmission of bacteria and viruses into the distribution system.

An early study by Hoerhammer (1911) suggested that the crustacean *Cyclops* could ingest large quantities of *Salmonella* bacteria and thus act as a reservoir for this pathogen. Chang *et al.* (1960*a*) reported that certain members of the nematode family are capable of ingesting significant numbers of pathogenic bacteria and viruses. Importantly, these ingested bacteria and viruses were shown to be protected against high levels of chlorine (80–90 mg/l), even though 90% of the nematodes were inactivated. As a result of investigations into coliform outbreaks in the San

Francisco water supply, Tracy *et al.* (1966) theorized that Crustacea such as *Daphnia* and *Cyclops* were responsible for harboring coliforms that eventually entered the drinking water supply. (It is important to note that chlorination was the only treatment used for this system.) Support for this theory is based on the common occurrence of these organisms in water reservoirs and their high tolerance to chlorine. To validate this theory, Tracy *et al.* (1966) collected plankton samples containing *Cyclops* and *Daphnia*, exposed them to chlorine, and blended them to release any associated coliforms. With the use of membrane filtration and multiple-tube fermentation techniques, it was determined that 8–54 coliforms/100 ml were contained within these organisms.

Work by Levy *et al.* (1984) indicated that high numbers of coliforms could be attached to the surface of the amphipod *Hyalella azteca*. Both *E. coli*, at 1.6×10^4 CFU/amphipod, and *E. cloacae*, at 1.4×10^3 CFU/amphipod, were found to be attached to this amphipod. After exposure to chlorine at 1 mg/l for 60 min, 2–15% of these attached coliforms were still viable, compared to the complete inactivation of unattached coliforms. Because of the disinfectant protection mediated by macroinvertebrate attachment, the authors posited that this phenomenon may be of significant epidemiological consequence. The work of Huq *et al.* (1983), in which copepods were demonstrated to serve as a vector of *Vibrio cholerae*, was cited as a case in point.

In contrast to the work of Levy *et al.* (1984), research by Haas, Khater & Wojtas (1985) indicated that organisms harboring bacteria may not necessarily afford protection from disinfectants. In this work, *Tetrahymena pyriformis* (strain W) was assayed for its sensitivity to free and combined chlorine. This organism was hypothesized to be an ideal model, as it is a vegetative, free-swimming organism that feeds on bacteria or other organic matter. Because this organism was shown to be six times more sensitive to monochloramine than either *E. coli* or *Salmonella typhi*, the ability of this protozoan to harbor and protect coliforms from disinfectants was concluded to be remote. It is important to note that these researchers determined viability by microscopic observation. King *et al.* (1988) suggested that Haas *et al.* (1985) may have incorrectly assumed that *T. pyriformis* was inactivated. In King's study, it was found that protozoa that appeared to be inactivated by microscopic analysis were viable following reinoculation into culture medium.

King *et al.* (1988) also found that coliforms including *E. coli*, *Citrobacter freundii*, *Enterobacter agglomerans*, *E. cloacae*, *Klebsiella oxytoca*, and *K. pneumoniae*, as well as the bacterial pathogens *S. typhimurium*, *Y. enterocolitica*, *S. sonnei*, *Legionella gormanii*, and *Campylobacter jejuni*, when contained in either *Acanthamoeba castellanii* or *T. pyriformis*, were more resistant to chlorine disinfection. When exposed to a free-chlorine residual of 1 mg/l (pH 7.0), freely suspended bacteria were inactivated within 1 min.

However, when ingested by protozoa and exposed to free chlorine at 2–10 mg/l, the coliforms and pathogenic bacteria increased their resistance to disinfection 30- to 120-fold. The authors indicated that even though bacteria are ingested by protozoa, they may survive digestion and presumably remain viable for up to 24 h. Shotts & Wooley (1990) suggested that these findings may explain the spontaneous occurrence of coliforms in chlorinated distribution systems. Presumably, viable coliforms could be released upon the death of the protozoa.

Recently, there has been considerable interest in the relationship between *Legionella* and protozoa (Rowbotham, 1980, 1983, 1986; Anand *et al.*, 1983; Fields *et al.*, 1986; Wadowsky *et al.*, 1988; Kilvington & Price, 1990). Rowbotham (1980) indicated that amoebae of the genera *Acanthamoeba* and *Naegleria*, commonly found in soil and freshwater environments, could harbor legionellae and act as the infectious reservoir of this organism. In later work, Rowbotham (1983) showed that the association of *Legionella* and amoebae was due to the ability of virulent legionellae actively to attack and multiply within suitable host amoebae. In work with the protozoan *T. pyriformis*, Fields *et al.* (1986) supported the observation of Rowbotham (1983) that only virulent *Legionella* could invade amoebae. Anand *et al.* (1983) confirmed previous work by Rowbotham (1980) and further indicated that intracellular multiplication also occurred at temperatures below 37 °C (e.g. 20 °C), which would be much more common in the environment. Anand *et al.* (1983) also suggested that *Legionella* contained within amoebae may be protected from chlorine used in potable water supplies.

Kilvington & Price (1990) presented evidence that legionellae contained within protozoans are indeed protected from disinfectants. In their study, viable *Legionella* could be recovered from infected cysts of *Acanthamoeba polyphaga* exposed to 50 mg/l chlorine (pH 7.2, 25 °C) for 18 h. This is significant, as free-living *Legionella* are readily inactivated by chlorine or chloramines. For example, more than 99% inactivation of *Legionella* occurred within 5 min contact time with free chlorine at 0.4 mg/l (pH 7.2, 25 °C) (Domingue *et al.*, 1988). Disinfection studies (Cunliffe, 1990) have demonstrated that 99% inactivation of *Legionella* occurred in 15 min when they were exposed to chloramines at 1 mg/l (pH 8.5, 30 °C). Clearly, *Legionella* harbored within protozoa are protected from potable water disinfectant agents.

The mediation of microbial protection against disinfectants by physicochemical agents has been described for both viruses and bacteria (Table 5.2). Tuberculation, particle association and association with protozoan vectors may all contribute to the protection of microorganisms in treated water systems. Importantly, these events commonly occur in potable water systems and may account for high levels of heterotrophic bacteria and even sporadic coliform episodes in disinfected systems.

Table 5.2. *Microbial resistance to disinfectants mediated by physical agents*

Observation	Reference
Bacteria	
Corrosion/tuberculation	
Heterotrophic bacteria association with tubercules protected from free chlorine at 1 mg/l	Tuovinen & Hsu (1982)
Tubercule material supplied growth of *Klebsiella pneumoniae*	Martin *et al.* (1982)
Particle association	
Coliforms associated with sewage solids resistant to free chlorine at 0.5 mg/l (60 min contact time)	Hoff (1978)
Fecal-associated *E. coli* and coliforms resistant to residual ozone at 0.014 mg/l (30 s contact time) at 5 NTU	Foster *et al.* (1980)
Attachment of heterotrophic bacteria to river water sediments at NTU > 13 provide protection from chlorine at 0.6 mg/l	LeChevallier *et al.* (1981)
Heterotrophic bacteria attached to particles of > 2 μm resistant to chlorine at 10 mg/l (2 min contact time)	Ridgway & Olson (1982)
Heterotrophic bacteria demonstrated increased resistance to free chlorine with increased turbidity	Herson *et al.* (1984)
Heterotrophic bacteria associated with granular activated carbon (GAC) particles resistant to chlorine at 1.6 mg/l (60 min contact time)	LeChevallier *et al.* (1984)
Coliforms associated with particles of > 7 μm were three times more resistant to chlorine (0.5 mg/l)	Berman *et al.* (1988)
Heterotrophic bacteria associated with GAC resistant to chlorine at 1.5 mg/l and chloramines at 1.5 mg/l (40 min contact time)	Stewart *et al.* (1990)
Vector association	
Bacteria ingested by nematodes protected from chlorine (80–90 mg/l)	Chang *et al.* (1960a)
Coliforms contained in *Cyclops* and *Daphnia* protected from chlorine	Tracy *et al.* (1966)
Attachment of *E. coli* and *E. cloacae* to *Hyalella azteca* provided protection from chlorine at 1 mg/l (60 min contact time)	Levy *et al.* (1984)
Coliforms and bacterial pathogens of the genera *Salmonella, Yersinia, Shigella, Legionella* and *Campylobacter*, when contained within *Acanthamoeba castellanii* or *Tetrahymena pyriformis*, were 30–120 times more resistant to chlorine	King *et al.* (1988)
Viable *Legionella* bacteria recovered from *Acanthamoeba polyphaga* exposed to chlorine at 50 mg/l (18 h contact time)	Kilvington & Price (1990)

Table 5.2. *Cont.*

Observation	Reference
Viruses	
Particle association	
Hepatitis virus in feces resistant to chlorine at 1.1 mg/l (30 min contact time)	Neefe *et al.* (1947)
Poliovirus and bacteriophage T7 associated with inorganic particles resistant to chlorine at 0.2 mg/l (10 min contact time)	Boardman (1976)
MS-2 bacteriophage twice as resistant to chlorine at 1 mg/l when attached to bentonite	Stagg *et al.* (1977)
Poliovirus associated with Hep-2 cells resistant to chlorine at 3 mg/l (50 min contact time)	Hoff (1978)
Poliovirus in fecal material four times as resistant to choramines	Hejkal *et al.* (1979)

Cell-mediated protection

Cell-mediated resistance is physiological in nature, requiring that the organism develop adaptive features to survive under adverse environmental conditions. This physiological process may be either passive or active, with the former as a constant feature of the cell and the latter as a newly acquired trait in response to adverse environmental conditions. Cell-mediated mechanisms of resistance to disinfectant agents are poorly understood compared to physicochemical protective factors. Only recently have studies been implemented to delineate the role of physiological adaptation of bacteria to the oxidizing effects of disinfectants. Most data on microbial resistance come from laboratory inactivation studies. These studies, however, are frequently based on bacteria grown in 'nutrient-rich' media. These conditions may not accurately reflect physiological characteristics developed by bacteria occurring in the environment, where nutrients are often limiting.

Recent studies have indicated the importance of growth conditions in modifying bacterial response to disinfectants through physiological alterations. Importantly, the growth conditions shown to enhance resistance are often those experienced by the organism in the natural aquatic environment. Consequently, experiments that simulate environmental growth conditions will produce data indicative of the effectiveness of disinfectant agents under field conditions.

Examples of cell-mediated resistance include (1) polymer or capsule production, which may act to limit diffusion of disinfectant agents into the cell or create localized demand conditions; (2) cellular aggregation, providing physical protection to internal cells; (3) cell wall and/or cell membrane alterations that result in reduced permeability of disinfectants; and (4) modification of sensitive target sites (e.g. enzymes). Because many of these physiological events are a function of the prevailing growth conditions, resistance to disinfectant agents may be primarily a result of the fortuitous development of mechanisms to cope with limiting nutrient conditions.

In order to study mechanisms of microbial resistance to disinfectants, many researchers have attempted to generate resistant strains under laboratory conditions (Farkas-Himsley, 1964; Bates, Shaffer & Sutherland, 1977; Haas & Morrison, 1981; Leyval et al., 1984). These studies are important because they provide information concerning the effects of growth conditions on mechanisms of cell-mediated disinfectant resistance.

Bates et al. (1977) demonstrated that laboratory strains of poliovirus acquired resistance after repeated exposure to chlorine. After 10 cycles of exposure to chlorine at 0.8 mg/l, the viruses were significantly more resistant than the original parent strain. The particular mechanism responsible for this increased resistance, however, was not determined. On the basis of subculturing work by Floyd, Johnson & Sharp (1976) and Sharp, Floyd & Johnson (1975, 1976), it appears that viral aggregation is not the mechanism of increased resistance. In a later study by Bates, Sutherland & Shaffer (1978), it was determined that disinfectant-treated viruses had an altered plaque morphology, which suggested that phenotypic changes had occurred.

Studies concerning the generation of disinfectant resistance for bacteria have been conducted by several investigators (Farkas-Himsley, 1964; Haas & Morrison, 1981; Leyval et al., 1984). Farkas-Himsley (1964) reported that bacteria isolated from and continuously subcultured in chlorinated water (swimming pool water) were more resistant than those grown in laboratory media. Interestingly, phage sensitivity testing indicated that the resistant strains had different surface properties from the chlorine-sensitive strains.

Haas & Morrison (1981) found that repeated exposure of E. coli to chlorine did not result in enhanced resistance but, rather, in greater sensitivity. However, the low concentration of chlorine (0.07 mg/l) used in this study may not have been sufficient to select for resistant strains. Furthermore, E. coli was recovered and continuously subcultured in nutrient broth and was maintained on nutrient slants for up to four months. These growth conditions probably did not promote selection for resistant strains.

In contrast to the results of Haas & Morrison (1981), Leyval et al. (1984) reported that E. coli could develop substantial resistance upon repeated exposure to chlorine. These authors indicated that cells continuously exposed to chlorine were 50 times more resistant than nonexposed cells. Importantly, no intermediate subculturing into nutrient media was used in this study. Indeed, the difference in subculturing may account for the difference in results between the work of Haas & Morrison (1981) and that of Leyval *et al.* (1984). Although the specific mechanism for enhanced resistance was not determined, it was observed that colonies of resistant strains possessed a more pronounced capsule than chlorine-sensitive strains.

The results of these studies suggest that growth conditions affect an organism's response to disinfectants. Importantly, simulating the original environment of isolation, or growth under low-nutrient conditions, seems to promote resistance. Moreover, resistant organisms tended to have altered physiological attributes compared to disinfectant-sensitive organisms. The following section summarizes studies concerned with the relationship between physiological characteristics and disinfectant resistance.

Capsule production and biofilm development

Many bacterial species are capable of producing an external layer composed predominantly of polysaccharides and, to a lesser extent, polypeptides or polysaccharide–protein complexes (Costerton & Irvin, 1981). This structure, commonly referred to as the capsule or slime layer, provides nutrient storage, protection from phagocytosis and adhesion to surfaces to enhance survival in the environment (Characklis, 1973*a*; Marshall, 1976; Costerton, Geesey & Cheng, 1978; Costerton & Irvin, 1981; Costerton, Irvin & Cheng, 1981). This structure has been termed 'the glycocalyx' by Costerton *et al.* (1978), and it has been redefined as any polysaccharide-containing bacterial surface that is distal to the surface of the outer membrane of gram-negative bacteria or to the surface of the peptidoglycan layer of gram-positive bacteria. The capsule, however, is generally considered to be directly associated with the cell membrane, where the glycocalyx structure includes any further extension of this polysaccharide material from the capsule. As the production and maintenance of this structure require cellular energy, the glycocalyx is a feature often found associated with environmental isolates, where it confers selective advantages (Costerton *et al.*, 1978). In fact, the occurrence of a glycocalyx on environmental isolates has been considered by Costerton *et al.* (1978) to represent the normal state of the bacterial organism. These advantages are often associated with enhancing nutrient exposure in a nutrient-poor environment. Bacteria attached to inert surfaces such as the pipe walls of distribution systems can extract nutrients and release waste products to the passing water. This topic has been reviewed by many authors (Characklis, 1973*b*; Allen *et al.*, 1980; Lee, O'Connor & Banergi, 1980; Tuovinen *et al.*, 1980; Olson, Ridgway & Means, 1981; Ridgway & Olson, 1981*a*; Tuovinen & Hsu, 1982; Nagy, 1984).

Many investigators have considered organisms involved in biofilm formation, or those capable of capsule formation and not associated with a biofilm, to be resistant to disinfectant agents (Characklis, 1973b; Ptak, Ginsburg & Willey, 1973; Earnhardt, 1980; Seyfried & Fraser, 1980; Ridgway & Olson, 1982; Reilly & Kippin, 1983; Costerton, 1984; Leyval et al., 1984; Olivieri et al., 1985b; LeChevallier et al., 1988a). Investigations have demonstrated that a pipe surface in contact with flowing water will undergo a series of physicochemical and biological transformations at the wall–liquid interface, resulting in biofilm formation (Characklis, 1973a,b, 1984a,b; Marshall, 1976; Trulear & Characklis, 1980). There seems to be an initial preconditioning phase during which compounds from the water bind to the surface of the solid. Some of these compounds may act as nutrients. Primary colonizers are most likely to be those organisms capable of attachment via polymer production to the nutrient-preconditioned pipe surface (Characklis, 1973a; Trulear & Characklis, 1980). The formation of this microbial biofilm undergoes primary and secondary microbial succession, eventually reaching a climax community (Alexander, 1971; McFeters, 1984; National Research Council, 1982). Organic and inorganic materials are trapped within the biofilm complex formed by the primary colonizers. This provides a suitable matrix for secondary colonizers that are less capable of attachment. The dynamics of this biofilm formation and the order of succession are functions of nutrient type and availability and physical factors, including temperature, pH, flow velocity, redox potential, turbidity, pipe diameter and material, concentration of disinfectant, and biological factors such as competing bacterial species and predatory organisms (Shindala & Chisholm, 1970; Geldreich et al., 1972; Victoreen, 1974; Hutchinson & Ridgway, 1977; Lee et al., 1980; Tuovinen et al., 1980; LeChevallier et al., 1981; Olson et al., 1981; National Research Council, 1982; Goshko et al., 1983a; Reilly & Kippin, 1983; Nagy, 1984). Once established, a biofilm can provide a continuous source of organisms to the distribution system (van der Wende, Characklis & Smith, 1989), of which many will be heavily encapsulated. It is important to note that in addition to physical forces that may disrupt biofilm-associated bacteria, changes in nutrient conditions (e.g. reduction of carbon or nitrogen) may also promote bacterial release into the distribution system (Delaquis et al., 1989).

Field investigations of chlorine-resistant bacteria provide circumstantial evidence of the protective advantages of capsule-producing species (Earnhardt, 1980; Ridgway & Olson, 1982; Reilly & Kippin, 1983). Earnhardt (1980) reported a coliform bloom incident occurring in the Muncie, Indiana, distribution system in which 14% of the samples collected had a mean value of 4 coliforms/100 ml. The free-chlorine residual of this system, which ranged from 0.5 to 2.5 mg/l, was increased to 4.5 mg/l, and later to 15 mg/l, to reduce coliform numbers. In spite of the high chlorine levels, coliforms were still isolated. The coliforms were identified as E. cloacae. Disinfection studies with this organism indicated that complete inactivation could be achieved with a chlorine concentration of only 0.5

mg/l. Earnhardt (1980) then grew these organisms under conditions that maximized capsule formation (procedure not described) and retested for chlorine sensitivity. Under these conditions, *E. cloacae* was found to survive even after exposure to free chlorine at 10 mg/l. While the experimental conditions were poorly described, this report provides important evidence of capsule-mediated protection to high levels of chlorine. Moreover, it was found that capsule production could be stimulated by growth conditions.

Reilly & Kippin (1983) reported that a variety of coliforms including *K. pneumoniae, Enterobacter aerogenes, E. cloacae* and *E. agglomerans* could survive in distribution water containing a free chlorine residual at 0.2 mg/l. These authors attributed survival of these strains to extensive capsule production. Unfortunately, definitive experimental proof of the role of capsule protection against chlorine was not provided. Even though these findings are observational in nature, this report does add support to the importance of capsule-producing strains in chlorine resistance.

Further experimental evidence for capsule protection against disinfection was provided by Ridgway & Olson (1982). In an analysis of the chlorine resistance patterns of bacteria from two drinking-water distribution systems, it was shown that bacteria attached to particles by polymer material survived chlorinated conditions. Furthermore, scanning electron photomicrographs revealed that there were also many bacterial aggregates. These aggregates and colonized particles were often surrounded by polymer material. The protective ability conferred by particle association or cellular aggregation, however, could not be distinguished from that provided by capsule production.

In an attempt to address the significance of biofilm formation in microbial resistance to chlorine, Olivieri *et al.* (1985*b*) isolated two strains of *K. pneumoniae* from a distribution system. These strains were grown in plate-count broth, resuspended and exposed to the same concentration of free chlorine as was used in the distribution system from which they were isolated (1.2 mg/l). Under these conditions, the strains were rapidly inactivated (5 log reduction in 5 min). The strains were then grown in a reduced-strength (not described) plate-count broth on a glass slide to promote biofilm formation. The intact slide was then exposed to free chlorine (final residual 1.3 mg/l) for 10 min. Strains not attached to glass slides were inactivated (4 log reduction) more readily then those strains attached to glass slides (only 2 log reduction) for the same exposure period. This work indicated that biofilm formation and stimulation by reduced-nutrient conditions may aid in resistance to disinfectant agents.

LeChevallier *et al.* (1988*a*) compared the resistance to chlorine and chloramines by a capsule-forming strain and a noncapsule strain of *K. pneumoniae*. To enhance the capsule quantity in the capsule-forming strain, this organism was grown in extracellular polysaccharide (EPS) agar. In spite of a 55-fold increase in capsule material, there was no difference in chlorine or chloramine resistance by either strain. However, when both strains were grown under low-nutrient conditions (1/10 000 strength EPS), resistance to chlorine increased two to three times for both strains, but no increase in chloramine resistance was observed. In additional experiments, *K. pneumoniae* was grown in the presence of EPS and glass microscope slides to induce biofilm formation. Under these conditions, resistance to chlorine increased 150-fold, while chloramine resistance merely doubled. When the organism was grown under low-nutrient conditions, biofilm formation on the microscope slide increased resistance 600-fold, indicating a multiplicative effect of resistance mechanisms. The results of this study indicate that capsule presence can promote resistance to disinfectants, but only under certain nutrient conditions. The authors posited that the capsule structure grown under low-nutrient conditions had changed in composition and was more colloidal in nature, thus increasing disinfectant resistance.

In related work, LeChevallier, Cawthon & Lee (1988*b*) evaluated the role of biofilm formation in promoting resistance to chlorine, chloramines and chlorine dioxide. In this study, a population of bacteria consisting of *Pseudomonas pickettii* (70%), *Moraxella* (18%) and *Pseudomonas paucimobilis* (12%) was grown in dechlorinated drinking water and used as seed inoculum for biofilm formation on GAC, glass slides, and metal coupons. These organisms, once attached, were found to be 150–3000 times more resistant to free chlorine (0.01–5 mg/l, pH 7.0) than were unattached cells. Chloramine disinfection studies indicated that attached bacteria were only 2–100 times more resistant than unattached bacteria. The chlorine dioxide results were similar to those with free chlorine. In addition, similar experiments using *K. pneumoniae* indicated that attachment to glass microscope slides increased resistance to free chlorine more than 125-fold compared to unattached *Klebsiella* organisms. Resistance to chloramines under the same conditions only doubled. The authors concluded that attachment via biofilm could substantially increase resistance to disinfectants. In addition, chloramines were concluded to have greater penetrating ability than free chlorine.

Studies not focusing on potable water have also indicated that the capsule may provide protection against disinfectants or antiseptics (Seyfried & Fraser, 1980; Berkelman *et al.*, 1984; Kolawole, 1984). Seyfried & Fraser (1980) examined the occurrence of *Pseudomonas aeruginosa* in chlorinated swimming pools. It was found that when the chlorine residual was high (e.g. 1.5 mg/l), pool isolates of *P. aeruginosa* had a much more pronounced capsule layer (as determined by colony appearance) than did *P. aeruginosa* isolates grown under low chlorine concentrations (e.g. 0.4 mg/l). The same *P. aeruginosa* strain was then grown in the presence of cetrimide to reduce production of capsular material. Upon exposure to free chlorine at 1 mg/l for 60 s, only 0.5 log of the capsule-producing strains were inactivated, compared to more than 4 log inactivation for the nonencapsulated strains. Although this study did not quantify the relationship between the amount of capsular material and chlorine resistance, it provided clear evidence of the capsules' role in disinfectant resistance.

Kolawole (1984) demonstrated that strains of *Staphylococcus aureus* capable of producing a large capsule were more resistant to disinfectants and antiseptic agents. *S. aureus* grown on *Staphylococcus* 110 medium were observed by phase-contrast microscopy to produce an extensive capsule layer. This capsule layer could be effectively removed by repeated washing in saline broth or by subculturing in brain–heart infusion broth. Strains prepared under capsule-promoting and capsule-reducing conditions were exposed to various disinfectants (e.g. Camel, Dettol, Hycolin, Izal and Savlon). In each case, after 2.5 min exposure to each of these disinfectants, the encapsulated strain was twice as resistant as the nonencapsulated strain. In a similar study, Berkelman *et al.* (1984) suggested that the resistance of *P. aeruginosa* to poloxamer-iodine was a result of its production of external polymer.

Aggregation

Microbial aggregation is defined as the combination of similar cells to form fairly stable, contiguous, multicellular associations (Calleja, 1984). The term 'aggregation' has also been referred to as adhesion, agglomeration, agglutination, association, autoagglutination, clumping, coagulation, or flocculation, and the resultant structure formed from this association of similar cells is referred to as an agglomerate, aggregate, aggregation, clump, film, floc, pellicle, or slime, among other descriptors (Calleja, 1984). Some researchers consider aggregation to include not only cellular clumping but also attachment to inert surfaces (Marshall, 1984). This review, however, will consider aggregation as the association of three or more similar cells to form a multicellular unit that can be observed under experimentally defined conditions.

Microbial aggregation of one or more of the following types can occur: (1) genetic, (2) physiological, (3) environmental, or (4) experimentally manipulated (Calleja, 1984). Genetically mediated aggregation often occurs for the purpose of exchanging genetic material, as in sexual conjugation in bacterial organisms. While the genetic information contained within the cell may ultimately control the processes involved in aggregation, other external environmental variables can also affect aggregation. Physiological structures such as fimbriae, external polymer compounds, or proteins located on the surface of the cell can aid in aggregation (Marshall, 1976; Calleja, 1984). Environmental conditions, including nutrient type and availability, ionic state of the surrounding solution, pH and temperature, can modify the potential to aggregate (Calleja, 1984; Marshall, 1984). Thus, the ability of a microorganism to form aggregates depends on the inter-action of many complex variables that may be internal or external to the cell and may act independently or in a concerted manner.

Olson & Stewart (1990) reported that bacterial aggregation enhanced resistance to chlorine, monochloramine and chlorine dioxide. Two isogenic strains of *Acinetobacter*, differing only in their ability to aggregate, were grown to stationary phase in peptone broth and tested for their disinfectant resistance. The Ct_{99} values for the nonaggregating strain exposed to chlorine dioxide, monochloramine and chlorine were >0.2, 8.4 and >0.2 mg min/l, respectively. By contrast, the Ct_{99} values for the aggregating strain were 6, 18 and 8 mg min/l for chlorine dioxide, monochloramine and chlorine, respectively. Importantly, when the aggregating strain was grown in the presence of a surfactant to promote disaggregation, it was observed that the aggregating strain of *Acinetobacter* disaggregated, resulting in increased sensitivity to disinfectants. Transmission electron microscopy indicated that the aggregating strain possessed significantly more fimbriae on its outer surface than did the nonaggregating strain, which may account for the difference in aggregation potential between these two strains.

The effects of bacterial aggregation on disinfectant resistance have also been observed by researchers conducting the 'use-dilution method for dis-infectant efficacy testing' in accordance with the Association of Official Analytical Chemists (Cole *et al.*, 1989). *P. aeruginosa* grown in peptone broth containing beef extract formed dense aggregates at least 10 μm in diameter. When exposed to a 1:256 dilution of a quaternary ammonium disinfectant (7.5% dodecyl dimethyl ammonium chloride, 5% *n*-alkyl di-methylbenzyl ammonium chloride), those test conditions in which aggre-gates were present consistently demonstrated surviving bacteria. However, when the aggregates were removed, few or no surviving bacteria were isolated. The authors attributed disinfectant resistance to the inability of the quaternary ammonium compound to penetrate into aggregated bacteria.

Wei & Chang (1975) developed a multi-Poisson distribution model for the analysis of disinfection data based on organism clumping. The impetus for the development of this model was based on consistent reports of nonlinear or nonexponential survival curves for both bacteria and viruses. Wei & Chang (1975) postulated that microbial inactivation by disinfectants was caused by random collisions between molecules of the disinfectant and the organism. As the number of molecules greatly outnumber the microorganisms present, the collision rate can be expressed as a Poisson distribution. Furthermore, as aggregates of microorganisms may exist in various sizes, the probability of collision involves more than one Poisson population.

Based on these assumptions, a mathematical multi-Poisson distribution model was developed to account for the effect of variations in clump size on inactivation curves. To test the validity of this model, the amoeba *Naegleria gruberi* was used because of its ability to form clumps of various sizes when grown under laboratory conditions. The clumped organisms were then exposed to iodine (2–5 mg/l). It was observed that as clump size increased, there was a concomitant decrease in the rate of inactivation. The results supported the multi-Poisson model and, more importantly, indicated the significance of aggregation in disinfectant resistance.

Studies concerning viral aggregation and disinfection resistance have also been reported in the literature (Galasso & Sharp, 1962; Berg, Chang & Harris, 1964; Poduska & Hershey, 1972; Katzenelson *et al.*, 1974; Young & Sharp, 1977; Scarpino, 1984). Scarpino (1984) reported that aggregated polioviruses were 1.7 times more resistant to chloramines than were mono-dispersed viruses. Katzenelson *et al.* (1974) found that sonicated polioviruses exposed to 0.1 mg/l ozone were appreciably more sensitive (a four-fold difference) than nonsonicated viruses. Because sonication did not affect virus viability, the authors attributed the difference in ozone sensitivity to clumping.

Poduska & Hershey (1972) examined viral inactivation data from other researchers and concluded that aggregation size could account for enhanced resistance and result in nonlinear inactivation. Young & Sharp (1977) observed the nonlinear inactivation kinetics of poliovirus exposed to bromine. Electron microscopy revealed that the virus suspension contained both monodispersed viruses and large viral aggregates. The authors concluded that the aggregated viruses accounted for the nonlinear phase of the inactivation curve. Young, Johnson & Sharp (1977) suggested that before aggregation could be a significant factor in resistance to disinfectants, the minimum aggregate size would have to be sufficient to surround a single virus completely. In the case of echovirus, it would require 16 virions to provide one layer of protection.

Cell wall and/or membrane alterations

Evidence indicating that microorganisms can acquire enhanced resistance to disinfectants by an alteration of their cellular envelope has been suggested by the work of several investigators (Farkas-Himsley, 1964; Hugo & Stretton, 1966; Gilbert & Brown, 1978, 1980; Chai, 1983). Farkas-Himsley (1964) observed that *E. coli* strains resistant to chlorine and bromine had different surface properties (as indicated by different phage sensitivity) than chlorine- and bromine-sensitive *E. coli* strains. Gilbert & Brown (1978) found that the cell envelope of strains of *P. aeruginosa* that were resistant to 3- and 4-chlorophenol had a higher content of phospholipid and lipopolysaccharide and a lower concentration of fatty acids than did nonresistant strains. On the basis of chlorophenol uptake studies, the authors concluded that the enhanced resistance was a function of decreased cell membrane permeability. Further studies by these researchers (Gilbert & Brown, 1980) indicated that 2-phenoxyethanol- and chlorhexidine-resistant *Bacillus megaterium* also had less permeable cell walls.

Hugo & Stretton (1966) observed that gram-positive organisms (*Bacillus subtilis, Streptococcus faecalis* and *Staphylococcus aureus*) with greater cell membrane lipid content were ten times more resistant to penicillin than when grown under conditions that resulted in less membrane lipid content. Chai (1983) found that the *E. coli* cells demonstrating the greatest resistance to heavy metals, colicins, detergents and dyes had altered outer membrane protein structure. He suggested that this altered envelope may have affected the permeability state of the cells.

In work with echovirus 1, Young *et al.* (1977) observed that exposure of this virus to chlorine resulted in nonlinear inactivation kinetics. Plaque titration techniques and observation of the virus preparation by electron microscopy revealed that the viruses were monodispersed and not aggregated. The authors suggested that conformational changes of the viral capsid, resulting in reduced chlorine penetration, were responsible for increased resistance to this disinfectant. In later studies, Young & Sharp (1985) confirmed that echovirus 1 resistance was related to conformational changes. It was demonstrated that prior to chlorine exposure, the viruses occurred in two distinct isoelectric forms. Following chlorine exposure, a third isoelectric form was observed. This fraction was subsequently determined to be infectious. The actual mechanism of resistance to chlorine was not determined in these studies.

Summary

Table 5.3 summarizes the categories of cell-mediated resistance to disinfectants. Capsule production, aggregation and altered permeability of the cellular envelope can affect an organism's sensitivity to a disinfectant. Moreover, growth conditions can alter these physiological structures. Consequently, responses to disinfectants by microorganisms should be described in terms of growth conditions, especially in instances where these conditions modify physiological features related to disinfection protection.

Effects of growth conditions on disinfection resistance

In general, microorganisms in potable water must be capable of growth under limited nutrient conditions. Adaptation features acquired by organisms grown under these conditions can render them phenotypically distinct

Table 5.3. *Microbial resistance to disinfectants by cell-mediated factors*

Observation	Reference
Bacteria	
Capsule	
Encapsulated *E. cloacae* recovered from potable water with residual of chlorine at 15 mg/l	Earnhardt (1980)
Encapsulated *P. aeruginosa* ten times more resistant to chlorine at 1.5 mg/l than nonencapsulated strains	Seyfried & Fraser (1980)
Bacteria isolated from chlorinated distribution system (1 mg/l) heavily encapsulated	Ridgway & Olson (1982)
Encapsulated *S. aureus* twice as resistant to clinical disinfectant agents (e.g. Dettol) compared to nonencapsulated strain	Kolawole (1984)
K. pneumoniae grown in biofilm twice as resistant to chlorine (1.3 ml/l) compared to nonbiofilm strain	Olivieri et al. (1985b)
Encapsulated strains of *K. pneumoniae, E. cloacae, E. aerogenes* and *E. agglomerans* isolated from chlorinated (residual of 0.2 mg/l) distribution system	Reilly & Kippin (1983)
K. pneumoniae grown under conditions promoting biofilm production resulted in 600-fold increase to resistance to chlorine and chloramines	LeChevallier et al. (1988a)
P. pickettii, Moraxella and *P. paucimobilis* grown under conditions promoting biofilm formation were 150–3000 times more resistant to chlorine and 2–100 times more resistant to chloramines	LeChevallier et al. (1988b)
Aggregation	
Bacteria isolated from chlorinated distribution system (1 mg/l) were aggregated	Ridgway & Olson (1982)
Aggregating strain of *Acinetobacter* was > 30, 2 and > 40 times more resistant to chlorine, chloramines and chlorine dioxide, respectively, than isogenic nonaggregating strain of *Acinetobacter*	Olson & Stewart (1990)
Alteration of cellular envelope	
Alteration of cellular envelope of *E. coli* resulted in increased resistance to chlorine and bromine	Farkas-Himsley (1964)
Increased phospholipid content of *P. aeruginosa* increased resistance to chloraphenol four-fold	Gilbert & Brown (1978)
Decrease in cell envelope permeability of *B. megaterium* resulted in seven- to ten-fold increase in resistance to chlorhexidine	Gilbert & Brown (1980)
Viruses	
Aggregation	
Aggregated polioviruses were four-fold more resistant to ozone (0.1 mg/l) than nonaggregated viruses	Katzenelson et al. (1974)
Aggregated polioviruses were approximately twice as resistant to bromine (10 μM) than monodispersed viruses	Young & Sharp (1977)
Aggregated polioviruses were two-fold more resistant to chloramines than nonaggregated poliovirus	Scarpino (1984)
Capsid alterations	
Capsid conformational state of monodispersed echovirus 1 was related to increased resistance to chlorine	Young & Sharp (1985)

from the same organism grown in nutrient-rich environments (Dawson, Humphrey & Marshall, 1981; Berg *et al.*, 1982; Kjelleberg *et al.*, 1983; Sterkenburg *et al.*, 1984). Many of these changes involve alterations of capsule presence or composition (Costerton *et al.*, 1978; LeChevallier *et al.*, 1988*a*; Stewart & Olson, 1992*b*), fimbriae (Calleja, 1984), outer-membrane proteins (Robinson & Tempest, 1973; Brown, Gilbert & Klemperer, 1979; Chai, 1983; Sterkenburg *et al.*, 1984; Williams, Brown & Lambert, 1984; Dalhoff, 1985), lipopolysaccharide components (Gill & Suisted, 1978; Brown *et al.*, 1979; Morse *et al.*, 1983; Dalhoff, 1985), cell-membrane lipids (Hugo & Stretton, 1966; Calcott & Petty, 1980; Dalhoff, 1985; Stewart & Olson, 1992*b*), and cell-wall components (Robinson & Tempest, 1973; Gilbert & Brown, 1980; Sterkenburg *et al.*, 1984). In addition, studies have also shown that low-nutrient conditions may induce the production of new proteins within the cell (Schultz, Latter & Matin, 1988), which may assist in resistance to compounds such as H_2O_2 (Jenkins, Schultz & Matin, 1988). Changes in membrane permeability seem to be a common link in many of these studies. It is believed that these changes in permeability are responsible for enhanced resistance to antimicrobial compounds (Hugo & Franklin, 1968; Brown & Melling, 1969; Roantree, Kuo & MacPhee, 1977; Gilbert & Brown, 1980; Yoshimura & Nikaido, 1982; Chai, 1983; Berg *et al.*, 1983). Movement of chemicals through the cellular envelope is controlled, in part, by binding, transport, or the formation of protein channels referred to as 'porins' (Nikaido & Nakae, 1979; Lugtenberg & Van Alphen, 1983; Nikaido & Vaara, 1985).

Differences in sensitivity to disinfectant agents between organisms grown in nutrient-rich environments and those grown in nutrient-limited environments may result from the disinfectants' decreased ability to pass through the envelope. For example, bacteria treated with ethylenediaminetetraacetic acid, which renders the cell more permeable to external agents, became more sensitive to antimicrobial agents (Leive, 1965; Brown & Melling, 1969). If cells grown under low-nutrient conditions have less protein available on the outer membrane for binding or facilitating transport, potential stressor compounds may not be able to gain entry into the cell. It is known that nonspecific diffusion for influx of nutrients or efflux of waste products occurs through porins (Lugtenberg & Van Alphen, 1983; Nikaido & Vaara, 1985). These channels generally allow passage of relatively small (e.g. less than 600 Da) hydrophilic molecules (Nikaido, 1976). Many antimicrobial compounds could conceivably enter the cell by this route. Interestingly, organisms with fewer sites, or with reduced activity at each site, have been shown to be more resistant to antibiotics and other antimicrobial agents (Hancock, 1984). Whether or not microorganisms resistant to potable water disinfectants have altered porin structures remains to be determined. The following paragraphs describe some of the key studies conducted in the past 40 years that illustrate the effects of antecedent growth conditions on an organism's response to disinfectant agents.

Early work by Milbauer & Grossowicz (1959) suggested that growth conditions affected bacterial resistance to chlorine. In these studies, *E. coli* cells grown on nutrient agar were approximately 5000 times more resistant to chlorine than cells grown on minimal agar. However, because important experimental variables (such as chlorine residuals) were not reported, it is difficult to conclude that nutrient-rich conditions did promote disinfectant resistance.

Favero & Drake (1966) reported the isolation of high numbers of several types of bacteria from swimming pools disinfected with chlorine and iodine. Examples of bacteria isolated included coliforms, enterococci, *Staphylococcus*, *Streptococcus salivarius*, *P. aeruginosa*, *P. alcaligenes*, and *Alcaligenes faecalis*. Laboratory studies indicated a substantial loss of resistance once these organisms were subcultured in trypticase soy broth or grown on standard-plate-count agar. For example, a strain of *P. alcaligenes* grown in sterile, dehalogenated swimming-pool water was ten times more resistant to iodine than it was when subcultured in trypticase soy broth.

Carson *et al.* (1972) indicated that a *P. aeruginosa* strain isolated from the distilled-water environment of a hospital's mist therapy unit was resistant to acetic acid, glutaraldehyde, chlorine dioxide and a quaternary ammonium compound. Importantly, after only one subculture on trypticase soy agar, this resistance was substantially reduced. These investigators also found that growth temperature and growth phase could alter disinfectant resistance. Naturally occurring *P. aeruginosa* cells grown at 25 °C were more resistant to acetic acid than when grown at 37 °C. Naturally occurring or subcultured cells grown to stationary phase were approximately twice as resistant to the disinfectants tested than were log-phase cells. This study demonstrated not only that nutrient conditions affected resistance, but also that growth phase and incubation temperature could alter cellular response to disinfectant agents. These authors did not speculate on the mechanism of resistance.

In a later study, Carson *et al.* (1978) examined isolates of *Mycobacterium chelonei* obtained from the peritoneal fluid of hospital patients, as well as American Type Culture Collection (ATCC) reference strains of *Mycobacterium fortuitum* and *M. chelonei*, for their sensitivity to formaldehyde, glutaldehyde and chlorine. All strains were grown in distilled water to stationary phase and then exposed to each disinfectant. It was observed that clinical strains were more resistant to the disinfectants than were the reference strains. Interestingly, even though the reference strains were grown in distilled water (low-nutrient conditions), they did not develop enhanced resistance. The authors concluded that longer adaptation periods under low-nutrient conditions may be required before resistance can be developed. This study also demonstrated the differences in the properties of environmental isolates and laboratory reference strains. Indeed, studies performed by Ward *et al.* (1984) demonstrated that strain variation in *E. coli* resulted in different sensitivities to chloramines.

Berg *et al.* (1982) examined the effect of antecedent growth conditions on the sensitivity of *E. coli* to chlorine dioxide (0.75 mg/l). The cells were grown in continuous culture at various nutrient dilution values. *E. coli* grown under the low-nutrient conditions were twice as resistant to chlorine dioxide as were cells grown under high-nutrient conditions. Berg *et al.* also determined that the type of limiting nutrient could affect resistance. Cells grown under decreased nutrient broth conditions were more (approximately 1 log) resistant than cells grown at an equivalent level of limiting glucose. Studies were also conducted to assess the effect of temperature on resistance. Cells grown at low temperatures (15 °C) were noted to be approximately 1 log more resistant than cells grown at higher temperatures (37 °C), which was consistent with the results of earlier work by Carson *et al.* (1972). Interestingly, these observations may suggest that the effect of low-temperature growth on disinfection resistance is similar for different bacterial genera.

Additional investigations were conducted by Berg *et al.* (1983) to study the effect of antecedent growth temperatures and oxygen tension on chlorine dioxide sensitivity. These researchers examined the hypothesis that the susceptibility of cells to disinfectant activity may depend on the temperature of assay rather than the temperature of growth. However, regardless of assay temperature (4, 23 or 37 °C), cells grown at 15 °C were still more resistant to chlorine dioxide than were cells grown at 37 °C. This observation indicated that cells do not undergo significant temperature shock during assay conditions and that the temperature of incubation prior to disinfectant exposure is indeed an important variable contributing to disinfection sensitivity.

Chai (1983) examined the response of *E. coli* isolated from bay-water and sewage to heavy metals, colicins, detergents, dyes and bacteriophages after growth in either rich laboratory medium (proteose peptone beef extract with 0.5% NaCl) or bay-water. *E. coli* was appreciably more sensitive to heavy metals (copper, mercury, lead, silver and cobalt), detergents and dyes (sodium dodecyl sulfate, Triton X-100, sodium sarcosinate, methylene blue and Zwittergent) when grown in bay-water than when grown in peptone. However, bay-water-grown cells were more resistant to colicins and bacteriophage attachment than were peptone-grown cells. To resolve the differences in sensitivities to these various agents, Chai (1983) examined the cellular envelope proteins of cells grown under bay-water and peptone conditions. Sodium dodecyl sulfate-gel electrophoresis revealed that cells grown in proteose peptone contained two or three major outer proteins (OmpA, OmpC, OmpF), whereas bay-water-grown cells lacked or had greatly reduced levels of each of these proteins. Chai posited that the resistance to bacteriophages and colicins was caused by a loss of binding proteins. The increased sensitivity to metals and dyes was attributed to differences in membrane permeability.

Harakeh *et al.* (1985) observed that growth of *Y. enterocolitica* and *K. pneumoniae* under reduced nutrient concentrations resulted in approximately a ten-fold increase in resistance to chlorine dioxide. In addition, it was demonstrated that this resistance was further increased when the cells were grown at lower temperatures (e.g. 15 °C). Although the mechanism of increased resistance was not determined, the authors speculated that alteration of cell envelope lipids resulted in decreased permeability of the cell to chlorine dioxide.

Pyle & McFeters (1989) studied the effects of bacterial inactivation by iodination. An 'iodine-resistant' *P. aeruginosa* species (isolated from a hospital povidone–iodine solution) was grown in brain–heart infusion, mineral salts medium, phosphate-buffered water, and reverse-osmosis water at room temperature. Growth under low-nutrient conditions (e.g. reverse-osmosis water and phosphate-buffered water) resulted in greater resistance to iodine than did growth under high-nutrient conditions (e.g. brain–heart infusion). In similar studies, Cargill *et al.* (1992) observed that growth under low-nutrient conditions (well-water growth) increased the resistance of *Legionella pneumophila* to iodine. This increased resistance was considered to have resulted from possible increased aggregation, protective extracellular material and/or other physiological changes under low-nutrient conditions.

In studies conducted by Stewart & Olson (1992*a*), the resistance of *K. pneumoniae* to chloramines was increased when cells were grown under low-nutrient conditions. The growth phase did not affect resistance to chloramines; however, there was a moderate increase in resistance to chloramines when the cells were grown at lower temperatures (e.g. 15 °C). The resistance of *K. pneumoniae* to chloramines was subsequently attributed by these investigators to multiple physiological factors (Stewart & Olson, 1992*b*). It was observed that chloramine-resistant *K. pneumoniae* cells that were highly aggregated (e.g. > 90 cells/aggregate) produced six times more capsule material and had a higher ratio of saturated fatty acids in the cellular envelope than did chloramine-sensitive *K. pneumoniae* cells. Importantly, upon exposure to chloramines, only 33% of the sulfhydryl groups in *K. pneumoniae* cells grown under low-nutrient conditions were oxidized, as compared with 80% oxidation of sulfhydryl groups in cells grown under high-nutrient conditions.

Evidence of the effect of nutrient growth conditions on resistance to disinfectant agents has also been observed in studies dealing with the development of resistant strains or the isolation of such strains (Farkas-Himsley, 1964; Leyval *et al.*, 1984; Shaffer, Metcalf & Sproul, 1980; Bates *et al.*, 1977). In each of these studies, organisms grown or subcultured under original isolation conditions were more resistant than strains subcultured in typical laboratory media containing elevated nutrient levels. Farkas-Himsley (1964) observed that *E. coli* maintained in its original swimming-pool water was five times more resistant to chlorine or bromine at 10 mg/l than were strains grown in laboratory media. Also, phage-typing tests determined that these resistant cells had different sensitivities from non-resistant cells, indicating that the cell surfaces had been altered.

Leyval *et al.* (1984) selected for chlorine-resistant *E. coli* strains by successive exposure of cells to increasing chlorine concentrations. This selection procedure resulted in strains that were 50 times more resistant to chlorine than was the original strain. Importantly, these strains were not subcultured but, rather, were maintained on the original recovery medium (R2A). Leyval *et al.* observed that colonies of the chlorine-resistant isolates had a significant mucoid appearance (indicative of a pronounced capsule layer). A similar study conducted by Haas & Morrison (1981) failed to create resistant *E. coli* strains upon repeated exposure to chlorine. However, in Haas & Morrison's methodology for generating resistant strains, cells were subcultured repeatedly in nutrient broth between exposure assays. This study, in contrast to that of Leyval *et al.* (1984), underscores the importance of growth conditions in the development of resistance features.

Clinical isolates of pathogenic bacteria have also been shown to lose resistance to antimicrobial compounds upon subculturing in nutrient-rich laboratory media (Bassett, Stokes & Thomas, 1970; Gilbert & Brown, 1978; Brown *et al.*, 1979; Gilbert & Brown, 1980; Anwar, Brown & Lambert, 1983; Cozens & Brown, 1983; Ombaka, Cozens & Brown, 1983; Costerton, 1984; Hancock, 1984; Brown & Williams, 1985; Dalhoff, 1985). These investigators have demonstrated that this loss of virulence or increased sensitivity to antimicrobial agents was a function of alterations occurring at the glycocalyx or cell envelope.

Hugo & Stretton (1966) illustrated that gram-positive bacteria (*B. subtilis*, *S. faecalis* and *S. aureus*) grown in nutrient broth containing glycerol were ten times more resistant to penicillin compounds than they were when grown in nutrient broth without glycerol. This resistance was concluded to be a result of the increased amount of lipid produced in the outer cell membrane (approximately 2% more) when cells were grown in the presence of high levels of glycerol. Glycerol-grown cells treated with lipase to reduce lipid content in the outer cell membrane were similar to the control cells in their sensitivity to antibiotics.

Bassett *et al.* (1970) found that a pathogenic strain of *Pseudomonas multivorans* could grow in a 1:30 dilution of Savlon solution (chlorhexidine 0.05%, cetrimide 0.5%). Subculturing this organism in 0.1% glucose broth resulted in a ten-fold increase in its sensitivity to Savlon. The original resistance level was restored by subculturing the isolate in a 1:30 dilution of Savlon in distilled water. The authors also reported that cells grown at higher temperatures (37 °C versus 30 °C) and those grown to late stationary phase were less resistant. Although the mechanism of cellular resistance was not determined, cell aggregation (not quantified) was observed.

Gilbert & Brown (1978) observed that *P. aeruginosa* was more resistant to 3-chlorophenol and 4-chlorophenol when grown under limiting conditions of magnesium or glucose. These limiting nutrient conditions promoted a higher lipopolysaccharide content. On the basis of proton uptake studies, Gilbert & Brown concluded that decreased nutrient conditions altered the phospholipid layer of the cell envelope, making the cell less permeable to both 3- and 4-chlorophenol.

Further work by Gilbert & Brown (1980) also demonstrated that antecedent growth conditions could alter disinfectant resistance in gram-positive bacteria. *B. megaterium* was grown under phosphate-, magnesium- and carbon (as glycerol)-limiting conditions and exposed to the disinfectant agents 2-phenoxyethanol and chlorhexidine. When the organism was grown in either magnesium- or carbon-limiting conditions, it became three times more resistant to 2-phenoxyethanol. However, cells grown under phosphate-limiting conditions showed little change in sensitivity to 2-phenoxyethanol, but showed a two-fold increase in resistance to chlorhexidine. On the basis of lysozyme and protoplast assays, Gilbert & Brown (1980) concluded that sensitivity to 2-phenoxyethanol and chlorhexidine was a function of cell-wall structure and composition. Cell-wall permeability tended to vary with respect to the type of limiting nutrient. It was further postulated that changes in the cytoplasmic membrane could contribute to altered sensitivity to disinfectant agents. In later work, Cozens & Brown (1983), investigating the effects of nutrient depletion of magnesium, carbon (as glucose), ammonium, phosphate and ferrous iron on the sensitivity of *Pseudomonas cepacia* to cetrimide, chlorhexidine and benzalkonium, determined that magnesium limitation was responsible for the greatest amount of resistance.

Anwar *et al.* (1983) examined the resistance of *P. cepacia* to serum-associated bactericidal agents (e.g. white blood cells). Cells grown under nutrient-limiting conditions gave the following order of decreasing resistance: magnesium > ammonium > phosphate > sulfate > iron > glucose. These findings illustrated that cellular resistance mediated by a single reduced nutrient could produce increased resistance against a broad range of stressor agents. Morse *et al.* (1983) also indicated that growth conditions could alter an organism's response to serum-associated bactericidal agents. *Neisseria gonorrhoeae* grown under low-nutrient conditions developed a two-fold increase in resistance to human serum. The surface hydrophobicity of these low-nutrient-grown cells increased 1.5 times compared to cells grown in rich media. The mechanism of increased resistance was not resolved; however, increased surface hydrophobicity may have enhanced the aggregation potential of the cells.

Altered resistance to disinfectants as a result of subculturing practices has also been demonstrated with enteric viruses. Bates *et al.* (1977) indicated that the resistance of poliovirus to chlorine was increased by subculturing surviving viruses exposed to chlorine at 0.6–0.8 mg/l (pH 7.0). After ten cycles of exposure, resistance was approximately doubled. The nature of this enhanced resistance was not discussed. Further work by these investigators (Bates *et al.*, 1978) revealed that chlorine-resistant viruses had different plaque morphologies, suggesting altered phenotypic properties of the virion.

Table 5.4 summarizes studies on altered resistance properties as a function of antecedent growth conditions. Many of these altered resistance properties involve changes at the cellular envelope. Indeed, it appears that this structure optimizes nutrient exchange and also provides protection from antimicrobial agents in the environment. Enhanced resistance may involve capsule production, changes in lipopolysaccharides, or outer membrane proteins. Physiological alterations resulting from nutrient status, however, are not limited to the outer membrane layers. Changes may also occur in the cytoplasmic membrane. This structure contains a significant number of metabolic enzymes, as well as protein and lipid components that regulate the permeability of the cell.

Summary and conclusions

The control of microorganisms in treated drinking-water supplies has been a major concern of water purveyors for many years. This concern is prompted by stricter regulations governing pathogen occurrence and permissible levels of indicator organisms in treated water supplies. Moreover, many organisms once thought to be harmless have been found to act as opportunistic pathogens or to promote the growth of primary pathogens. Consequently, the need to provide safe drinking water is still a critical necessity in the water industry. In spite of significant advances in water treatment, microorganisms – including coliforms and pathogens – are still isolated from drinking-water systems that maintain a disinfectant residual. Historically, the presence of a disinfectant has been considered sufficient to control the growth of microorganisms. However, microorganisms can grow, and even flourish, in treated drinking-water environments, which may contain a residual concentration of disinfectant. Numerous physical and physicochemical factors have been found to be responsible for the persistence of these organisms in distribution systems. It is therefore essential to understand the dynamic environment encountered by organisms in

Table 5.4. *Effects of antecedent growth conditions on bacterial resistance to disinfectants*

Organism	Growth condition	Disinfectant	Response	Reference
Escherichia coli	Swimming-pool water	Chlorine or bromide at 10 mg/l	Five times more resistant than cells grown on nutrient agar	Farkas-Himsley (1964)
Pseudomonas alcaligenes	Swimming-pool water	Iodine at 0.1–0.6 mg/l	Ten times more resistant than strains grown in trypticase soy broth	Favero & Drake (1966)
Pseudomonas aeruginosa	Distilled water	Chlorine dioxide	100 times more resistant than strains grown in trypticase soy broth	Carson et al. (1972)
Escherichia coli	Nutrient broth; 0.02–0.6 dilution	Chlorine dioxide at 0.75 mg/l	Twice as resistant as cells grown in undiluted nutrient broth	Berg et al. (1982)
Klebsiella pneumoniae	Glucose minimal media	Chlorine dioxide at 0.25 mg/l	Approximately ten times more resistant as glucose concentration decreased by a factor of 6	Harakeh et al. (1985)
Yersinia enterocolitica	Glucose minimal media	Chlorine dioxide at 0.25 mg/l	Approximately ten times more resistant as glucose concentration decreased by a factor of 3	Harakeh et al. (1985)
Flavobacterium	Drinking water	Chlorine at 0.45–0.75 mg/l	>200 times more resistant than when grown in R2A broth	Wolfe & Olson (1985)
Klebsiella pneumoniae	Extracellular polysaccharide agar (EPS)	Chlorine	600 times more resistant when grown in 1:10000 dilution EPS and attached to microscope slide	LeChevallier et al. (1988a)

drinking-water distribution systems in order to devise effective control strategies.

Physiological alterations occurring in response to changing environmental conditions may modify an organism's sensitivity to a myriad of antimicrobial agents. For example, organisms grown in pure culture, under nutrient-rich conditions and optimal growth temperatures, have dramatically different physiological characteristics from those they have when grown under environmental conditions. Interestingly, the adaptive features developed by environmental organisms to cope with nutrient-poor conditions often increase their resistance to antimicrobial agents. This is likely to be merely a fortuitous event for the organism. Unfortunately, the complex interaction of many factors occurring under environmental conditions complicates the task of determining which variables actually stimulate resistance mechanisms. The goal of this review, however, has been to identify conditions promoting resistance and, where possible, to note the physiological features contributing to disinfectant resistance.

It is hoped that a better understanding of microbial resistance to disinfectant agents will provide critical information for the future control of these organisms in drinking water.

Acknowledgments

The authors gratefully acknowledge the assistance of Roy L. Wolfe, Associate Director of Water Quality at the Metropolitan Water District of Southern California, who reviewed the original manuscript, and Peggy Kimball, technical editor at Metropolitan's Water Quality Laboratory, who accomplished the detail work needed to complete the chapter.

References

Adair, F. W., Geftic, S. G. & Gelzer, J. (1969). Resistance of *Pseudomonas* to quaternary ammonium compounds. *Applied Microbiology*, **18**, 299–302.
Alexander, M. (1971). *Microbial Ecology*. New York: John Wiley and Sons.
Allen, M. J., Taylor, R. H. & Geldreich, E. E. (1980). The occurrence of microorganisms in water main encrustations. *Research and Technology*, **72**, 614–25.
Alvarez, M. E. & O'Brien, R. T. (1982). Mechanisms of inactivation of poliovirus by chlorine dioxide and iodine. *Applied and Environmental Microbiology*, **44**, 1064–71.
American Public Health Association (APHA) (1989). *Standard Methods for the Examination of Water and Wastewater*, 17th edn. Washington, DC: APHA, American Water Works Association, and Water Pollution Control Federation.
Anand, C. M., Skinner, A. R., Malic, A. & Kurtz, J. B. (1983). Interaction of *L. pneumophila* and a free living amoeba (*Acanthamoeba palestinensis*). *Cambridge Journal of Hygiene*, **91**, 167–78.

Anderson, R. L., Berkelman, R. L., Mackel, D. C., Davis, B. J., Holland, B. W. & Martone, W. J. (1984). Investigations into the survival of *Pseudomonas aeruginosa* in poloxamer-iodine. *Applied and Environmental Microbiology*, **47**, 757–62.

Anwar, H., Brown, M. R. W. & Lambert, P. A. (1983). Effect of nutrient depletion on sensitivity of *Pseudomonas cepacia* to phagocytosis and serum bactericidal activity at different temperatures. *Journal of General Microbiology*, **129**, 2021–7.

Atwood, K. C. & Norman, A. (1949). On the interpretation of multihit survival curves. *Proceedings of the National Academy of Sciences (USA)*, **35**, 696–709.

Baker, J. C. (1926). Chlorine in sewage and waste disposal. *Canadian Journal of Engineering, Water and Sewage*, **50**, 127–34.

Bassett, D. C. J., Stokes, K. J. & Thomas, W. R. G. (1970). Wound infection with *Pseudomonas multivorans*. *Lancet*, **i**, 1188–91.

Bates, R. C., Shaffer, P. T. & Sutherland, S. M. (1977). Development of poliovirus having increased resistance to chlorine inactivation. *Applied and Environmental Microbiology*, **34**, 849–53.

Bates, R. C., Sutherland, S. M. & Shaffer, P. T. B. (1978). Development of resistant polioviorus by repetitive sublethal exposure to chlorine. In *Water Chlorination: Environmental Impact and Health Effects*, vol. 2, ed. R. L. Jolley, H. Gorchev & D. H. Hamilton Jr, pp. 471–82. Ann Arbor, Michigan: Ann Arbor Science Publishers.

Bekhtereva, M. N. & Krainova, O. A. (1975). Changes in the activity of enzyme systems in *Bacillus anthracoides* spores during germination and due to the action of calcium hypochlorite. *Microbiology*, **44**, 712–15. [Translated from *Mikrobiologiya* (Russia), **44**, 791–5 (1975).]

Berg, G., Chang, S. L. & Harris, E. K. (1964). Devitalization of microorganisms by iodine. 1. Dynamics of the devitalization of enteroviruses by elemental iodine. *Virology*, **22**, 469–81.

Berg, J. D., Matin, A. & Roberts, P. V. (1982). Effect of antecedent growth conditions on sensitivity of *Escherichia coli* to chlorine dioxide. *Applied and Environmental Microbiology*, **44**, 814–19.

Berg, J. D., Matin, A. & Roberts, P. V. (1983). Growth of disinfection-resistant bacteria and simulation of natural aquatic environments in the chemostat. In *Water Chlorination: Environmental Impact and Health Effects*, vol. 4, book 2, ed. R. L. Jolley, W. A. Brungs, J. A. Cotruvo, R. B. Cumming, J. S. Mattice & V. A. Jacobs, pp. 1137–48. Ann Arbor, Michigan: Ann Arbor Science Publishers.

Berg, J. D., Roberts, P. V. & Matin, A. (1986). Effect of chlorine dioxide on selected membrane functions of *Escherichia coli*. *Journal of Applied Bacteriology*, **60**, 213–20.

Berkelman, R. L., Anderson, R. L., Davis, B. J., Highsmith, A. K., Petersen, N. J., Bond, W. W., Cook, E. H., Mackel, D. C., Favero, M. S. & Martone, W. J. (1984). Intrinsic bacterial contamination of a commercial iodophor solution: investigation of the implicated manufacturing plant. *Applied and Environmental Microbiology*, **47**, 752–6.

Berman, D., Rice, E. W. & Hoff, J. C. (1988). Inactivation of particle-associated coliforms by chlorine and monochloramine. *Applied and Environmental Microbiology*, **54**, 507–12.

Bernarde, M. A., Snow, W. B. & Olivieri, V. P. (1967). Kinetics and mechanisms of bacterial disinfection by chlorine dioxide. *Applied Microbiology*, **15**, 257–65.

Boardman, G. D. (1976). Protection of waterborne viruses by virtue of their affiliation with particulate matter. Doctoral thesis, University of Maine, Orono.

Bocharov, D. A. (1970). Content of nucleic acids in bacterial suspension after influence on it of chlorinated and alkaline solutions. *Trudy Vsesoiuznvi Nauchno-Issledovatel'skii Institut Veterinarnoi Sanitarii* (Soviet Union), **34**, 242–52.

Bocharov, D. A. & Kulikovsky, A. V. (1971). Structural and biochemical changes of bacteria after the action of some chlorine-containing preparations. Report 1. *Trudy Vsesoiuznvi Nauchno-Issledovatel'skii Institut Veterinarnoi Sanitarii* (Soviet Union), **38**, 165–70.

Boring, J. R., Martin, W. T. & Elliot, L. M. (1971). Isolation of *Salmonella typhimurium* from municipal water, Riverside, California. *American Journal of Epidemiology*, **93**, 49.

Boyle, W. C. (1963). Studies on the biochemistry of disinfection by monochloramine. Doctoral thesis, California Institute of Technology, Pasadena.

Brown, M. R. W., Gilbert, P. & Klemperer, R. M. M. (1979). Influence of the bacterial cell envelope on combined antibiotic action. In *Antibiotic Interactions*, ed. J. D. Williams. San Diego, California: Academic Press.

Brown, M. R. W. & Melling, J. (1969). Role of divalent cations in the action of polymyxin B and EDTA on *Pseudomonas aeruginosa*. *Journal of General Microbiology*, **59**, 263–74.

Brown, M. R. W. & Williams, P. (1985). The influence of environment on envelope properties affecting survival of bacteria in infections. *Annual Review of Microbiology*, **39**, 527–56.

Butterfield, C. T. (1948a). Bactericidal properties of chloramines and free chlorine in water. *Public Health Report*, **63**, 934–40.

Butterfield, C. T. (1948b). Bactericidal properties of free and combined available chlorine. *Journal, American Water Works Association*, **40**, 1305–12.

Butterfield, C. T. & Wattie, E. (1946). Influence of pH and temperature on the survival of coliforms and enteric pathogens when exposed to chloramines. *Public Health Report*, **61**, 157–92.

Butterfield, C. T., Wattie, E., Mergregion, S. & Chambers, C. W. (1943). Influence of pH and temperature on the survival of coliforms and enteric pathogens when exposed to free chlorine. *Public Health Report*, **58**, 1837–66.

Calcott, P. H. & Petty, R. S. (1980). Phenotypic variability of lipids of *Escherichia coli* grown in chemostat culture. *FEMS Microbiology Letters* (Amsterdam), **7**, 23–7.

Calleja, G. B. (1984). *Microbial Aggregation*. Boca Raton, Florida: CRC Press.

Camper, A. K. & McFeters, G. A. (1979). Chlorine injury and the enumeration of waterborne coliform bacteria. *Applied and Environmental Microbiology*, **37**, 633–41.

Cargill, K. L., Pyle, B. H., Sauer, R. L. & McFeters, G. A. (1992). Effects of culture conditions and biofilm formation on the iodine susceptibility of *Legionella pneumophila*. *Canadian Journal of Microbiology*, **38**, 423–9.

Carson, L. A., Favero, M. S., Bond, W. W. & Petersen, N. J. (1972). Factors affecting comparative resistance of naturally occurring and subcultured *Pseudomonas aeruginosa* to disinfectants. *Applied Microbiology*, **23**, 863–9.

Carson, L. A., Petersen, N. J., Favero, M. S. & Aguero, S. M. (1978). Growth characteristics of atypical mycobacteria in water and their comparative resistance to disinfectants. *Applied and Environmental Microbiology*, **36**, 839–46.

Chai, T. J. (1983). Characteristics of *Escherichia coli* grown in bay water as compared with rich medium. *Applied and Environmental Microbiology*, **45**, 1316–23.

Chang, S. L. (1944). Destruction of micro-organisms. *Journal, American Water Works Association*, **36**, 1192–207.

Chang, S. L., Berg, G., Clarke, N. A. & Kabler, P. W. (1960a). Survival and protection against chlorination of human enteric pathogens in free living nematodes isolated from water supplies. *American Journal of Tropical Medicine and Hygiene*, **9**, 366–71.

Chang, S. L., Woodward, R. L. & Kabler, P. W. (1960b). Survey of free living nematode worms in a city water supply. *Journal, American Water Works Association*, **52**, 613–18.

Characklis, W. G. (1973a). Attached microbial growths I. Attachment and growth. *Water Research*, **7**, 1113–27.

Characklis, W. G. (1973b). Attached microbial growths II. Frictional resistance due to microbial slimes. *Water Research*, **7**, 1249–58.

Characklis, W. G. (1984a). Biofilm development and its consequences. In *Microbial Adhesion and Aggregation*, ed. K. C. Marshall. Berlin: Springer-Verlag.

Characklis, W. G. (1984b). Biofilm development: a process analysis. In *Microbial Adhesion and Aggregation*, ed. K. C. Marshall. Berlin: Springer-Verlag.

Chick, H. (1908). Investigations of the laws of disinfection. *Journal of Hygiene*, **8**, 92–160.

Cole, E. C., Rutala, W. A., Carson, J. L. & Alfano, E. M. (1989). *Pseudomonas* pellicle in disinfectant testing: electron microscopy, pellicle removal, and effect on test results. *Applied and Environmental Microbiology*, **55**, 511–13.

Collins, H. & Selleck, R. (1972). Process kinetics of wastewater chlorination. Sanitary Engineering Research Laboratory (SERL) Report No. 72–5. Berkeley, California: University of California.

Costerton, J. W. (1984). The formation of biocide-resistant biofilms in industrial, natural and medical systems. *Developments in Industrial Microbiology*, **25**, 363–72.

Costerton, J. W., Geesey, G. G. & Cheng, K. J. (1978). How bacteria stick. *Scientific American*, **238**, 86–95.

Costerton, J. W. & Irvin, R. T. (1981). The bacterial glycocalyx in nature and disease. *Annual Review of Microbiology*, **35**, 299–324.

Costerton, J. W., Irvin, R. T. & Cheng, K. J. (1981). The role of bacterial surface structures in pathogenesis. *Critical Reviews in Microbiology*, **8**, 303–38.

Cozens, R. M. & Brown, M. R. W. (1983). Effect of nutrient depletion on the sensitivity of *Pseudomonas cepacia* to antimicrobial agents. *Journal of Pharmaceutical Sciences*, **72**, 1363–5.

Cunliffe, D. A. (1990). Inactivation of *Legionella pneumophila* by monochloramine. *Journal of Applied Bacteriology*, **68**, 453–9.

Dalhoff, A. (1985). Differences between bacteria grown *in vitro* and *in vivo*. *Journal of Antimicrobial Chemotherapy*, **15**, 175–95.

Dawson, M. P., Humphrey, B. A. & Marshall, K. C. (1981). Adhesion: a tactic in the survival strategy of a marine vibrio during starvation. *Current Microbiology*, **6**, 195–9.

Delaquis, P. J., Caldwell, D. E., Lawrence, J. R. & McCurdy, A. R. (1989). Detachment of *Pseudomonas fluorescens* from biofilms on glass surfaces in response to nutrient stress. *Microbial Ecology*, **18**, 199–210.

Dennis, W. H. (1977). The mode of action of chlorine on f2 bacterial virus during disinfection. Doctoral thesis, Johns Hopkins University, Baltimore.

Dennis, W. H., Olivieri, V. P. & Kruse, C. W. (1979a). The reaction of nucleotides with aqueous hypochlorous acid. *Water Research*, **13**, 357–62.

Dennis, W. H., Olivieri, V. P. & Kruse, C. W. (1979b). Mechanism of disinfection: incorporation of Cl-36 into f2 virus. *Water Research*, **13**, 363–9.

Domingue, E. L., Tyndall, R. L., Mayberry, W. R. & Pancorbo, O. C. (1988). Effects of three oxidizing biocides on *Legionella pneumophila* Serogroup 1. *Applied and Environmental Microbiology*, **54**, 741–7.

Earnhardt, K. B. Jr (1980). Chlorine resistant coliform – the Muncie, Indiana experience. In *Proceedings, AWWA Water Quality Technology Conference*, pp. 371–6. Denver, Colorado: American Water Works Association.

Fair, G. M., Morris, J. C. & Chang, S. L. (1947). The dynamics of water chlorination. *Journal of the New England Water Works Association*, **61**, 285.

Fair, G. M., Morris, J. C., Chang, S. L., Weil, I. & Burden, R. J. (1948). The behavior of chlorine as a water disinfectant. *Journal, American Water Works Association*, **40**, 1051–61.

Farkas-Himsley, H. (1964). Killing of chlorine-resistant bacteria by chlorine–bromine solutions. *Applied Microbiology*, **12**, 1–6.

Favero, M. S., Carson, L. S., Bond, W. W. & Petersen, N. J. (1971). *Pseudomonas aeruginosa*: growth in distilled water from hospitals. *Science*, **173**, 836–8.

Favero, M. S. & Drake, C. H. (1966). Factors influencing the occurrence of high numbers of iodine-resistant bacteria in iodinated swimming pools. *Applied Microbiology*, **14**, 627–34.

Fetner, R. H. (1962). Chromosome breakage in *Vicia faba* by monochloramine. *Nature*, **196**, 1122–3.

Fields, B. S., Barbaree, J. M., Shotts, E. B., Feeley, J. C., Morrill, W. E., Sanden, G. N. & Dykstra, M. J. (1986). Comparison of guinea pig and protozoan models for determining virulence of *Legionella* species. *Infection and Immunity*, **53**, 553–9.

Floyd, R., Johnson, J. D. & Sharp, D. G. (1976). Inactivation by bromine of single poliovirus particles in water. *Applied and Environmental Microbiology*, **31**, 298–303.

Foster, D. M., Emerson, M. A., Buck, C. E., Walsh, D. S. & Sproul, O. J. (1980). Ozone inactivation of cell- and fecal-associated viruses and bacteria. *Journal, Water Pollution Control Federation*, **52**, 2174–84.

Friberg, L. (1956). Quantitative studies on the reaction of chlorine with bacteria in water disinfection. *Acta Pathologica Microbiologica Scandinavica* (Copenhagen), **38**, 135–44.

Friberg, L. (1957). Further quantitative studies on the reaction of chlorine with bacteria in water disinfection. *Acta Pathologica Microbiologica Scandinavica* (Copenhagen), **40**, 67–80.

Fujioka, R. S., Tenno, K. M. & Loh, P. C. (1983). Mechanism of chloramine inactivation of poliovirus: a concern for regulators? In *Water Chlorination: Environmental Impact and Health Effects*, vol. 4, book 2, ed. R. L. Jolley, W. A.

Brungs, J. A. Cotruvo, R. B. Cumming, J. S. Mattice & V. A. Jacobs, pp. 1067–76. Ann Arbor, Michigan: Ann Arbor Science Publishers.

Galasso, G. L. & Sharp, D. G. (1962). Virus particle aggregation and the plaque forming unit. *Journal of Immunology*, **88**, 339.

Gard, S. (1957). Chemical inactivation of viruses. In *CIBA Foundation Symposium on the Nature of Viruses* (Amsterdam), p. 123. Boston: Little, Brown.

Geldreich, E. E., Nash, H. D., Reasoner, D. J. & Taylor, R. H. (1972). The necessity of controlling bacterial populations in potable waters: community water supply. *Journal, American Water Works Association*, **64**, 596–602.

Gilbert, P. & Brown, M. R. W. (1978). Influence of growth rate and nutrient limitation on the gross cellular composition of *Pseudomonas aeruginosa* and its resistance to 3- and 4-chlorophenol. *Journal of Bacteriology*, **133**, 1066–72.

Gilbert, P. & Brown, M. R. W. (1980). Cell wall-mediated changes in sensitivity of *Bacillus megaterium* to chlorhexidine and 2-phenoxyethanol, associated with growth rate and nutrient limitation. *Journal of Applied Bacteriology*, **48**, 223–30.

Gill, C. O. & Suisted, J. R. (1978). The effects of temperature and growth rate on the proportion of unsaturated fatty acids in bacterial lipids. *Journal of General Microbiology*, **104**, 31–6.

Goshko, M. A., Minnigh, H. A., Pipes, W. O. & Christian, R. R. (1983a). Relationships between standard plate counts and other parameters in water distribution systems. *Journal, American Water Works Association*, **75**, 568–71.

Goshko, M. A., Pipes, W. O. & Christian, R. R. (1983b). Coliform occurrence and chlorine residual in small water distribution systems. *Journal, American Water Works Association*, **75**, 371–82.

Gould, J. P., Richards, J. T. & Miles, M. G. (1984). The kinetics and primary products of uracil chlorination. *Water Research*, **18**, 205–12.

Green, D. E. & Stumpf, P. K. (1946). The mode of action of chlorine. *Journal, American Water Works Association*, **38**, 1301–5.

Haas, C. N. & Engelbrecht, R. S. (1980a). Physiological alterations of vegetative microorganisms resulting from chlorination. *Journal, Water Pollution Control Federation*, **52**, 1976–89.

Haas, C. N. & Engelbrecht, R. S. (1980b). Chlorine dynamics during inactivation of coliforms, acid-fast bacteria and yeasts. *Water Research*, **14**, 1749–57.

Haas, C. N. & Karra, S. B. (1984a). Kinetics of microbial inactivation by chlorine. I. Review of results in demand-free systems. *Water Research*, **18**, 1443–9.

Haas, C. N. & Karra, S. B. (1984b). Kinetics of microbial inactivation by chlorine. II. Kinetics in the presence of chlorine demand. *Water Research*, **18**, 1451–4.

Haas, C. N., Khater, K. M. & Wojtas, A. T. (1985). Sensitivity of vegetative protozoa to free and combined chlorine. In *Water Chlorination: Chemistry, Environmental Impact and Health Effects*, vol. 5, ed. R. L. Jolley, R. J. Bull, W. P. Davis, S. Katz, M. H. Roberts Jr & V. A. Jacobs, pp. 667–80. Chelsea, Michigan: Lewis Publishers.

Haas, C. N. & Morrison, E. C. (1981). Repeated exposure of *Escherichia coli* to free chlorine: production of strains possessing altered sensitivity. *Water, Air and Soil Pollution*, **16**, 233–42.

Hamelin, C. & Chung, Y. S. (1978). Role of the POL, REC, and DNA gene products in the repair of lesions produced in *E. coli* by ozone. *Studia* (Berlin), **68**, 229.

Hancock, R. E. W. (1984). Alterations in outer membrane permeability. *Annual Review of Microbiology*, **38**, 237–64.

Harakeh, M. S., Berg, J. D., Hoff, J. C. & Matin, A. (1985). Susceptibility of chemostat-grown *Yersinia enterocolitica* and *Klebsiella pneumoniae* to chlorine dioxide. *Applied and Environmental Microbiology*, **49**, 69–72.

Hejkal, T. W., Wellings, F. M., Larock, P. A. & Lewis, A. L. (1979). Survival of poliovirus within organic solids during chlorination. *Applied and Environmental Microbiology*, **38**, 114–18.

Herson, D. S., Marshall, D. R. & Victoreen, H. T. (1984). Bacterial persistence in the distribution system. In *Proceedings, AWWA Water Quality Technology Conference*, pp. 309–22. Denver, Colorado: American Water Works Association.

Hess, S., Diachishin, A. & DeFalco, P. Jr (1953). Bactericidal effects of sewage chlorination: theoretical aspects. *Sewage and Industrial Wastes*, **25**, 909.

Hoerhammer, C. (1911). Studies of the resistance of lower crustaceans to bacteria in water. *Archive für Hygiene und Bakteriologie* (Munich), **73**, 183.

Hoff, J. C. (1978). The relationship of turbidity to disinfection of potable water. In *Evaluation of Microbiological Standards for Drinking Water*, ed. C. W. Hendricks, EPA-57/014-78-006. Washington, DC: Office of Drinking Water, Environmental Protection Agency.

Hoff, J. C. & Geldreich, E. E. (1978). Effects of turbidity and other factors on the inactivation of viruses by chlorine. *Proceedings, AWWA 1978 Annual Conference*, Part 2, Session 35–1c, pp. 1–11. Denver, Colorado: American Water Works Association.

Hoff, J. C. & Geldreich, E. E. (1981). Comparison of the biocidal efficiency of alternative disinfectants. *Journal, American Water Works Association*, **73**, 40–4.

Holwerda, K. (1928). On the control and degree of reliability of the chlorination process of drinking water in connection with the chloramine procedure and the chlorination of ammoniacal water. *Mededeelingen van den Dienst der Volksgezondheid in Nederlandsch-Indie* (D.W.I.), **17**, 151–297.

Hom, L. W. (1972). Kinetics of chlorine disinfection in an ecosystem. *Journal of the Sanitary Engineering Division: Proceedings of the American Society of Civil Engineers* (ASCE), SA1, pp. 183–94.

Hoyano, Y., Bacon, V., Summons, R. E., Pereira, W. E., Halpern, B. & Duffield, A. M. (1973). The reaction of aqueous hypochlorous acid with pyrimidine and purine bases. *Biochemical and Biophysical Research Communications*, **53**, 1195–9.

Hudson, L. D., Hankins, J. W. & Battaglia, M. (1983). Coliforms in a water distribution system: a remedial approach. *Journal, American Water Works Association*, **75**, 564–8.

Hugo, W. B. & Franklin, I. (1968). Cellular lipid and the antistaphylococcal activity of phenols. *Journal of General Microbiology*, **52**, 365–73.

Hugo, W. B. & Stretton, R. G. (1966). Effect of cellular lipid on the sensitivity of some gram-positive bacteria to penicillins. *Nature*, **209**, 940.

Huq, A., Small, E. B., West, P. A., Hug, M. I., Rahman, R. & Colwell, R. R. (1983). Ecological relationships between *Vibrio cholerae* and planktonic crustacean copepods. *Applied and Environmental Microbiology*, **45**, 275–83.

Hutchinson, M. & Ridgway, J. W. (1977). Microbiological aspects of drinking water supplies. In *Aquatic Microbiology*, ed. F. A. Skinner & J. M. Shewan, pp. 179–218. New York: Academic Press.

Ingols, R. S. (1958). The effect of monochloramine and chromate on bacterial chromosomes. *Public Works*, **89**, 105–6.

Ingols, R. S. & Ridenour, G. M. (1948). Chemical properties of chlorine dioxide in water treatment. *Journal, American Water Works Association*, **40**, 1207–27.

Ingols, R. S., Wyckoff, H. A., Kethley, T. W., Hodgden, H. W., Fincher, E. L., Hildebrand, J. C. & Mandel, J. E. (1953). Bactericidal studies of chlorine. *Industrial Engineering and Chemistry*, **45**, 996–1000.

Jacangelo, J. G. & Olivieri, V. P. (1985). Aspects of the mode of action of monochloramine. In *Water Chlorination: Chemistry, Environmental Impact and Health Effects*, vol. 5, ed. R. L. Jolley, R. J. Bull, W. P. Davis, S. Katz, M. H. Roberts Jr & V. A. Jacobs, pp. 575–86. Chelsea, Michigan: Lewis Publishers.

Jacangelo, J. G., Olivieri, V. P. & Kawata, K. (1987). Oxidation of sulfhydryl groups by monochloramine. *Water Research*, **21**, 1339–44.

Jacangelo, J. G., Olivieri, V. P. & Kawata, K. (1991). Investigating the mechanism of inactivation of *Escherichia coli* B by monochloramine. *Journal, American Water Works Association*, **83**, 80–7.

Jenkins, D. E., Schultz, J. E. & Matin, A. (1988). Starvation-induced cross protection against heat or H_2O_2 challenge in *Escherichia coli*. *Journal of Bacteriology*, **170**, 3910–14.

Kaminski, J. J., Huycke, M. M., Selk, S. H., Bodor, N. & Higuchi, T. (1976). *N*-halo derivatives V: comparative antimicrobial activity of soft *N*-chloramine systems. *Journal of Pharmaceutical Sciences*, **65**, 1737–42.

Katzenelson, E., Kletter, B. & Shuval, H. I. (1974). Inactivation kinetics of virus and bacteria in water by use of ozone. *Journal, American Water Works Association*, **66**, 725–9.

Kilvington, S. & Price, J. (1990). Survival of *Legionella pneumophila* within cysts of *Acanthamoeba polyphaga* following chlorine exposure. *Journal of Applied Bacteriology*, **68**, 519–25.

Kim, C. K., Gentile, D. M. & Sproul, O. J. (1980). Mechanism of ozone inactivation of bacteriophage f2. *Applied and Environmental Microbiology*, **39**, 210–18.

Kimball, A. W. (1953). The fitting of multi-hit survival curves. *Biometrics*, **9**, 201–11.

King, C. H., Shotts, E. B., Wooley, R. E. & Porter, K. G. (1988). Survival of coliforms and bacterial pathogens within protozoa during chlorination. *Applied and Environmental Microbiology*, **54**, 3023–33.

Kjelleberg, S., Humphrey, B. A. & Marshall, K. C. (1983). Initial phases of starvation and activity of bacteria at surfaces. *Applied and Environmental Microbiology*, **46**, 978–84.

Knox, W. E., Stumpf, P. K., Green, D. E. & Auerbach, V. H. (1948). The inhibition of sulfhydryl enzymes as the basis of the bactericidal action of chlorine. *Journal of Bacteriology*, **55**, 451–8.

Kolawole, D. O. (1984). Resistance mechanisms of mucoid-grown *Staphylococcus aureus* to the antibacterial action of some disinfectants and antiseptics. *FEMS Microbiology Letters* (Amsterdam), **25**, 205–9.

Kulikovsky, A., Pankratz, H. S. & Sadoff, H. L. (1975). Ultrastructural and chemical changes in spores of *Bacillus cereus* after action of disinfectants. *Journal of Applied Bacteriology*, **38**, 39–46.

Lamka, K. G., LeChevallier, M. W. & Seidler, R. J. (1980). Bacterial contamination of drinking water supplies in a modern rural neighborhood. *Applied and Environmental Microbiology*, **39**, 734–8.

LeChevallier, M. W., Cawthon, C. D. & Lee, R. G. (1988a). Factors promoting survival of bacteria in chlorinated water supplies. *Applied and Environmental Microbiology*, **54**, 649–54.

LeChevallier, M. W., Cawthon, C. D. & Lee, R. G. (1988b). Inactivation of biofilm bacteria. *Applied and Environmental Microbiology*, **54**, 2492–9.

LeChevallier, M. W., Evans, T. M. & Seidler, R. J. (1981). Effect of turbidity on chlorination efficiency and bacterial persistence in drinking water. *Applied and Environmental Microbiology*, **42**, 159–67.

LeChevallier, M. W., Hassenauer, T. S., Camper, A. K. & McFeters, G. A. (1984). Disinfection of bacteria attached to granular activated carbon. *Applied and Environmental Microbiology*, **48**, 918–23.

LeChevallier, M. W., Seidler, R. J. & Evans, T. M. (1980). Enumeration and characterization of standard plate count bacteria in chlorinated and raw water supplies. *Applied and Environmental Microbiology*, **40**, 922–30.

Lee, S. H., O'Connor, J. T. & Banergi, S. K. (1980). Biologically mediated corrosion and its effects on water quality in the distribution system. *Journal, American Water Works Association*, **72**, 636–45.

Leive, L. (1965). Release of lipopolysaccharide by EDTA treatment of *E. coli*. *Biochemical and Biophysical Research Communications*, **21**, 290–6.

Levy, R. V., Cheetham, R. D., Davis, J., Winer, G. & Hart, F. L. (1984). Novel method for studying the public health significance of macroinvertebrates occurring in potable water. *Applied and Environmental Microbiology*, **47**, 889–94.

Leyval, C., Arx, C., Block, J. C. & Rizet, M. (1984). *Escherichia coli* resistance to chlorine after successive chlorination. *Environmental Technology Letters*, **5**, 359–64.

Lugtenberg, B. & Van Alphen, L. (1983). Molecular architecture and functioning of the outer membrane of *Escherichia coli* and other gram-negative bacteria. *Biochimica et Biophysica Acta*, **737**, 51–115.

Mackenthun, K. M. & Keup, L. E. (1970). Biological problems encountered in water supplies. *Journal, American Water Works Association*, **62**, 520–5.

Maki, J. S., LaCroix, S. J., Hopkins, B. S. & Staley, J. T. (1986). Recovery and diversity of heterotrophic bacteria from chlorinated drinking waters. *Applied and Environmental Microbiology*, **51**, 1047–55.

Marshall, K. C. (1976). *Interfaces in Microbial Ecology*. Cambridge, Massachusetts: Harvard University Press.

Marshall, K. C., ed. (1984). *Microbial Adhesion and Aggregation: Report of the Dahlem Workshop*, Life Sciences Research Reports, vol. 31. Berlin: Springer-Verlag.

Martin, R. S., Gates, W. H., Tobin, R. S., Grantham, D., Sumarah, R., Wolfe, P. & Forestall, P. (1982). Factors affecting coliform bacteria growth in distribution systems. *Journal, American Water Works Association*, **74**, 34–7.

McFeters, G. A. (1984). Biofilm development and its consequences. In *Microbial Adhesion and Aggregation: Report of the Dahlem Workshop*, Life Sciences Research Reports, vol. 31, ed. K. C. Marshall. Berlin: Springer-Verlag.

McFeters, G. A. & Camper, A. K. (1983). Enumeration of indicator bacteria exposed to chlorine. *Advances in Applied Microbiology*, **29**, 177–93.

Means, E. G., Hanami, L., Ridgway, H. F. & Olson, B. H. (1981). Evaluating mediums and plating techniques for enumerating bacteria in water distribution systems. *Journal, American Water Works Association*, **58**, 585–90.

Milbauer, R. & Grossowicz, N. (1959). Reactivation of chlorine-inactivated *Escherichia coli*. *Applied Microbiology*, **7**, 67–70.

Moats, W. A. (1971). Kinetics of thermal death of bacteria. *Journal of Bacteriology*, **105**, 165–71.

Montgomery, James M., Consulting Engineers, Inc. (1985). Disinfection. In *Water Treatment Principles and Design*, pp. 262–83. New York: John Wiley & Sons.

Morris, J. C. (1966). The future of chlorination. *Journal, American Water Works Association*, **58**, 1475–82.

Morse, S. A., Mintz, C. S., Sarafian, S. K., Bartenstein, L., Bertram, M. & Apicella, M. A. (1983). Effect of dilution rate on lipopolysaccharide and serum resistance of *Neisseria gonorrhoeae* grown in continuous culture. *Infection and Immunity*, **41**, 74–82.

Nagy, L. A. (1984). An experimental analysis of the microbiological dynamics of a drinking water distribution system. Doctoral thesis, University of California, Irvine.

National Research Council (1977). *Drinking Water and Health*, vol. 1. Washington, DC: National Academy of Sciences.

National Research Council (1980). *Drinking Water and Health*, vol. 2. Washington, DC: National Academy Press.

National Research Council (1982). *Drinking Water and Health*, vol. 4. Washington, DC: National Academy Press.

National Research Council (1987). *Drinking Water and Health: Disinfectants and Disinfectant By-Products*, vol. 7. Washington, DC: National Academy Press.

Neefe, J. R., Baty, J. B., Reinhold, J. G. & Stokes, J. (1947). Inactivation of the virus of infectious hepatitis in drinking water. *American Journal of Public Health*, **37**, 365–72.

Nikaido, H. (1976). Outer membrane of *Salmonella*. Transmembrane diffusion of some hydrophobic molecules. *Biochimica et Biophysica Acta*, **433**, 118–32.

Nikaido, H. & Nakae, T. (1979). The outer membrane of gram-negative bacteria. *Advances in Microbial Physiology*, **20**, 164–250.

Nikaido, H. & Vaara, M. (1985). Molecular basis of bacterial outer membrane permeability. *Microbiology Reviews*, **49**, 1–32.

Noss, C. I., Dennis, W. H. & Olivieri, V. P. (1983). Reactivity of chlorine dioxide with nucleic acids and proteins. In *Water Chlorination: Environmental Impact and Health Effects*, vol. 4, book 2, ed. R. L. Jolley, W. A. Brungs, J. A. Cotruvo, R. B. Cumming, J. S. Mattice & V. A. Jacobs, pp. 1077–86. Ann Arbor, Michigan: Ann Arbor Science Publishers.

O'Connor, J. T., Hash, L. & Edwards, A. B. (1975). Deterioration of water quality in distribution systems. *Journal, American Water Works Association*, **67**, 113–16.

Olivieri, V. P. (1974). The mode of action of chlorine on f2 bacterial virus. Doctoral thesis, Johns Hopkins University, Baltimore.

Olivieri, V. P., Bakalian, A. E., Bossung, K. W. & Lowther, E. D. (1985*a*). Recurrent coliforms in water distribution systems in the presence of free residual chlorine. In *Water Chlorination: Chemistry, Environmental Impact and Health Effects*, vol. 5, ed. R. L. Jolley, R. J. Bull, W. P. Davis, S. Katz, M. H. Roberts Jr & V. A. Jacobs, pp. 651–66. Chelsea, Michigan: Lewis Publishers.

Olivieri, V. P., Dennis, W. H., Snead, M. C., Richfield, D. T. & Kruse, C. W. (1980). Reaction of chlorine and chloramines with nucleic acids under disinfection conditions. In *Water Chlorination: Environmental Impact and Health Effects*, vol. 3, ed. R. L. Jolley, W. A. Brungs & R. B. Cumming, pp. 651–63. Ann Arbor, Michigan: Ann Arbor Science Publishers.

Olivieri, V. P., Hauchman, F. S., Noss, C. I. & Vasl, R. (1985*b*). Mode of action of chlorine dioxide on selected viruses. In *Water Chlorination: Chemistry, Environmental Impact and Health Effects*, vol. 5, ed. R. L. Jolley, R. J. Bull, W. P. Davis, S. Katz, M. H. Roberts Jr & V. A. Jacobs, pp. 619–34. Chelsea, Michigan: Lewis Publishers.

Olivieri, V. P., Krusé, C. W., Hsu, T. C., Griffiths, A. C. & Kawata, K. (1975). The comparative mode of action of chlorine, bromine, and iodine on f2 bacterial virus. In *Disinfection, Water and Wastewater*, ed. J. D. Johnson. Ann Arbor, Michigan: Ann Arbor Science Publishers.

Olson, B. H. & Nagy, L. (1984). Microbiology of potable water. In *Advances in Applied Microbiology*, **30**, 73–132.

Olson, B. H., Ridgway, H. F. & Means, E. G. (1981). Bacterial colonization of mortar-lined and galvanized iron water distribution mains. In *Proceedings, AWWA 1981 Annual Conference*, part II, pp. 1027–39. Denver, Colorado: American Water Works Association.

Olson, B. H. & Stewart, M. H. (1990). Factors that change bacterial resistance to disinfection. In *Water Chlorination: Chemistry, Environmental Impact and Health Effects*, vol. 6, ed. R. L. Jolley, L. W. Condie, J. D. Johnson, S. Katz, R. A. Minear, J. S. Mattice & V. A. Jacobs, pp. 885–904. Chelsea, Michigan: Lewis Publishers.

Ombaka, E. A., Cozens, R. M. & Brown, M. R. W. (1983). Influence of nutrient limitation of growth on stability and production of virulence factors of mucoid and nonmucoid strains of *Pseudomonas aeruginosa*. *Reviews in Infectious Diseases*, **5**, 880–8.

Patton, W., Bacon, V., Duffield, A. M., Halpern, B., Hoyano, Y., Pereira, W. & Lederberg, J. (1972). The reaction of aqueous hypochlorous acid with cytosine. *Biochemical and Biophysical Research Communications*, **48**, 880–4.

Poduska, R. A. & Hershey, D. (1972). Model for virus inactivation by chlorination. *Journal, Water Pollution Control Federation*, **44**, 738–45.

Prat, R. C. & Cier, A. (1968). Effects of sodium hypochlorite, ozone, and ionizing radiation on the pyrimidine constituents of *Escherichia coli*. *Annales de L'Institut Pasteur*, **114**, 595–607.

Prokop, A. & Humphrey, A. E. (1970). Kinetics of disinfection. In *Disinfection*, ed. M. A. Bernarde, pp. 61–83. New York: Marcel Dekker.

Ptak, D. J., Ginsburg, W. & Willey, F. (1973). Identification and incidence of *Klebsiella* in chlorinated water supplies. *Journal, American Water Works Association*, **65**, 604–8.

Pyle, B. H. & McFeters, G. A. (1989). Iodine sensitivity of bacteria isolated from iodinated water systems. *Canadian Journal of Microbiology*, **35**, 520–3.

Reasoner, D. J. & Geldreich, E. E. (1985). A new medium for the enumeration and subculture of bacteria from potable water. *Applied and Environmental Microbiology*, **49**, 1-7.

Reilly, J. K. & Kippin, J. S. (1983). Relationship of bacterial counts with turbidity and free chlorine in two distribution systems. *Journal, American Water Works Association*, **75**, 309-12.

Ridgway, H. F. & Olson, B. H. (1981*a*). Scanning electron microscope evidence for bacterial colonization of a drinking-water distribution system. *Applied and Environmental Microbiology*, **41**, 274-87.

Ridgway, H. F. & Olson, B. H. (1981*b*). Iron bacteria in drinking-water distribution systems: elemental analysis of *Gallionella* stalks, using X-ray energy-dispersive microanalysis. *Applied and Environmental Microbiology*, **41**, 288-97.

Ridgway, H. F. & Olson, B. H. (1982). Chlorine resistance patterns of bacteria from two drinking-water distribution systems. *Applied and Environmental Microbiology*, **44**, 972-87.

Roantree, R. J., Kuo, T. T. & MacPhee, D. G. (1977). The effect of defined lipopolysaccharide core defects upon antibiotic resistances of *Salmonella typhimurium*. *Journal of General Microbiology*, **103**, 223-34.

Robinson, A. & Tempest, D. W. (1973). Phenotypic variability of the envelope proteins of *Klebsiella aerogenes*. *Journal of General Microbiology*, **78**, 361-70.

Roller, S. D., Olivieri, V. P. & Kawata, K. (1980). Mode of bacterial inactivation by chlorine dioxide. *Water Research*, **14**, 635-41.

Rosenkranz, H. S. (1973). Sodium hypochlorite and sodium perborate: preferential inhibitors of DNA polymerase-deficient bacteria. *Mutation Research*, **21**, 171-4.

Rowbotham, T. J. (1980). Preliminary report on the pathogenicity of *Legionella pneumophila* for freshwater and soil amoebae. *Journal of Clinical Pathology*, **33**, 1179-83.

Rowbotham, T. J. (1983). Isolation of *Legionella pneumophila* from clinical specimens via amoebae, and the interaction of those and other isolates with amoebae. *Journal of Clinical Pathology*, **36**, 978-86.

Rowbotham, T. J. (1986). Current views on the relationships between amoebae, *Legionella* and man. *Israel Journal of Medical Sciences*, **22**, 678-89.

Roy, D., Wong, K. Y., Engelbrecht, R. S. & Chian, E. S. K. (1981). Mechanism of enteroviral inactivation by ozone. *Applied and Environmental Microbiology*, **41**, 718-23.

Scarpino, P. V. (1984). *Effect of Particulates on Disinfection of Enteroviruses in Water by Chloramines*, EPA-600/S2084-094, pp. 1-6. Cincinnati, Ohio: Environmental Protection Agency.

Scarpino, P. V., Cronier, S., Zink, M. L. & Brigano, F. A. O. (1977). Effect of particulates on disinfection of enteroviruses and coliform bacteria in water by chlorine dioxide. In *Proceedings, AWWA Water Quality Technology Conference*, Session 2B-3, pp. 1-11. Denver, Colorado: American Water Works Association.

Schultz, J. E., Latter, G. I. & Matin, A. (1988). Differential regulation by cyclic AMP of starvation protein synthesis in *Escherichia coli*. *Journal of Bacteriology*, **170**, 3903-9.

Scott, D. B. & Lesher, E. C. (1963). Effect of ozone on survival and permeability of *Escherichia coli*. *Journal of Bacteriology*, **85**, 567-76.

Selleck, R. E., Saunier, B. M. & Collins, H. F. (1978). Kinetics of bacterial deactivation with chlorine. *Journal of the Environmental Engineering Division*: *Proceedings of the ASCE*, **104**, 1197–212.

Severin, B. F., Suidan, M. T. & Engelbrecht, R. S. (1983). Kinetic modeling of U.V. disinfection of water. *Water Research*, **17**, 1669–78.

Seyfried, P. L. & Fraser, D. J. (1980). Persistence of *Pseudomonas aeruginosa* in chlorinated swimming pools. *Canadian Journal of Microbiology*, **26**, 350–5.

Shaffer, P. T. B., Metcalf, T. G. & Sproul, O. J. (1980). Chlorine resistance of poliovirus isolants recovered from drinking water. *Applied and Environmental Microbiology*, **40**, 115–21.

Sharp, D. G., Floyd, R. & Johnson, J. D. (1975). Nature of the surviving plaque-forming unit of reovirus in water containing bromine. *Applied Microbiology*, **29**, 94–101.

Sharp, D. G., Floyd, R. & Johnson, J. D. (1976). Initial fast reaction of bromine on reovirus in turbulent flowing water. *Applied and Environmental Microbiology*, **31**, 173–81.

Shih, K. L. & Lederberg, J. (1976). Chloramine mutagenesis in *Bacillus subtilis*. *Science*, **181**, 463–4.

Shih, K. L. & Lederberg, J. (1979). Effects of chloramine on *Bacillus subtilis* deoxyribonucleic acid. *Journal of Bacteriology*, **125**, 934–45.

Shindala, A. & Chisholm, C. H. (1970). Water quality changes in the distribution system. *Water and Waste Engineering*, **7**, 35–7.

Shinriki, N., Ishizaki, K., Yoshizaki, T., Miura, K. & Ueda, T. (1988). Mechanism of inactivation of tobacco mosaic virus with ozone. *Water Research*, **22**, 933–8.

Shotts, E. G. & Wooley, R. E. (1990). *Protozoan Sources of Spontaneous Coliform Occurrence in Chlorinated Drinking Water*, EPA/600/S2–89/019, Project Summary. Cincinnati, Ohio: US Environmental Protection Agency.

Small, I. C. & Greaves, G. F. (1968). A survey of animals in distribution systems. *Water Treatment and Examination*, **19**, 150–83.

Sproul, O. J., Pfister, R. M. & Kim, C. K. (1982). The mechanism of ozone inactivation of water borne viruses. *Water Science and Technology*, **14**, 303–14.

Stagg, C. H., Wallis, C. & Ward, C. H. (1977). Inactivation of clay-associated bacteriophage MS-2 by chlorine. *Applied and Environmental Microbiology*, **33**, 385–91.

Sterkenburg, A., Vlegels, E. & Wouters, J. T. M. (1984). Influence of nutrient limitation and growth rate on the outer membrane proteins of *Klebsiella aerogenes* NCTC 418. *Journal of General Microbiology*, **130**, 2347–55.

Stewart, M. H. & Olson, B. H. (1986). Mechanisms of bacterial resistance to inorganic chloramines. In *Proceedings, AWWA Water Quality Technology Conference*, pp. 577–90. Denver, Colorado: American Water Works Association.

Stewart, M. H. & Olson, B. H. (1992a). Impact of growth conditions on resistance of *Klebsiella pneumoniae* to chloramines. *Applied and Environmental Microbiology*, **58**, 2649–53.

Stewart, M. H. & Olson, B. H. (1992b). Physiological studies of chloramine resistance developed by *Klebsiella pneumoniae* under low-nutrient growth conditions. *Applied and Environmental Microbiology*, **58**, 2918–27.

Stewart, M. H., Wolfe, R. L. & Means, E. G. (1990). Assessment of bacteriological activity associated with granular activated carbon treatment of drinking water. *Applied and Environmental Microbiology*, **56**, 3822–9.

Tracy, H. W., Camarena, V. M. & Wing, F. (1966). Coliform persistence in highly chlorinated waters. *Journal, American Water Works Association*, **58**, 1151–9.

Trulear, M. G. & Characklis, W. G. (1980). Dynamics of biofilm processes. In *Proceedings, WPCF 53rd Annual Conference*. Washington, DC: Water Pollution Control Federation.

Tuovinen, O. H., Button, K. S., Vuorinen, A., Carlson, L., Mair, D. M. & Yut, L. A. (1980). Bacterial, chemical, and mineralogical characteristics of tubercles in distribution pipelines. *Journal, American Water Works Association*, **72**, 626–35.

Tuovinen, O. H. & Hsu, J. C. (1982). Aerobic and anerobic microorganisms in tubercles of the Colombus, Ohio, water distribution system. *Applied and Environmental Microbiology*, **44**, 761–4.

US Environmental Protection Agency (1975). National interim primary drinking water regulations. *Federal Register*, **40** (248), 59566–87.

US Environmental Protection Agency (1983). *Assessment of Microbiology and Turbidity Standards for Drinking Water*, EPA 570/9/83/001. Washington, DC: Office of Drinking Water, Environmental Protection Agency.

van der Wende, E., Characklis, W. G. & Smith, D. B. (1989). Biofilms and bacterial drinking water quality. *Water Research*, **23**, 1313–22.

Venkobachar, C., Iyengar, L. & Prabhakara Rao, A. V. S. (1975). Mechanism of disinfection. *Water Research*, **9**, 119–24.

Venkobachar, C., Iyengar, L. & Prabhakara Rao, A. V. S. (1977). Mechanism of disinfection: effect of chlorine on cell membrane functions. *Water Research*, **11**, 727–9.

Venosa, A. D. (1972). Ozone as a water and wastewater disinfectant: a literature review. In *Ozone in Water and Wastewater Treatment*, ed. F. L. Evans III, pp. 82–100. Ann Arbor, Michigan: Ann Arbor Science Publishers.

Victoreen, H. T. (1969). Soil bacteria and color problem in distribution systems. *Journal, American Water Works Association*, **61**, 429–31.

Victoreen, H. T. (1974). Control of water quality in transmission and distribution mains. *Journal, American Water Works Association*, **66**, 369–70.

Victoreen, H. T. (1984). Controlling corrosion by controlling bacterial growth. *Journal, American Water Works Association*, **76**, 87–9.

Wadowsky, R. M., Butler, L. J., Cook, M. K., Verma, S. M., Paul, M. A., Fields, B. S., Keleti, G., Sykora, J. L. & Yee, R. B. (1988). Growth-supporting activity for *Legionella pneumophila* in tap water cultures and implication of *Hartmannellid* amoebae as growth factors. *Applied and Environmental Microbiology*, **54**, 2677–82.

Ward, N. R., Wolfe, R. L., Justice, C. A. & Olson, B. H. (1984). Effect of pH, application technique, and chlorine-to-nitrogen ratio on disinfectant activity of inorganic chloramines with pure culture bacteria. *Applied and Environmental Microbiology*, **48**, 508–14.

Watson, H. E. (1908). A note on the variation of the rate of disinfection with change in the concentration of the disinfectant. *Journal of Hygiene*, **8**, 536.

Wei, J. H. & Chang, S. L. (1975). A multi-Poisson distribution model for treating disinfection data. In *Disinfection: Water and Wastewater*, ed. J. D. Johnson, pp. 11–48. Ann Arbor, Michigan: Ann Arbor Science Publishers.

White, G. C. (1972). *Handbook of Chlorination*. New York: Van Nostrand Reinhold.

White, G. C. (1975). Disinfection: the last line of defense for potable water. *Journal, American Water Works Association*, **67**, 410–13.

Wierenga, J. T. (1985). Recovery of coliforms in the presence of a free chlorine residual. *Journal, American Water Works Association*, **77**, 83–8.

Williams, P., Brown, W. R. W. & Lambert, P. A. (1984). Effect of iron deprivation on the production of siderophores and outer membrane proteins in *Klebsiella aerogenes*. *Journal of General Microbiology*, **130**, 2357–65.

Wolfe, R. L. & Olson, B. H. (1985). Inability of laboratory models to accurately predict field performance of disinfectants. In *Water Chlorination: Chemistry, Environmental Impact and Health Effects*, vol. 5, ed. R. L. Jolley, R. J. Bull, W. P. Davis, S. Katz, M. H. Roberts Jr & V. A. Jacobs, pp. 555–73. Chelsea, Michigan: Lewis Publishers.

Wolfe, R. L., Stewart, M. H., Liang, S. & McGuire, M. J. (1989). Disinfection of model indicator organisms in a drinking water pilot plant by using PEROXONE. *Applied and Environmental Microbiology*, **55**, 2230–41.

Wolfe, R. L., Ward, N. R. & Olson, B. H. (1985). Inactivation of heterotrophic bacterial populations in finished drinking water by chlorine and chloramines. *Water Research*, **19**, 1393–403.

Yoshimura, F. & Nikaido, H. (1982). Permeability of *Pseudomonas aeruginosa* outer membrane to hydrophilic solutes. *Journal of Bacteriology*, **152**, 636–42.

Young, D. C., Johnson, J. D. & Sharp, D. G. (1977). The complex reaction kinetics of ECHO-1 virus with chlorine in water (39965). *Proceedings, Society for Experimental Biology and Medicine*, **156**, 496–9.

Young, D. C. & Sharp, D. G. (1977). Poliovirus aggregates and their survival in water. *Applied and Environmental Microbiology*, **33**, 168–77.

Young, D. C. & Sharp, D. G. (1985). Virion conformational forms and the complex inactivation kinetics of echovirus by chlorine in water. *Applied and Environmental Microbiology*, **49**, 359–64.

Chapter 6

Preventing foodborne infectious disease

CHRISTON J. HURST

Introduction

Diseases associated with the consumption of foods can be either chronic or acute in nature. Chronic diseases sometimes represent dietary deficiencies, in which case blame lies with the general diet rather than with the food itself. At other times, chronic food-associated disease represents over-consumption of heavy metals or normally healthful nutrients such as fat-soluble vitamins. Acute diseases associated with food are usually caused by either chemical toxins or microorganisms.

Plants and plant products may contain natural chemical toxins that are produced by those plants. Diseases that are due to the ingestion of natural plant toxins may be avoided by treating the plant material to leach out or neutralize the toxic compounds, growing the plant under specified conditions so that the toxins are not produced, eating only nontoxic parts of the plant, or simply not eating the plant. Plants can also biologically accumulate chemical contaminants such as heavy metals and radioactive elements from the water or soil that is used to sustain their growth. There also are times when plants and plant products may contain exogenous toxic chemical contaminants from human sources. Field and orchard crops may become coated with chemical contaminants such as fungicides, herbicides, insecticides and rodenticides. These disease agents represent contaminants variously acquired from the environment prior to harvesting, or during storage and handling after the harvest. Herbicides and some insecticides are applied prior to harvesting to increase crop yields. Other insecticide compounds and fungicides may be applied after harvesting to prevent degradation of the crop by fungi, and consumption of the crop by invertibrate animals. Rodenticides are occasionally found in foods, and these usually represent inadvertent contamination acquired at some point after harvesting.

Animals that are consumed as food likewise can become contaminated by chemicals acquired from the environment. This contamination represents bioaccumulation that results when the animals ingest contaminated food or water (Eastaugh & Shepherd, 1989). Some of the chemical contami-nants are toxins of natural biological origin, such as those contained in ingested dinoflagellates (Eastaugh & Shepherd, 1989). Other chemicals, such as insecticides, polychlorinated biphenyls and organometal com-pounds, tend to represent contaminants that humans have released into the environment. Some of the ingested toxins are concentrated in a stepwise manner through the food chain. This can result in dangerously high toxin concentrations within the bodies of longlived animals, particularly those carnivorous animals that occupy the top of the food chain. Both plant and animal products can acquire chemical contaminants from processing equipment or packaging materials as a result of inadequate care during handling and preparation.

Despite the disease hazards represented by chemical contaminants, the causative agents of foodborne disease outbreaks are usually microbiologi-cal in nature, representing either microbially produced toxins or infectious agents. Microbial contaminants, as is the case with chemical contaminants, can be acquired either from the environment, or during the subsequent steps involved with food processing, packaging and handling that occur prior to consumption. Unlike chemical contaminants, these microbial con-taminants may be capable of replicating within the food, which can increase the disease hazard associated with the contaminated foods over the course of time.

The focus of this book is infectious disease. Therefore, the remainder of this chapter will be restricted to the topic of microbial contaminants in food, and will primarily address those diseases caused by microbial contaminants that infect human consumers. Water is considered to be a food item in some cases (Cliver, 1990; Warburton et al., 1992). However, for the purpose of this book, infectious disease hazards associated with the consumption of water are discussed separately, in Chapter 4.

Outbreaks of foodborne infectious disease

Foodborne disease outbreaks tend to be smaller in size, but more numer-ous, than waterborne disease outbreaks (Tulchinsky et al., 1988; Cliver, 1990). A recent review by Cliver suggests that in the United States, bac-terial contaminants cause approximately 66% of foodborne disease out-breaks, accounting for 92% of the cases of food-associated illness (Cliver, 1990). The corresponding estimates for viral contaminants were that they caused 4% of outbreaks and 5% of illness cases. Other causes of foodborne disease, including parasites and chemicals, together account for the remain-ing 30% of outbreaks and 3% of illness cases. Foodborne infectious disease represents a substantial economic cost to society (Roberts, 1988), and it is important to recognize that the associated microbial contaminants cause disease even in the absence of recognized foodborne outbreaks (Hedberg *et al.*, 1993).

Many of the individual foodborne pathogens are characteristically associated with specific types of foods (Table 6.1). The predominant bacterial organisms associated with foodborne infectious disease seem to be *Salmonella*, *Shigella* and *Campylobacter*. *Salmonella* tends to be associated with red meats, poultry (Roberts, 1988) and eggs (Ching-Lee *et al.*, 1991; Ewert *et al.*, 1993; Hedberg *et al.*, 1993). *Campylobacter* seems to be associated with milk (Wegmüller, Lüthy & Candrian, 1993) and likewise with poultry (Pearson *et al.*, 1993; Tauxe *et al.*, 1988). *Listeria* is associated with milk and other dairy products. *Vibrio* is often associated with two categories of shellfish: the crustaceans and bivalve molluscs (Davis & Sizemore, 1982; Grimes, 1991; Ragazzoni *et al.*, 1991; Roman *et al.*, 1991). *Escherichia* tends to be associated with ground beef (Centers for Disease Control, 1993; Le Saux *et al.*, 1993; Turney *et al.*, 1994). The predominant viruses associated with foodborne diseases are hepatitis A virus, which is the sole member of the *Hepatovirus* genus, and *Calicivirus*, a genus most often represented by Norwalk and Norwalk-like organisms. These viruses are characteristically associated with illnesses acquired from ingesting bivalve molluscs (O'Mahony *et al.*, 1983; Morse *et al.*, 1986; Cliver, 1990). Caliciviruses may also be associated with ice (Khan *et al.*, 1994), an association

Table 6.1 *Examples of characteristic associations between foods and infectious diseases*

Food	Illness	Causative microorganism
Crops (vegetables)	Enteritis	*Salmonella, Vibrio*
Dairy products	Enteritis	*Brucella, Campylobacter, Listeria*
Eggs	Enteritis	*Campylobacter, Salmonella*
Fish	Sparganosis	*Spirometra*
Poultry	Enteritis	*Campylobacter, Salmonella*
Red meats	Enteritis	*Escherichia*
	Taeniasis	*Taenia*
	Trichinosis	*Trichinella*
	Tularemia	*Francisella*
Shellfish (crustacean)	Enteritis	*Vibrio*
Shellfish (molluscan)	Enteritis	*Vibrio*
	Gastroenteritis	*Calicivirus*
	Hepatitis	*Hepatovirus*
Ice	Gastroenteritis	*Calicivirus*

Note: These characteristic associations generally represent microbial contamination that occurs either before harvesting or else when food is processed prior to marketing. The diseases associated with *Staphylococcus* and *Clostridium* are not listed in this table, because they generally represent intoxications rather than infections.

that probably results from using virally contaminated water to produce the ice. Helminths, such as *Taenia* and *Trichinella*, are usually associated with meat. These characteristic associations tend to represent contamination acquired either prior to harvesting, or else during processing operations performed before the food is marketed.

Non-characteristic associations between microorganisms and foods have also been noticed, and these tend to represent contamination that is acquired during the course of food handling after marketing. Some examples of these non-characteristic associations are *Salmonella* in orange juice (Birkhead *et al.*, 1993), *Giardia* on raw sliced vegetables (Mintz *et al.*, 1993) or in noodle salads (Petersen, Cartter & Hadler, 1988), *Streptococcus* in macaroni and cheese (Farley *et al.*, 1993) and *Calicivirus* on fresh-cut fruit (Herwaldt *et al.*, 1994).

Detection of causative microorganisms in foods

The existence of foodborne infectious disease outbreaks suggests that the associated foods were contaminated with pathogenic microorganisms at the time of their consumption. The significance of microbially contaminated foods as a cause of morbidity and mortality has resulted in the development of techniques for detecting pathogenic microorganisms in foods. Bacteria are usually detected by growth on selective media (Davis & Sizemore, 1982; Ching-Lee *et al.*, 1991; Luby *et al.*, 1993). This process may be supplemented by more specific tests, including those that use nucleic acid hybridization probes (Miceli, Watkins & Rippey, 1993; Taylor *et al.*, 1993) and immunofluorescence assays (Pearson *et al.*, 1993). Parasites are usually detected by direct microscopic examination. Viruses can be detected by cultivation in susceptible hosts. The polymerase chain reaction nucleic acid amplification technique has successfully been used to aid biochemical detection of both bacteria (Bej *et al.*, 1994; Wegmüller *et al.*, 1993) and viruses (Atmar *et al.*, 1993; Herwaldt *et al.*, 1994).

The sources of microbial contaminants in foods

Foods can become contaminated with microorganisms at many different steps between the time of harvesting and eventual consumption. Some contamination is intentional, with certain bacteria and fungi deliberately added to food so that through their associated metabolic processes, these microorganisms will impart desired characteristics to the food. These characteristics include modifications of food coloration, texture, aroma and taste. The gases released by respiration of yeasts cause bread to rise. Yeasts can ferment sugars to produce ethanol, which is a desired component of many beverages. Bacteria in the same beverages can then convert the ethanol and other compounds to acetic acid, thereby producing vinegar. Usually, it is the unintentional contamination of foods with microorganisms that is of public health concern. This contamination may result in chemical spoilage of the food, the production of microbial toxins, and infections of humans or animals that consume the food. It is these categories of microorganisms, the unintentional contaminants, that will be addressed below.

Contamination acquired from the environment prior to harvesting

Microbial contamination can occur before harvesting, the very first link in the chain of events. As indicated above, some of these microbial contaminants are hazardous because they infect consumers, whereas other microbes primarily are hazardous because of the toxins that they produce. Among those microbial contaminants that represent infectious disease hazards are the bacterial genera *Vibrio*, *Salmonella* and *Campylobacter*, and the viral genera *Calicivirus* and *Hepatovirus*. *Vibrio* species naturally colonize the chitinous shells of crustaceans (Grimes, 1991) and can cause illness if the crustaceans are not adequately cooked prior to their being ingested (Davis & Sizemore, 1982; Ragazzoni *et al.*, 1991). There is also some evidence that vibrios may infect the interior of the crustaceans (Davis & Sizemore, 1982). Vibrios can also occur as contaminants in bivalve molluscs (Miceli *et al.*, 1993), with this contamination presumably resulting from the molluscs having ingested the bacteria during the course of filter feeding. In fact, during the course of filter feeding, bivalve molluscs can become contaminated with any enteric pathogens present in the shellfish-growing waters, including such viral contaminants as *Calicivirus* and *Hepatovirus* (Metcalf, 1978; Atmar *et al.*, 1993). *Salmonella* can be detected in bivalve molluscs (Bej *et al.*, 1994), although it is not certain whether this contamination is acquired prior to harvesting, or during processing and handling. Additional examples of microbial contamination that occurs prior to harvesting are represented by poultry which may be infected with *Campylobacter* (Pearson *et al.*, 1993) and eggs which may contain *Salmonella* (Ching-Lee *et al.*, 1991).

The microorganisms listed above cause illness by infecting the consumer. Some other microbial contaminants are notable for producing toxins which, when the food is ingested, cause diseases that represent intoxications. Dinoflagellate toxins can be accumulated by fish and shellfish prior to harvesting (Eastaugh & Shepherd, 1989). Additional examples of environmentally acquired microbial toxins in foods are fungal toxins, such as aflatoxins, caused by the growth of fungi on plant materials either prior to harvesting or during storage. Bacterial toxins can result from growth of microorganisms on improperly stored foods (Eastaugh & Shepherd, 1989; Allen & Baron, 1991). These intoxications usually do not represent infectious disease, but nevertheless are sufficiently notable that one bacterial example, *Staphylococcus* food poisoning, is among the top three causes of foodborne illness. *Clostridium botulinum* is a bacterium that causes disease by two different mechanisms. During the growth of this organism in foods it produces a toxin which, when ingested, results in an intoxication. This organism normally cannot grow within the human intestines beyond the age of infancy. However, growth of *Clostridium botulinum* within the intestines of infants can result in the intoxication known as infant botulism. Viruses cannot produce toxins, and it is generally presumed that protozoans also do not produce toxins.

Contamination acquired during processing and production prior to marketing

Many foods are processed prior to marketing. In some cases this can involve procedures as simple as washing and sorting, followed by packaging. Of course, processing can involve far more complex operations, including the addition of multiple ingredients from different sources, and cooking. Each one of the steps involved, and each one of the ingredients added, represents an opportunity for microbial contamination. Contaminants can come from food contact surfaces including processing equipment such as vats and plumbing. Other sources of contamination are water used in washing operations (Taylor *et al.*, 1993), aerosols which may fall into the food as it is being processed and packaged, and the packaging material itself (von Bockelmann, 1991). Cross contamination can occur between subsequently processed lots of raw ingredients. Cross contamination can also occur between raw and cooked products if these should come into contact with one another, either directly, or through the use of common processing or packaging equipment. Assuring sanitation at the processing step is particularly important, in light of the fact that some microbial contaminants such as bacteria and fungi can grow in the foods following packaging, and the distribution and marketing process for some foods may involve very long periods during which that growth can occur. A few recent examples of contamination during processing are bacterial contamination of meat (Luby *et al.*, 1993), frozen fresh foods (Taylor *et al.*, 1993) and dairy products (Wegmüller *et al.*, 1993).

There are a variety of legal regulations that have been established to protect the quality of food products, as reviewed by Walker & LaGrange (1991). Perhaps the most important of these are regulations pertaining to sanitation and that guarantee the microbial quality not only of the basic material used in a food, but even of additives as simple as potable water (Walker & LaGrange, 1991). Preventing microbial contamination requires close attention to sanitation practices. Good sanitation can ensure that food products have a greater health safety level and may increase the shelf life of the product.

Contamination acquired during handling and preparation following marketing

Handlers who are involved in the preparation of food represent a major source of contaminants. Again, the largest factor involved is one of proper sanitation. The infectious microbial contaminants associated with improper handling during food preparation tend to represent microorganisms that are present in the feces of infected individuals, and for which the transmission route involves ingestion of fecally contaminated materials. An example of bacterial disease caused by enteric microorganisms that were introduced during handling and preparation is the recent report of *Vibrio* in a reconstituted beverage (Birkhead *et al.*, 1993). Recent examples of enteric protozoan contamination of foods that occured during handling and preparation are *Giardia* in raw sliced vegetables (Grabowski *et al.*, 1989; Mintz *et al.*, 1993) and in a pasta salad (Petersen *et al.*, 1988). Protozoan contamination of the vegetables in one of those two outbreaks (Grabowski *et al.*, 1989), and of the pasta salad (Petersen *et al.*, 1988), presumably represented cross contamination contributed by water that had been used to prepare the food. An example of enteric viral contamination that may have originated during handling is represented by *Calicivirus* associated with fresh cut fruit (Herwaldt *et al.*, 1994). Handlers also can contaminate food through contact of the food with infected wounds, resulting in outbreaks of diseases that normally are not considered to be enteric in nature. An example of this route of contamination is an outbreak of *Streptococcus* pharyngitis attributed to prepared macaroni and cheese (Farley *et al.*, 1993). It should also be recognized that the predominating types of contaminating microorganisms may change during the chain of events between the time of harvesting and final cooking prior to consumption. With fish used as an example: at the time of harvesting, the predominant bacterial and fungal contaminants are environmental organisms, many of which prefer to grow at refrigeration temperatures. During processing by humans, human enteric microflora are added, and these prefer to grow at 35–40 °C.

Preserving the existing quality of foods

The preservation of the quality of foods is a subject that can be divided into two principle components. The first of these is the reduction of the existing level of pathogenic microbial contaminants, both in the food substance and in packaging materials, hopefully while the addition of newly introduced microbial contaminants is limited (von Bockelmann, 1991; Walker & LaGrange, 1991). The second of these is the prevention of subsequent growth of the remaining microorganisms before the food can be consumed (Foegeding & Busta, 1991).

Reducing the levels of microbial contaminants

In the case of molluscan shellfish, pathogenic microorganisms that the shellfish acquire from the environment during their natural filter feeding can be purged prior to marketing by either relaying or depuration (Morse et al., 1986; Richards, 1988; De Leon & Gerba, 1990; Power & Collins, 1990). Relaying involves moving the shellfish to environmental waters that are known to be uncontaminated. Depuration involves placing the shellfish into an artificial environment of flowing, usually recirculated, water that may be disinfected by a process such as ultraviolet (UV) irradiation. In both instances, the shellfish are then kept in the clean water for a length of time adequate for them to expel their load of microbial contaminants.

Reducing the existing level of microbial contaminants in other types of food, and in the materials used for packaging, can be accomplished by many different approaches during the processing that occurs prior to marketing. The simplest of these is washing both the food (Ching-Lee et al., 1991) and the packaging materials with water that contains sanitizing agents in the form of chemical disinfectants such as chlorine and chlorine compounds (Walker & LaGrange, 1991) or hydrogen peroxide (von Bockelmann, 1991).

All organisms can be destroyed by the application of sufficient heat. Applying heat until the food reaches an adequately high temperature, and maintaining that temperature for an appropriate period of time, is one of the most effective methods for reducing the level of microorganisms in foods. Pasteurization is a term used to describe the application of heat for a specific period of time to destroy pathogenic bacteria. It is normally performed when the products are processed, prior to marketing. Pasteurization is often used to treat dairy products (Tauxe et al., 1988; Centers for Disease Control, 1993). Retorting is another heating technique that is used in food processing. Retorting is commonly used in canning operations, during which externally applied heat generates steam pressure inside sealed containers of food. This process simulates the practice of autoclaving that is used in laboratory sterilization operations.

Adequately cooking food prior to consumption, by using either heat or microwave radiation, can be effective in reducing foodborne illness. An example of this is represented by salmonellosis associated with the consumption of eggs (Ewert *et al.*, 1993). In fact, the extent to which eggs are cooked prior to consumption has been found to be statistically associated with the likelihood of acquiring illness (Hedberg *et al.*, 1993). Outbreaks of bacterial disease are also associated with meat that has been inadequately cooked prior to consumption (Centers for Disease Control, 1993; Le Saux *et al.*, 1993; Turney *et al.*, 1994). Consumption of inadequately cooked crustacean shellfish is one of the characteristic associations with foodborne cholera (Ragazzoni *et al.*, 1991; Roman *et al.*, 1991). Consumption of raw or inadequately cooked bivalve molluscan shellfish is characteristically associated with viral hepatitis and gastroenteritis (Morse *et al.*, 1986; De Leon & Gerba, 1990).

The sensitivity of various groups of microorganisms to the effects of heating varies tremendously. This is a particular problem with regard to *Bacillus* and *Clostridium*, two genera of bacteria that produce endospores. Although the vegetative cells of these bacteria can be killed fairly easily by heating, their spores are relatively resistant to the effects of heating (von Bockelmann, 1991). After cooked food is allowed to cool, any spores that survived the heat treatment can germinate into new vegetative bacterial cells that may subsequently grow in the food (Allen & Baron, 1991; Luby *et al.*, 1993). Because of this resistance the spore-producing organism *Clostridium perfringens* ranks third behind *Salmonella* and *Staphylococcus aureus* as the most common cause of food poisoning in the United States (Allen & Baron, 1991). The subject of thermal inactivation of microorganisms will be discussed more extensively in Chapter 12 of this book.

Alternative techniques for destroying pathogenic microorganisms in food include freezing, irradiation, dehydration and chemical addition. Freezing can be effective for destroying protozoans (Petersen *et al.*, 1988), although it cannot be considered to be protective against viruses (Khan *et al.*, 1994) and bacteria (Taylor *et al.*, 1993). Gamma irradiation is used during the processing of some foods, including a potential application for shellfish (Harewood, Rippey & Montesalvo, 1994). Other types of radiations including microwave (Welt *et al.*, 1994), UV and electron beams (von Bockelmann, 1991) may have beneficial roles by reducing the levels of contaminating microorganisms in foods during processing.

Some microorganisms will be destroyed by drying, as discussed in Chapter 1. The role of drying is, however, more important from the standpoint of its effectiveness at preventing microbial growth (Foegeding & Busta, 1991). Drying reduces the water activity in foods and is achieved either by vaporization of liquid water, or by freeze drying, which involves the sublimation of ice from frozen foods. Chemical supplements also can be effective at destroying microbial contaminants. These supplements are many in number, and include such common additives as salt and sugar, which act by changing osmolarity and reducing water activity, and smoke, which is a complex mixture of chemicals. Enzymes, such as lysozyme and lactoperoxidase, are sometimes added to food because of their anti-microbial potential. The review prepared by Foegeding & Busta (1991) presents an excellent in-depth discussion of the chemical supplements that are used to reduce the levels of microbial contaminants in foods.

Preventing the growth of microbial contaminants

Bacteria, fungi and perhaps algae are groups of pathogenic microbial con-taminants whose environmental needs are sufficiently simple that they can potentially replicate within stored food. Microbial pathogens which, for the most part, cannot replicate in food are protozoa, helminths and viruses. Water activity, temperature and the presence of competing microorganisms are factors that play a major role in regulating the growth of microorgan-isms in food. Other factors that are slightly less important, although still significant in terms of regulating the ability of microorganisms to grow in foods, include pH, the presence of inhibiting chemicals and the availability of suitable electron acceptors to support metabolic processes. In the case of aerobic microbial growth, the electron acceptor is usually oxygen. Other inorganic and organic molecules serve as electron acceptors during the growth of microorganisms under anaerobic conditions.

Control of water activity is probably the oldest technique known for preventing the growth of microorganisms in foods, and it can effectively be accomplished by many different approaches. Drying is the simplest technique for long-term storage of foods, and will inhibit the growth of all microorganisms below a critical moisture level. Drying offers the benefit that the food can be stored either at ambient temperature or, for longer storage, under refrigeration. Drying is commercially important for preserv-ing many whole, uncooked materials such as meats, fruits and vegetables. Drying is also used for preserving many powdered foods such as eggs, milk and other assorted dairy products; flour and flour-based products such as pasta and cereals; plus sugar and sugar-containing products such as flavored beverage mixes. Entire prepackaged dried meals are available to use in settings that necessitate lightweight transportation and extended storage. Drying, followed by maintaining adequate dryness, can protect the micro-biological quality of food for decades. However, drying will not prevent the chemical oxidation of food, which, over the course of time, can render some foods unpalatable. Other means for controling microbial growth through reduction of water activity include increasing the level of highly soluble compounds such as salt and sugar. This approach may not be effective for as long a period of time as drying, because some bacteria can grow in salt at very low moisture levels and fungi can grow in highly sugared foods, including corn syrups.

Temperature control can be accomplished by freezing, which inhibits all growth and severely reduces the rate of chemical oxidation. Freezing potentially can preserve food for centuries. Unfortunately, maintaining food in a frozen condition is costly, and freezing can produce undersirable changes in the texture of some foods. Storage of food at temperatures above freezing has contributed to the knowledge that temperature has a very clear impact upon microbial growth rates (Allen & Baron, 1991; Ewert *et al.*, 1993; Farley *et al.*, 1993; Islam, Hasan & Khan, 1993). This relationship has proven itself well suited to the use of mathematical modeling, with microbial growth being a function of time and temperature (Zanoni *et al.*, 1993; Zwietering *et al.*, 1994*a,b*). Storing foods at temperatures that allow the uncontrolled growth of microbial pathogens is termed 'thermal abuse', and represents a major cause of illness. Time and salinity also constitute an important relationship in terms of microbial growth (Kaspar & Tamplin, 1993), as does the combination of temperature and pH (Abdul-Raouf, Beuchat & Ammar, 1993).

There are numerous types of supplements that can be added to foods to inhibit the growth of undesired microorganisms. The list of chemical supplements includes, of course, those solid compounds such as salt and sugar, which reduce water activity (Foegeding & Busta, 1991), as discussed above; acidulants such as acetic acid, citric acid and lactic acid, which lower the pH (Foegeding & Busta, 1991; Abdul-Raouf *et al.*, 1993); nitrite, which may have its effect through inhibiting heme-based microbial enzymes (Foegeding & Busta, 1991); and gases such as carbon dioxide and nitrogen, which largely are inert and act by displacing oxygen. Carbon dioxide also can act as an acidulant when solubilized into aqueous liquids, and when solubilized under pressure, it has a powerful inhibitory effect upon microbial growth (Foegeding & Busta, 1991). Other gases used for their antimicrobial effects include ozone, sulfur dioxide, which may be added in the form of sulfites, epoxides and chlorine (Foegeding & Busta, 1991). Table 6.2 lists examples of chemicals that are used as food preservatives.

Many, and indeed quite possibly most, bacteria and fungi produce and release antimicrobial compounds that are antagonistic to the growth of other microorganisms. This is a common occurrence in microbial ecology, and represents evolutionarily developed mechanisms that favor the ability of organisms to compete for available nutrients. Some of these antimicrobial compounds have been developed commercially as antibiotics. Others include siderophores, which act by binding iron (Gram, 1993), which is essential for the growth of organisms, and bacteriocins (Winkowski, Crandall & Montville, 1993), which produce cell death. One of the newest

Table 6.2 *Examples of chemicals used as food preservatives*

Reducers of water activity
Salt
Sugar

Enzymes
Lysozyme
Lactoperoxidase

Acidulants
Acetic acid
Dehydroacetic acid
Halogenacetic acids
Adipic acid
Ascorbic acid
Benzoic acid
Boric acid
Citric acid
Formic acid
Hydrochloric acid
Lactic acid
Phosphoric acid
Propionic acid
Salicylic acid
Sorbic acid
Succinic acid
Tartaric acid

Gases
Carbon dioxide
Chlorine
Epoxides
Nitrogen
Ozone
Sulfur dioxide

Others
Ethanol
Hydrogen peroxide
Nitrates
Nitrites
Phosphates
Plant oils
Smoke

developments in food microbiology is research on the application of microbial competition as a means of preserving the healthful quality of foods, a process termed 'biopreservation'. The goal of biopreservation is to seed nonpathogenic strains of organisms such as *Lactobacillus* into food, and allow the seeded organisms to grow, thereby inhibiting subsequent growth of pathogenic organisms such as *Listeria* (Winkowski *et al.*, 1993).

Modeling the risk of acquiring infectious disease from consumption of foods

Figure 6.1 presents one format that can be used to estimate the risk of acquiring infectious disease from the consumption of pathogenic microorganisms in foods. This is a compartmental model, which takes into consideration the various concepts introduced earlier in this chapter. Foods may have an initial concentration of pathogenic microorganisms at the time of harvest. These can be reduced in number during processing, although pathogenic contaminants may inadvertently be added at this stage. Bacterial and fungal contaminants can increase in number by replicating within foods during the storage time involved in marketing. Viruses and parasites that infect humans generally die-off during food storage, because they are not able to grow and replicate in the stored food. Following marketing, cooking food for a sufficient length of time at an adequately high temperature will reduce the level of microbial contaminants, although surviving bacterial and fungal organisms may regrow in the cooked foods, depending upon the subsequent holding temperature and duration of storage. Reheating foods to a sufficiently high temperature for an adequate length of time will reduce the level of microorganisms. This model allows for the fact that foods can be consumed at many different points, represented by the circles in the figure. The rectangles represent the initial microbial contamination that exists prior to harvesting, and the level of microbial load in foods after each of the different steps from harvesting to consumption. If the numbers of microorganisms expected to be present in food at the time of consumption can be estimated, per gram, decagram, hectogram or kilogram, then these values can be fed into (pun intended!) the infectious disease risk equations (Equations 6.1–6.4).

$$\text{Number of organisms ingested} = \begin{bmatrix} \text{Number of} \\ \text{organisms} \\ \text{per unit} \\ \text{weight or} \\ \text{volume of} \\ \text{food} \end{bmatrix} \times \begin{bmatrix} \text{Weight or} \\ \text{volume,} \\ \text{in units,} \\ \text{of food} \\ \text{ingested} \end{bmatrix} \tag{6.1}$$

$$\text{Risk of infection} = \begin{bmatrix} \text{Number of} \\ \text{organisms} \\ \text{ingested} \end{bmatrix} \times \begin{bmatrix} \text{Probability} \\ \text{of infection} \\ \text{per ingested} \\ \text{organism} \end{bmatrix} \tag{6.2}$$

$$\text{Risk of illness} = \begin{bmatrix} \text{Number of} \\ \text{organisms} \\ \text{ingested} \end{bmatrix} \times \begin{bmatrix} \text{Probability} \\ \text{of infection} \\ \text{per ingested} \\ \text{organism} \end{bmatrix} \times \begin{bmatrix} \text{Probability} \\ \text{of illness} \\ \text{per incidence} \\ \text{of infection} \end{bmatrix} \tag{6.3}$$

$$\text{Risk of death} = \begin{bmatrix} \text{Number of} \\ \text{organisms} \\ \text{ingested} \end{bmatrix} \times \begin{bmatrix} \text{Probability} \\ \text{of infection} \\ \text{per ingested} \\ \text{organism} \end{bmatrix} \times \begin{bmatrix} \text{Probability} \\ \text{of illness} \\ \text{per incidence} \\ \text{of infection} \end{bmatrix} \times \begin{bmatrix} \text{Probability} \\ \text{of death} \\ \text{per incid-} \\ \text{ence of} \\ \text{illness} \end{bmatrix} \tag{6.4}$$

Table 6.3. *Estimated risk of infection, illness and death from microbial contaminants in food*

Level of organisms per 100 g or per 100 ml	Risk of infection	Risk of illness	Risk of death
Bacterial			
0.01	1.0×10^{-6}	4.9×10^{-7}	8.8×10^{-9}
0.1	1.0×10^{-5}	4.9×10^{-6}	8.8×10^{-8}
1.0	1.0×10^{-4}	4.9×10^{-5}	8.8×10^{-7}
10.0	1.0×10^{-3}(0.1%)	4.9×10^{-4}	8.8×10^{-6}
Protozoan			
0.01	6.7×10^{-4}	2.9×10^{-4}	5.9×10^{-8}
0.1	6.7×10^{-3}	2.9×10^{-3}	5.9×10^{-7}
1.0	6.7×10^{-2}	2.9×10^{-2}	5.9×10^{-6}
10.0	6.7×10^{-1} (67%)	2.9×10^{-1} (29%)	5.9×10^{-5}
Viral			
0.01	5.0×10^{-3}	2.9×10^{-3}	2.7×10^{-5}
0.1	5.0×10^{-2}	2.9×10^{-2}	2.7×10^{-4}
1.0	5.0×10^{-1}	2.9×10^{-1}	2.7×10^{-3}
10.0	1 (100%)	1 (100%)	2.7×10^{-2} (2.7%)

Note: These calculations are based upon an assumption that the quantity of food ingested was 100 g or 100 ml. The listed values were derived by substituting values for probability of infection per ingested organism, probability of illness per incidence of infection and probability of death per incidence of illness from Table 4.8 into Equations 6.1–6.4.

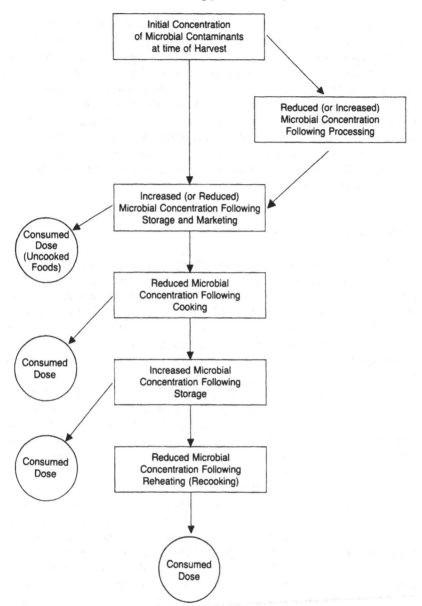

Figure 6.1   Diagrammatic tree, showing the points at which the microbial level in foods changes, and how these changes relate to the acquisition of infection by ingestion of food.

These equations utilize the values from Table 4.8 which define the probability of infection per ingested organism, along with the associated risks of infection leading to illness, and illness progressing to death. These health risks can also be calculated per meal, or over the course of any desired time period, by estimating the amount of contaminated food that would be likely to be ingested per meal, week, month, or year. Such risk calculations are valuable, because foodborne illness represents a significant cause of morbidity (Cliver, 1990) and societal cost (Roberts, 1988). Table 6.3 presents an example of the use of these equations to estimate the infectious disease risks associated with consuming foods that contain microbial pathogens.

Summary

Food is essential for survival, but unfortunately the ingestion of pathogenic microorganisms that may exist as contaminants in foods can result in infectious disease. Contamination with pathogenic microorganisms can occur prior to harvesting, so that the food is processed in a contaminated state. Microbial contaminants can also enter food during the handling and processing that occurs prior to marketing. Likewise, following marketing, foods can become microbially contaminated during the handling and preparation that occurs prior to consumption. A number of different treatment techniques can be used to reduce the potential of foods to cause infectious disease. These include the washing of foods, heating, freezing, irradiation and chemical additions. The various categories of foodborne pathogenic microorganisms differ with respect to their sensitivity to each of these treatment techniques. As an example, parasites are extremely susceptible to freezing, while viruses and bacteria are able to survive freezing in foods.

Bacterial and fungal contaminants can grow within foods. This potential for microbial replication presents a problem that necessitates care both during the time period between which food is harvested and processed, as well as the time period between processing and eventual consumption. Controlling the storage temperature of foods, either by refrigeration or freezing, is extremely useful for restricting microbial growth. However, this technique can be utilized only in areas of the world where ice or some suitable energy source is available; in some regions refrigerated storage is simply too costly. Reducing the water activity in foods by drying them is an approach that is far less expensive than refrigeration and is universally available. Other techniques used for reducing the water activity of foods and thereby controlling the growth of microorganisms during food storage are the addition of highly soluble solids such as salt and sugar. Additional inhibitory compounds, including acidulants and gases, can also be added to foods to restrict microbial growth. One of the newer methods for controling the growth of pathogenic microorganisms in foods is biopreservation, which relies upon seeding nonpathogenic microorganisms into the foods. This process is based upon the expectation that these seeded microorganisms will multiply in the food during storage, and that they will in turn produce compounds that are biologically active in preventing the growth of other, unwanted microorganisms. Lastly, it is possible to model the risk of acquiring infectious disease from foods, and this chapter presented a basic compartment modeling approach that can be used for making such risk assessments.

References

Abdul-Raouf, U. M., Beuchat, L. R. & Ammar, M. S. (1993). Survival and growth of *Escherichia coli* 0157:H7 in ground, roasted beef as affected by pH, acidulants, and temperature. *Applied and Environmental Microbiology*, **59**, 2364–8.

Allen, S. D. & Baron, E. J. (1991). *Clostridium*. In *Manual of Clinical Microbiology*, 5th edn, ed. A. Balows, W. J. Hausler Jr, K. L. Herrmann, H. D. Isenberg & H. J. Shadomy, pp. 505–21. Washington: The American Society for Microbiology.

Atmar, R. L., Metcalf, T. G., Neill, F. H. & Estes, M. K. (1993). Detection of enteric viruses in oysters by using the polymerase chain reaction. *Applied and Environmental Microbiology*, **59**, 631–5.

Bej, A. K., Mahbubani, M. H., Boyce, M. J. & Atlas, R. M. (1994). Detection of *Salmonella* spp. in oysters by PCR. *Applied and Environmental Microbiology*, **60**, 368–73.

Birkhead, G. S., Morse, D. L., Levine, W. C., Fudala, J. K., Kondracki, S. F., Chang, H.-G., Shayegani, M., Novick, L. & Blake, P. A. (1993). Typhoid fever at a resort hotel in New York: a large outbreak with an unusual vehicle. *Journal of Infectious Diseases*, **167**, 1228–32.

Centers for Disease Control (1993). Preliminary report: foodborne outbreak of *Escherichia coli* 0157:H7 infections from hamburgers – Western United States, 1993. *Morbidity and Mortality Weekly Report*, **42**, 85–6.

Ching-Lee, M. R., Katz, A. R., Sasaki, D. M. & Minette, H. P. (1991). Salmonella egg survey in Hawaii: evidence for routine bacterial surveillance. *American Journal of Public Health*, **81**, 764–6.

Cliver, D. O. (1990). *Report of WHO Collaborating Center on Food Virology Literature Review – 1990*. Madison, Wisconsin: Food Research Institute, University of Wisconsin.

Davis, J. W. & Sizemore, R. K. (1982). Incidence of *Vibrio* species associated with blue crabs (*Callinectes sapidus*) collected from Galveston Bay, Texas. *Applied and Environmental Microbiology*, **43**, 1092–7.

De Leon, R. & Gerba, C. P. (1990). Viral disease transmission by seafood. In *Food Contamination from Environmental Sources*, ed. J. O. Nriagu & M. S. Simmons, pp. 639–62. New York: John Wiley & Sons.

Eastaugh, J. & Shepherd, S. (1989). Infectious and toxic syndromes from fish and shellfish consumption. *Archives of Internal Medicine*, **149**, 1735–40.

Ewert, D., Bendaña, N., Tormey, M., Kilman, L., Mascola, L., Gresham, L. S.,

Ginsberg, M. M., Tanner, P. A., Bartzen, M. E., Hunt, S., Marks, R. S., Peter, C. R., Mohle-Boetani, J., Fenstersheib, M., Gans, J., Coy, K., Liska, S., Abbott, S., Bryant, R., Barrett, L., Reilly, K., Wang, M., Werner, S. B., Jackson, R. J. & Rutherford, G. W. (1993). Outbreaks of *Salmonella enteritidis* gastroenteritis – California, 1993. *Morbidity and Mortality Weekly Report*, **42**, 793–7.

Farley, T. A., Wilson, S. A., Mahoney, F., Kelso, K. Y., Johnson, D. R. & Kaplan, E. L. (1993). Direct inoculation of food as the cause of an outbreak of group A streptococcal pharyngitis. *Journal of Infectious Diseases*, **167**, 1232–5.

Foegeding, P. M. & Busta, F. F. (1991). Chemical food preservatives. In *Disinfection, Sterilization and Preservation*, 4th edn, ed. S. S. Block, pp. 802–32. Philadelphia: Lea & Febiger.

Grabowski, D. J., Tiggs, K. J., Senke, H. W., Salas, A. J., Powers, C. M., Knott, J. A. & Sewell, C. M. (1989). Common-source outbreak of giardiasis – New Mexico. *Morbidity and Mortality Weekly Report*, **38**, 405–7.

Gram, L. (1993). Inhibitory effect against pathogenic and spoilage bacteria of *Pseudomonas* strains isolated from spoiled and fresh fish. *Applied and Environmental Microbiology*, **59**, 2197–203.

Grimes, D. J. (1991). Ecology of estuarine bacteria capable of causing human disease: a review. *Estuaries*, **14**, 345–60.

Harewood, P., Rippey, S. & Montesalvo, M. (1994). Effect of gamma irradiation on shelf life and bacterial and viral loads in hard-shelled clams (*Mercenaria mercenaria*). *Applied and Environmental Microbiology*, **60**, 2666–70.

Hedberg, C. W., David, M. J., White, K. E., MacDonald, K. L. & Osterholm, M. T. (1993). Role of egg consumption in sporadic *Salmonella enteritidis* and *Salmonella typhimurium* infections in Minnesota. *Journal of Infectious Diseases*, **167**, 107–11.

Herwaldt, B. L., Lew, J. F., Moe, C. L., Lewis, D. C., Humphrey, C. D., Monroe, S. S., Pon, E. W. & Glass, R. I. (1994). Characterization of a variant strain of Norwalk virus from a foodborne outbreak of gastroenteritis on a cruise ship in Hawaii. *Journal of Clinical Microbiology*, **32**, 861–6.

Islam, M. S., Hasan, M. K. & Khan, S. I. (1993). Growth and survival of *Shigella flexneri* in common Bangladeshi foods under various conditions of time and temperature. *Applied and Environmental Microbiology*, **59**, 652–4.

Kaspar, C. W. & Tamplin, M. L. (1993). Effects of temperature and salinity on the survival of *Vibrio vulnificus* in seawater and shellfish. *Applied and Environmental Microbiology*, **59**, 2425–9.

Khan, A. S., Moe, C. L., Glass, R. I., Monroe, S. S., Estes, M. K., Chapman, L. E., Jiang, X., Humphrey, C., Pon, E., Iskander, J. K. & Schonberger, L. B. (1994). Norwalk virus-associated gastroenteritis traced to ice consumption aboard a cruise ship in Hawaii: comparison and application of molecular method-based assays. *Journal of Clinical Microbiology*, **32**, 318–22.

Le Saux, N., Spika, J. S., Friesen, B., Johnson, I., Melnychuck, D., Anderson, C., Dion, R., Rahman, M. & Tostowaryk, W. (1993). Ground beef consumption in noncommercial settings is a risk factor for sporadic *Escherichia coli* 0157:H7 infection in Canada. *Journal of Infectious Diseases*, **167**, 500–2.

Luby, S., Jones, J., Dowda, H., Kramer, J. & Horan, J. (1993). A large outbreak of gastroenteritis caused by diarrheal toxin-producing *Bacillus cereus*. *Journal of Infectious Diseases*, **167**, 1452–5.

Metcalf, T. G. (1978). Indicators of viruses in shellfish. In *Indicators of Viruses in Water and Food*, ed. G. Berg, pp. 383–415. Ann Arbor, Michigan: Ann Arbor Science Publishers.

Miceli, G. A., Watkins, W. D. & Rippey, S. R. (1993). Direct plating procedure for enumerating *Vibrio vulnificus* in oysters (*Crassostrea virginica*). *Applied and Environmental Microbiology*, **59**, 3519–24.

Mintz, E. D., Hudson-Wragg, M., Mshar, P., Cartter, M. L. & Hadler, J. L. (1993). Foodborne giardiasis in a corporate office setting. *Journal of Infectious Diseases*, **167**, 250–3.

Morse, D. L., Guzewich, J. J., Hanrahan, J. P., Stricof, R., Shayegani, M., Deibel, R., Grabau, J. C., Nowak, N. A., Herrmann, J. E., Cukor, G. & Blacklow, N. R. (1986). Widespread outbreaks of clam- and oyster-associated gastroenteritis role of Norwalk virus. *New England Journal of Medicine*, **314**, 678–81.

O'Mahony, M. C., Gooch, C. D., Smyth, D. A., Thrussell, A. J., Bartlett, C. L. R. & Noah, N. D. (1983). Epidemic hepatitis A from cockles. *Lancet*, **i**, 518–20.

Pearson, A. D., Greenwood, M., Healing, T. D., Rollins, D., Shahamat, M., Donaldson, J. & Colwell, R. R. (1993). Colonization of broiler chickens by waterborne *Campylobacter jejuni*. *Applied and Environmental Microbiology*, **59**, 987–96.

Petersen, L. R., Cartter, M. L. & Hadler, J. L. (1988). A food-borne outbreak of *Giardia lamblia*. *Journal of Infectious Diseases*, **157**, 846–8.

Power, U. F. & Collins, J. K. (1990). Tissue distribution of a coliphage and *Escherichia coli* in mussels after contamination and depuration. *Applied and Environmental Microbiology*, **56**, 803–7.

Ragazzoni, H., Mertz, K., Finelli, L., Genese, G., Dunston, F. J., Russell, B., Riley, W., Feller, E., Ares, M. B., Fernandez, M., Sfakianaki, E., Simmons, J. A. & Hopkins, R. S. (1991). Cholera – New Jersey and Florida. *Morbidity and Mortality Weekly Report*, **40**, 287–9.

Richards, G. P. (1988). Microbial purification of shellfish: a review of depuration and relaying. *Journal of Food Protection*, **51**, 218–51.

Roberts, T. (1988). Salmonellosis control: estimated economic costs. *Poultry Science*, **67**, 936–43.

Roman, R., Middleton, M., Cato, S., Bell, E., Ong, K. R., Gruenewald, R., Ferguson, A. & Ramon, A. (1991). Cholera – New York, 1991. *Morbidity and Mortality Weekly Report*, **40**, 516–18.

Tauxe, R. V., Hargrett-Bean, N., Patton, C. M. & Wachsmuth, I. K. (1988). *Campylobacter* isolates in the United States, 1982–1986. *Morbidity and Mortality Weekly Report*, **37** (SS-2), 1–13.

Taylor, J. L., Tuttle, J., Pramukul, T., O'Brien, K., Barrett, T. J., Jolbitado, B., Lim, Y. L., Vugia, D., Morris, J. G. Jr, Tauxe, R. V. & Dwyer, D. M. (1993). An outbreak of cholera in Maryland associated with imported commercial frozen fresh coconut milk. *Journal of Infectious Diseases*, **167**, 1330–5.

Tulchinsky, T. H., Levine, I., Abrookin, R. & Halperin, R. (1988). Waterborne enteric disease outbreaks in Israel, 1976–1985. *Israeli Journal of Medical Sciences*, **24**, 644–51.

Turney, C., Green-Smith, M., Shipp, M., Mordhorst, C., Whittingslow, C., Brawley, L., Koppel, D., Bridges, E., Davis, G., Voss, J., Lee, R., Jay, M., Abbott, S., Bryant, R., Reilly, K., Werner, S. B., Barrett, L., Jackson, R. J., Rutherford, G. W. III. & Lior, H. (1994). *Escherichia coli* 0157:H7 outbreak linked to home-cooked hamburger – California, July 1993. *Morbidity and Mortality Weekly Report*, **43**, 213–16.

von Bockelmann, B. (1991). Aseptic packaging. In *Disinfection, Sterilization and Preservation*, 4th edn, ed. S. S. Block, pp. 833–45. Philadelphia: Lea & Febiger.

Walker, H. W. & LaGrange, W. S. (1991). Sanitation in food manufacturing operations. In *Disinfection, Sterilization and Preservation*, 4th edn, ed. S. S. Block, pp. 791–801. Philadelphia: Lea & Febiger.

Warburton, D. W., Dodds, K. L., Burke, R., Johnston, M. A. & Laffey, P. J. (1992). A review of the microbiological quality of bottled water sold in Canada between 1981 and 1989. *Canadian Journal of Microbiology*, **38**, 12–19.

Wegmüller, B., Lüthy, J. & Candrian, U. (1993). Direct polymerase chain reaction detection of *Campylobacter jejuni* and *Campylobacter coli* in raw milk and dairy products. *Applied and Environmental Microbiology*, **59**, 2161–5.

Welt, B. A., Tong, C. H., Rossen, J. L. & Lund, D. B. (1994). Effect of microwave radiation on inactivation of *Clostridium sporogenes* (PA 3679) spores. *Applied and Environmental Microbiology*, **60**, 482–8.

Winkowski, K., Crandall, A. D. & Montville, T. J. (1993). Inhibition of *Listeria monocytogenes* by *Lactobacillus bravaricus* MN in beef systems at refrigeration temperatures. *Applied and Environmental Microbiology*, **59**, 2552–7.

Zanoni, B., Garzaroli, C., Anselmi, S. & Rondinini, G. (1993). Modeling the growth of *Enterococcus faecium* in bologna sausage. *Applied and Environmental Microbiology*, **59**, 3411–17.

Zwietering, M. H., Cuppers, H. G. A. M., de Wit, J. C. & van 't Riet, K. (1994a). Evaluation of data transformations and validation of a model for the effect of temperature on bacterial growth. *Applied and Environmental Microbiology*, **60**, 195–203.

Zwietering, M. H., de Wit, J. C., Cuppers, H. G. A. M. & van 't Riet, K. (1994b). Modeling of bacterial growth with shifts in temperature. *Applied and Environmental Microbiology*, **60**, 204–13.

PART 3   PREVENTING DISEASE TRANSMISSION BY AEROSOLS, SURFACES AND MEDICAL DEVICES

Disinfection of microbial aerosols

SCOTT CLARK and
PASQUALE SCARPINO

Introduction

Microbial aerosols of pathogenic organisms can serve as infective agents
either through direct inhalation or through pathways involving the deposition of microbes onto food substances or other surfaces, which eventually
contaminate objects entering an individual's body. This chapter will deal
with methods to disinfect air that contains microbial aerosols, resulting in
a reduction in the airborne concentration of the microbial aerosols, and
hopefully a decrease in their infectivity. The disinfection of surfaces that
may have been contaminated by microbial aerosols, and the ultraviolet
disinfection of liquids that may be a reservoir, supplying microbes to the
air, are covered in other chapters. Other means for disinfecting liquids that
contain microbial populations are adequately covered in numerous other
documents.

A comprehensive plan for controlling microbial aerosols must contain
a variety of measures to reduce the source(s) of the microbes, to make
conditions less favorable for their survival and to reduce their airborne
levels by use of other engineering controls (e.g. ventilation, or the covering
of vessels containing microbes). Work practice techniques (e.g. type of
clothing worn, removing work areas from locations of highest exposure)
are also important to control both the concentration and infectivity of
microbial aerosols. The types of staff clothing worn in an ultraclean
operating room with down-flow air enclosure has been found to affect
bacterial air contamination, being able to reduce contamination from the
skin of persons attending the operation to below 1 colony forming unit
(CFU) per cubic meter (Sanzen, Carlsson & Walder, 1990). In one study
in which cotton gowns were used under two types of operating gowns
(exhaust and nonwoven), use of the exhaust gowns was found to result in
lower airborne bacterial levels than use of the non-woven disposable
gowns. In another study involving the wearing of cotton and synthetic

scrub gowns under nonwoven operating gowns, the use of the synthetic gowns was found to be associated with airborne contamination levels lower than 1 CFU/m^3, and in nine of 20 operations using cotton scrub suits the air concentration exceeded 1 CFU/m^3.

Disinfection of microbial aerosols is generally recommended only as a supplement to other engineering controls. Some reports on the efficiency of microbial aerosol disinfection are expressed as the equivalent effect of a certain number of air changes per hour. Under certain circumstances personal protective equipment such as respirators must also be a part of the overall program to protect human health.

Source control measures have considerable potential for indirectly reducing levels of microbial aerosols. Such measures include limiting the exposure to the air of sources of microbially contaminated materials, reducing the opportunity for microbial growth, preventing the entry of aerosols into the environment of concern and disinfecting surfaces and liquids contaminated with microbes. This chapter, however, focuses on measures relating directly to controlling the airborne microbes once the aerosolization has occurred.

Microbial aerosol disinfection methods

There are three general methods to reduce airborne concentrations of microbes: use of ultraviolet lamps for germicidal irradiation (UVGI), chemical disinfection and air filtration through high-efficiency particulate filters (HEPA). There has been little work done in the area of chemical disinfection, because of the dangers involved in releasing such chemical agents into the air. Although technically not a true disinfection mechanism, HEPA filtration has virtually equivalent results and is discussed in this chapter as it can also be effective in removing non-microbial contaminants. Microbial disinfection methods may have undesirable side effects, which must be considered. UV radiation, for example, has a number of adverse effects. Short-term UV over-exposure can cause erythema and keratoconjunctivitis (Sterenborg, van der Putte & van der Leun (1988). Broadspectrum UV radiation has been associated with increased risk for squamous and basal cell carcinomas of the skin. Some studies have indicated that UV radiation can increase replication of the human immunodeficiency virus (Zmudzka & Beer, 1990).

Applications of microbial aerosol disinfection

The disinfection of microbial aerosols has application in a wide variety of situations such as those involving food (processing, distribution, sales,

preparation and consumption), healthcare (surgical wards, dental treatment areas, laboratories, laundries, supply and preparation areas, waiting
rooms), waste processing industries (wastewater collection and treatment,
composting and other sludge treatment and distribution), agricultural facilities (poultry and swine confinements) and veterinary facilities.

Ultraviolet inactivation

Ultraviolet radiation is defined as the portion of the electromagnetic spectrum with wavelengths between 100 and 400 nm. The use of UV radiation
to inactivate microorganisms is currently known as UVGI. The use of
ultraviolet light to inactivate microorganisms has been a topic of research
for many years (Wells & Wells, 1936; Sharp, 1940; Wells, 1942, 1955; Riley
& O'Grady, 1961; Riley *et al.*, 1962; Riley, Knight & Middlebrook, 1976).
Many of these studies documented the effectiveness of UVGI for the
control of tuberculosis in controlled settings. Several factors are thought
to be responsible for the lack of widespread use of UVGI: subsequent
experiments in schools did not duplicate the earlier findings, the use of
other treatments showed promise for the control of tuberculosis and concerns were raised over the safety of UV radiation (Macher, 1993). There
has been a renewed interest in the use of UVGI, because of the increased
concern for the threat of tuberculosis, especially in compromised populations, and the identification of multi-drug-resistant strains.

The United States Centers for Disease Control (CDC) recommends
UVGI as a supplement to other tuberculosis control measures in situations
where there is the need for the killing or inactivation of tubercle bacteria.
Much of the material in this section has been taken from the latest CDC
recommendation (Centers for Disease Control, 1994).

Techniques used

There are three types of applications used for UVGI: duct irradiation,
self-contained air-cleaning units and upper-room air irradiation. In duct-
work applications, the UV lamps are placed within duct-work systems that
are designed to remove air from rooms and treat it with UV radiation
before returning it to the rooms. Incorporating the UV lamps within the
ductwork, if properly installed and maintained, can restrict human
exposure to only those situations where maintenance operations occur. A
number of commercially available units now incorporate UV; lamps in
self-contained portable systems can be placed in areas where UVGI is
needed. (These units may also contain HEPA filtration.) In upper-air radiation, shielded lamps are placed in upper areas of rooms and regular room

circulation, such as by convection, is used to ensure that room air circulates and comes into contact with the radiation. The lamps may be either suspended from the ceilings or attached to the walls.

Laboratory studies

Using an experimental air-conditioning system incorporating both HEPA and UVGI, Nakamura (1987) found that the germicidal efficiency of spiral UV lamps was above 99.99% for *Bacillus subtilis* spores with an exposure time of 0.5 s.

Scarpino *et al.* (1994) reported an inactivation of more than 99% of *Escherichia coli, Pseudomonas fluorescens, Serratia marcescens* and *Micrococcus luteus*, using a lamp-powered UVGI unit containing four lamps.

Salie *et al.* (1995) evaluated a prototype ceiling fan equipped with an enclosed UV lamp, using three test organisms in a laboratory setting: *E. coli, Sarcinia lutea* and *B. subtilis*. The residence time in the UV-equipped fan assembly was 26 ms and the power intensity of the lamp was 64×10^6 $\mu W/cm^2$. For geometric mean upstream air concentrations of 35.1, 1.76 and 2.46×10^6 CFU/m^3, reductions of 73.8, 3.8 and 8.6%, respectively, were observed. Based on a room 6.1 m (20 ft) × 6.1 m (20 ft) × 3.05 m (10 ft) in size, and assuming perfect mixing, and a fan equipped with two such lamps, 30 min would be required for all of the air in the room to pass through the blade assembly a single time.

Field studies

The effectiveness of UVGI is often measured in terms of equivalent number of room air changes per hour (ACH) required to achieve the same reductions. Riley *et al.* (1976) found that UVGI was the equivalent of 10 ACH in reducing BCG (*Mycobacterium boris* strain bacille Calmette–Guérin) aerosolized in a room. Kethley & Branch (1972) reported UVGI to be equivalent to 39 ACH in reducing *Serratia marcescens* aerosolized in another room. Macher *et al.* (1992) determined the effectiveness of 254-nm ultraviolet radiation from four 15-W wall-mounted germicidal lamps in inactivating airborne microorganisms in an outpatient waiting room. A reduction of an estimated 14–19% in culturable airborne bacteria was observed.

Adverse effects

Exposure to ultraviolet radiation can have undesirable effects on humans and recommendations for worker exposure limits have been developed

(National Institute for Occupational Safety and Health, 1972). Short-term over-exposure to UV radiation can cause erythema (reddening of the skin), photokeratitis (inflammation of the cornea) and conjunctivitis (inflammation of the conjunctiva) (National Institute for Occupational Safety and Health, 1972). Broad-spectrum UV radiation has been associated with squamous and basal cell carcinomas of the skin and has led to UV-C (100–290 nm) being classified as 'probably carcinogenic to humans' by the International Agency for Research on Cancer (1992). The National Institute for Occupational Safety and Health (NIOSH) recommended exposure limit (REL) is intended to protect workers from the acute effects of UV exposure. However, those photosensitive may not be protected. The NIOSH REL is wavelength-dependent, because different wavelengths have different effects. The recommended exposure times for selected values of effective irradiation is shown in Table. 7.1. Properly trained individuals should be engaged to ensure that appropriate safety precautions are in place to protect workers and others from over-exposure.

Table 7.1 *Maximum recommended exposure limits for selected values of effective irradiation*

Recommended exposure time per day	Effective irradiation[a] (μW/cm^2)
8 h	0.1
4 h	0.2
2 h	0.4
1 h	0.8
30 min	1.7
15 min	3.3
10 min	5.0
5 min	10.0
1 min	50.0
30 s	100.0

[a] Relative to the effectiveness of 270 nm, the wavelength of maximum ocular sensitivity.
Source: National Institute for Occupational Safety and Health

HEPA filtration

HEPA filters are air cleaning devices that have been documented to have a minimum removal efficiency of 99.97% of particles >0.3 μm in diameter. Many pathogenic bioaerosols are of a size range that should make them candidates for removal by HEPA filters. *Aspergillus* spores (1.5–6 μm) have been demonstrated to be removed by HEPA filtration (Opel *et al.*, 1986; Sherentz *et al.*, 1987). Since *Mycobacterium tuberculosis* droplet nuclei are probably 1–5 μm in diameter (Centers for Disease Control, 1994), they are likely candidates for removal by HEPA filtration, but this has not yet been demonstrated.

Types of installation

There are a number of effective ways that HEPA filters can be placed in ventilation systems. They can be used to cleanse air that is to be recirculated into rooms (Woods, 1989), or they can be used before air is exhausted to the outside of the contaminated area where the microbes were generated, or as a combination of both. The HEPA filters can be installed within duct work or located elsewhere in a room, e.g. mounted on the wall or free-standing. Modular HEPA filtration units are available for each of these types of installation. Many modular units are self-contained with mechanisms to pull air through the unit and recirculate it back into the room. These units have the advantage of being portable and being able to be used to supplement existing systems when needed to control microbial air quality. Disadvantages include the fact that their effective operation may be adversely affected by room occupants or by inappropriate location of the filtration units.

Use of HEPA filtration for exhaust air

Air exhausted from healthcare facilities, laboratories, etc. may need to be disinfected or 'cleaned' to prevent pathogen exposure to persons who come in contact with this air in exterior areas near the exhaust, in other interior areas where the exhausted air may be used or by direct contact with the contaminated filter. In some instances the filtered air may be exhausted from one area for subsequent use in other presumably uncontaminated indoor areas of a facility. Sometimes the exhausted air is passed to a heat recovery device such as a heat wheel before recirculation. In these cases the HEPA filtration device should be installed upstream of the heat recovery device in order to reduce the potential for contamination.

Use of HEPA filtration for recirculation within a room

The Centers for Disease Control (1994) suggests two scenarios for use of HEPA filtration within a room. One involves exhausting the air into a duct and passing that air through HEPA filters before returning it to the room. The other involves a ceiling-mounted HEPA unit. The appropriate positioning of these HEPA units with respect to a patient's bed will result in contaminated air being pulled away from the patient and cleaned air moving in the direction of the patient's breathing zone. Portable HEPA filtration units may be useful in some situations, but their uses are subject to several limitations. The effectiveness of the operation of the portable unit may be influenced by configurations of the room, placement of the units, and the fact that the actions of room occupants may compromise the efficiency of the units. They have not been evaluated for their effectiveness in tuberculosis infection control programs (Centers for Disease Control, 1994).

A ventilation system capable of removing 95% of all air particles larger than 0.3 μm has been found to be effective in reducing airborne bacteria and dust particles in a calf nursery (Hillman, Gebremedhin & Warner, 1992).

Maintenance of HEPA filters

Careful installation and appropriate routine maintenance are essential if the benefits of HEPA filtration are to be realized (Woods & Rask, 1988). Timely replacement of the HEPA filters is necessary and they should be handled with care during their disposal, since the filters contain the pathogenic microorganisms that were contained in the aerosols removed. The manufacturers' guidelines should be carefully followed. When using HEPA filtration units, placing prefilters upstream of the HEPA filters will greatly extend their useful life.

References

Centers for Disease Control (1994) Guidelines for preventing the transmission of *Mycobacterium tuberculosis* in health care facilities. *Morbidity and Mortality Weekly Report*, **43**, 1–113.

Hillman, P., Gebremedhin, K. & Warner, R. (1992) Ventilation system to minimize airborne bacteria, dust, humidity, and ammonia in calf nurseries. *Journal of Dairy Science*, **75**, 1305–13.

International Agency for Research on Cancer (1992). *Monographs on the Evaluation of Carcinogenic Risks to Humans: Solar and Ultraviolet Radiation*, vol. 55. Lyon, France: World Health International Agency for Research on Cancer.

Kethley, T. W. & Branch, K. (1972). Ultraviolet lamps for room air disinfection: effect of sampling location and particle size of bacterial aerosol. *Archives of Environmental Health*, **235**, 205–14.

Macher, J. M. (1993) The use of germicidal lamps to control tuberculosis in healthcare facilities. *Infection Control and Hospital Epidemiology*, **14**, 723–9.

Macher, J. M., Aleantils, L. E., Chang, Y.-L. & Liu, K.-S. (1992). Effect of ultraviolet germicidal lamps on airborne microorganisms in an outpatient waiting room. *Journal of Applied Occupational and Environmental Hygiene*, **7**, 505–13.

Nakamura, H. (1987). Sterilization efficiency of ultraviolet irradiation on microbial aerosols under dynamic airflow by experimental air conditioning systems. *Bulletin of Tokyo Medical Dentistry University* **34**, 25–40.

National Institute for Occupational Safety and Health (1972). *Criteria for a Recommended Standard. Occupational Exposure to Ultraviolet Radiation*, Publication No. HSM 73–11009. Cincinnati, Ohio: Department of Health and Human Services.

Opel, S. M., Asp, A. A. Cannady, P. B., Morse, P. & Mammer, P. G. (1986). Efficacy of infection control measures during a nosocomial outbreak of disseminated *Aspergillus* associated with hospital construction. *Journal of Infectious Diseases*, **153**, 63–7.

Riley, R. L., Knight, M. & Middlebrook, G. (1976). Ultraviolet susceptibility of BCG and virulent tubercle bacilli. *American Review of Respiratory Diseases*, **113**, 413–18.

Riley, R. L., Mills, C. C., O'Grady, F., Sultan, L. U., Wittstadt, F. & Shivpuri, D. N. (1962). Infectiousness of air from a tuberculosis ward. *American Review of Respiratory Diseases*, **85**, 511–25.

Riley, R. L. & O'Grady, F. (1961). *Airborne Infection*. New York: MacMillan.

Salie, F., Scarpino, P., Clark, S. & Willeke, K. (1995). Laboratory evaluation of airborne microbial reduction by an ultraviolet light positioned in a modified hollow ceiling fan blade. *American Industrial Hygiene Association Journal*, **56**, 987–92.

Sanzen, L., Carlsson, A. S. & Walder, M. (1990). Air contamination during total hip arthroplasty in an ultraclean air enclosure using different types of staff clothing. *Journal of Arthroplasty*, **6**, 127–30.

Scarpino, P. V., Stoeckel, D. M. & Jensen, P. A. (1994). Ability of a fan-powered UVGI unit to inactivate selected airborne bacteria (abstract). *Conference of the Society of Occupational and Environmental Health, Preventing TB in the Workplace: Principles and Practices for Controlling Transmission*, Rockville, Maryland, December.

Sharp, D. G. (1940). The effects of ultraviolet light on bacteria suspended in air. *Journal of Bacteriology*, **39**, 535–47.

Sherertz, R. J., Belani, A., Kramer, B. S., Elfenbein, G. J., Weiner, R. S., Sullivan, M. L., Thomas, R. G. & Samsa, G. P. (1987). Impact of air filtration on nosocomial *Aspergillus* infections. *American Journal of Medicine* **83**, 709–18.

Sterenborg, H. J., van der Putte, S. C. & van der Leun, J. C. (1988). The dose–response relationship of tumorigenesis by ultraviolet radiation at 254 nm. *Photochemistry and Photobiology*, **47**, 245–53.

Wells, W. F. (1942). Radiation disinfection of air. *Archives of Physical Therapy*, 143–8.

Wells, W. F. (1955). *Airborne Contagion and Air Hygiene.* Cambridge, Massachusetts: Harvard University Press.

Wells, W. F. & Wells, M. W. (1936). Airborne infection. *Journal of the American Medical Association,* **107**, 1805–9.

Woods, J. E. (1989). Cost avoidance and productivity in owning and operating buildings. *Journal of Occupational Medicine,* **4**, 753–70.

Woods, J. E. & Rask, C. R. (1988). Heating ventilation, air conditioning systems; the enlightening approach to methods of control. In *Architectural Design and Indoor Microbial Pollution,* ed. R. B. Kandsin, pp. 123–63. New York: Oxford University Press.

Zmudzka, B. Z. & Beer, J. (1990). Activation of human immunodeficiency virus by ultraviolet radiation. *Photochemistry and Photobiology,* **52**, 1153–62.

Chapter 8

Transmission of viral infections through animate and inanimate surfaces and infection control through chemical disinfection

SYED A. SATTAR and V. SUSAN SPRINGTHORPE

Introduction

Modeling the transmission of viral diseases through animate or inanimate surfaces is extremely complex, unless individuals are isolated within a confined area to control their movements. Although this may be the case in certain animal husbandry situations, in general humans, and most domestic animals of economic importance, roam freely and make many contacts with surfaces in their environments and communities. Unlike exposure to water and food, which is often relatively easily defined, and to air, which is continuous, surface contacts are frequent but often random and may differ markedly between different individuals in a population. Moreover, there is the regular interaction, and the resulting transfer of infectious materials, between animate and inanimate surfaces. For the purposes of this chapter, the term 'inanimate' or 'environmental' surface will include both porous and non-porous objects and fomites. The term 'animate' surface, as used here, will refer to human hands, even though other body surfaces may also act as vehicles for viruses. Although the discussion in this chapter is restricted almost entirely to viral disease and its transmission in humans, the principles of models to describe transmission of viral diseases in animals are directly analogous.

Certain aspects of the information necessary to model virus transmission and its interruption by chemical disinfection have been recently reviewed. These include the chemical disinfection of virus-contaminated surfaces (Springthorpe & Sattar, 1990), survival and inactivation of individual viruses such as HIV (Sattar & Springthorpe, 1991; Sattar *et al.*, 1994*b*) and rotaviruses (Ansari, Springthorpe & Sattar, 1991*b*) and the methodology for testing the virucidal activity of topical antiseptics (Sattar & Springthorpe, 1994). As far as we know, the only comprehensive review on the survival of human pathogenic viruses on fomites was published nearly 15 years ago (England, 1981), and more recent information is given here to augment the data given by England (1981) on virus survival. This chapter, therefore, places particular emphasis on those aspects of the topic not covered in the earlier reviews. The disinfection of medical instruments and other devices is being dealt with in another chapter of this book.

Viruses and the infectious disease burden

The eradication of smallpox and the successful control of many other infectious diseases during the past two decades raised expectations of a general and lasting victory against microbial pathogens. This prompted consideration of an 'epidemiological transformation', in which the emphasis would shift from infectious diseases to degenerative and non-communicable diseases. Whereas the initial sense of optimism about strategies against infectious agents has not proven to be entirely misplaced, it has been tempered with the realization that (1) anticipated breakthroughs in vaccination and chemotherapy against many infectious diseases have not materialized; (2) many common chemotherapeutic agents have been rendered ineffective, due to overuse and abuse; (c) a number of diseases that had been successfully controlled have re-emerged, due to changing demographics and disruptions in disease surveillance systems; and (d) several 'new' infectious agents have emerged as important human pathogens as a result of many factors, including changes in our life styles. Therefore, even today, infectious diseases continue to exert a heavy burden on human health and may continue to do so for the foreseeable future, especially in view of the increasing human population and population densities in various parts of the world.

Globally, infectious diseases are estimated to cause at least one-third of all deaths annually (World Health Organization, 1992). In the United States, they account for at least 740 million nonfatal symptomatic cases and more than 90 000 fatalities every year, making microbial agents the fourth most frequent cause of death (McGinnis & Foege, 1993). Estimates in the United States also show that infections account for 25% of all visits to physicians and that antimicrobials are the second most prescribed class of drugs; the resulting healthcare expenses and other economic losses are believed to cost no less than $120 billion every year (Centers for Disease Control and Prevention, 1994).

Viruses are important among infectious agents affecting humans and, paradoxically, their relative significance as agents of disease has been increasing with the advent of antibiotics and vaccines against many common bacterial diseases. Moreover, viruses figure very prominently in the list of human pathogens discovered in the past 25 years (Lederberg, Shope & Oaks, 1992). Some examples of these newly discovered viruses are the human immunodeficiency viruses (HIV), rotaviruses and enterovirus 70. The impact of HIV on human health is too well-known to be reiterated here. Human rotaviruses, discovered just over 20 years ago, are now recognized as among the most important causes of acute gastroenteritis (Ho et al., 1988; Hoshino & Kapikian, 1994). They are responsible for over 80 million episodes of diarrhea and nearly 900 000 deaths annually in the world (Centers for Disease Control and Prevention, 1993). Each year in the USA, rotaviruses are estimated to result in 3 million cases of acute diarrhea, 500 000 visits to the doctor and 75–125 deaths (Centers for Disease Control, 1992). Enterovirus 70 was first discovered in 1969 and it has spread around the globe at unsurpassed speed, causing two pandemics with an estimated total of 80 million clinical cases of acute hemorrhagic conjunctivitis (Wadia et al., 1983). At the present time, poliomyelitis is the only viral disease that is targeted for eradication, but mumps and rubella are considered potentially eradicable by the International Task Force for Disease Eradication (Centers for Disease Control and Prevention, 1993).

Costs due to influenza, hepatitis B and rotaviral gastroenteritis alone account for more than 15% of the annual losses mentioned above (Centers for Disease Control and Prevention, 1994). Even rhinoviruses, which generally cause relatively mild infections, are estimated to result in substantial economic losses in the USA, due to millions of work- and school-days lost every year (National Health Survey, 1991). It should also be noted here that current estimates of the economic burden due to viral infections do not take into account the damage due to delayed sequelae, as exemplified by the post-polio syndrome (Aston, 1992). Nor do they consider the likely impact of viruses in heart (Friman & Fohlman, 1993) and rheumatic (Naides, 1994) diseases, or other chronic disorders where the etiological role of viruses is less well defined. In addition, enormous economic consequences result from viral disease in all types of animal husbandry, from cattle feed lots to fish farming (Springthorpe & Sattar, 1990).

Information needed to model virus transmission

The first step to initiate infection of a susceptible host is the deposition of the required dose of the etiological agent at its portal of entry. Since virus particles are discharged from an infected host in various body secretions and excretions, transmission of virus infections requires direct or indirect contact with such contamination. Figure 8.1 shows how spread of human pathogenic viruses can occur in nature. Since viruses are obligate parasites, they remain in a state of suspended animation when outside living hosts, and the capacity of a given virus to spread from host to host is determined by, among other things, its ability to retain its infectivity during transit to the susceptible host. Investigations of single-source outbreaks have clearly established the role of water, food and air as vehicles for a number of viruses. On the other hand, very little systematic work has been carried out to assess the relative importance of animate and inanimate surfaces as vehicles for viruses. The nature and frequency of interactions with these vehicles in many settings make it difficult retrospectively to determine their role as virus vehicles in outbreak investigations.

The model for stages involved in transmission of infectious agents including viruses, and for which specific steps or submodels could be designed, is shown in Figure 8.2. Discharge or shedding of virus generally begins prior to the onset of clinical symptoms and lasts for several days after recovery. A further complication for modeling virus disease transmission is the large number of individuals infected with viruses who do not become sick; such 'subclinical cases', which generally remain unrecognized and can be as high as 99.9% of the total individuals infected (Gerba & Rose, 1993), can also discharge viruses into their surroundings. In fact, subclinically infected adults can initiate outbreaks of rotaviral disease in infants and

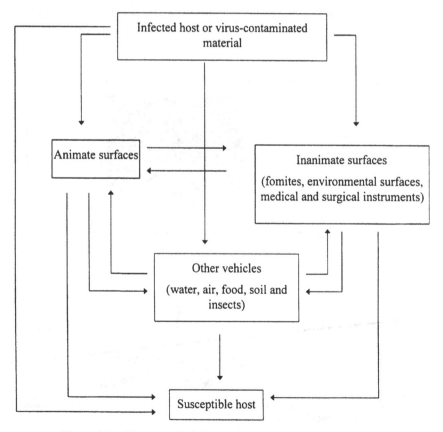

Figure 8.1   The spread of virus infections in nature.

children (Holdaway *et al.*, 1982). Conversely, asymptomatic cases of hepatitis A in children can be the source of virus for symptomatic disease in their adult contacts (Hadler & McFarland, 1986). For prevention of nosocomial viral infections, nursing personnel frequently cite asymptomatic cases as being more of a problem than those where the disease is overt, because the perceived need for infection control is absent. The instances of recognized viral infections therefore represent only the tip of the iceberg in terms of potential for virus transmission.

Virus discharge

During infection, the actual amount of virus discharged varies considerably, depending on the type of infecting agent and the stage of the infection. For example, in the acute phase of the *Rhinovirus* cold, the nasal discharge may contain up to 1.6×10^3 infectious units of virus/ml (Fox *et al.*, 1985), whereas at the peak of rotaviral gastroenteritis, every gram of feces may have 10^7 to 10^8 infectious units of virus (Ward *et al.*, 1984). Obviously, virus discharge cannot be modeled, but the levels of individual viruses in

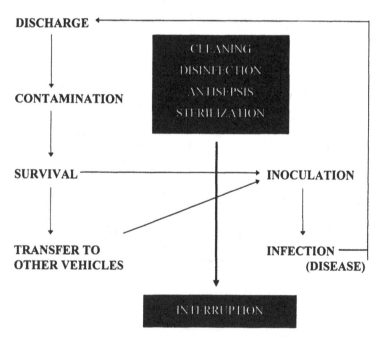

Figure 8.2   Interruption of the spread of infectious agents through the use of chemical germicides.

particular body secretions or excretions can be measured directly and the values obtained can be used in the model.

Direct virus contamination of surfaces and objects

The nature and extent of surface contamination will depend on the site of infection, the degree and nature of the discharge medium from the host, as well as the personal habits of the infected individual and the hygienic facilities available. Hands are among the most obvious surfaces to become contaminated, whether it is from nose-blowing or diaper changing. The degree of contamination can vary widely. For example, some enteric viral infections (e.g. *Rotavirus*) can produce a profuse and almost explosive diarrhea which may be difficult to contain. Addressing such an infection in a toilet-trained child in a household setting could be quite different from a similarly infected child either in a developing country with limited hygienic facilities or in an institution for the mentally retarded, where hygiene awareness of the patients may be very low.

Virus survival on inanimate surfaces

Survival of the virus on the contaminated surface is obviously the key to the potential of that surface to act as a source of the virus. Just how long can viruses survive outside the host? A simple answer to this question is not possible, because virus survival is governed by a variety of factors including environmental or climatic conditions (relative humidity and air movement, temperature, visible and ultraviolet (UV) light exposure), the nature of the matrix (secretion/excretion) in which the virus is suspended and the nature of the surface on which the virus is deposited. Of these, the most important is probably the fluid or semi-solid matrix in which the virus is discharged and which may protect it. Moreover, in nature, viruses may be embedded as aggregates or clumps in cellular material that also protects them (Springthorpe & Sattar, 1990). Elevated temperatures are generally detrimental to virus survival, and colder temperatures favor survival; some viruses, especially those with a lipid envelope, are more readily inactivated by freezing. UV and visible light exposure is also generally detrimental, although its effects may be mitigated by the exact conditions of exposure. Each virus reacts in a particular way to a given relative humidity level (Buckland & Tyrrell, 1962; Sattar *et al.*, 1986; Ansari *et al.*, 1991b), and therefore survival can be affected quite differently in different climatic conditions (Sattar *et al.*, 1986).

Two specific phases during virus survival should be considered. Many viruses may lose much of their infectivity during the drying process (Abad,

Pintò & Bosch, 1994), but others can be dried without significant loss of infectious titer under the right conditions (Sattar *et al.*, 1986; Mbithi *et al.*, 1991). After drying, biological decay continues, but usually at a slower rate. During laboratory studies of modeling it may be important to distinguish these two phases. Virus decay on surfaces is not usually linear (Abad *et al.*, 1994) and, although clumping or different populations of viruses may play a role (Mbithi *et al.*, 1991), the observed kinetics could also be the resultant of two first-order inactivation curves that intersect. Although generalizations can be dangerous, enveloped viruses usually survive better in dryer conditions and lower relative humidities than can be tolerated by non-enveloped viruses (Buckland & Tyrrell, 1962; Sattar & Ijaz, 1987). A few viruses, for example enterovirus 70 (Sattar *et al.*, 1988), may become totally inactivated during drying, which may indicate the importance of some critical water in virus integrity. For this virus, climate is extremely important as the pandemics it causes tend to be restricted mainly to humid or coastal regions (Sattar *et al.*, 1988). Other enteroviruses, like poliovirus, which is often considered as a model virus, are also affected by drying. However, for this virus group, as for others, the presence of organic matter in body secretions and excretions can be significantly protective during the drying process (Springthorpe & Sattar, 1990). Even HIV, contrary to popular belief, can survive for several hours to days once dried on a surface (Sattar & Springthorpe, 1991).

The importance of surface composition and microtopography may be difficult to determine, because survival is usually only assessed by recovery of infectious virus from the contaminated surface, and the data obtained may be confounded with lack of virus recovery. The only way to resolve this issue is to make use of virus which has been labelled with a physical tracer, such as a radioisotope (Sattar *et al.*, 1986). Preparation of such a labelled virus will inevitably require extensive purification of the virus. Therefore, if the virus is to be used in modeling applied to a natural setting, it should be resuspended in the natural organic matter in which it is normally discharged.

Table 8.1 summarizes the available information on the survival of human pathogenic viruses on inanimate surfaces and objects. However, the wide variations in test methodology, amounts of infectious virus in the inoculum and reporting of virus survival data make direct comparisons between results of these studies difficult. Many studies also fail to specify clearly the nature of the virus-suspending medium and the exact conditions under which the experimentally-contaminated surfaces were held. Another important source of variation is the method by which virus is recovered from the contaminated carrier. Most studies also tend to use new and clean surfaces and objects for such experimentation and the extent of virus survival on them may or may not be representative of what happens under field conditions, where most such materials normally carry residues of soil and cleaning/disinfecting chemicals. Because these are experimental studies, it is inevitable that the virus strains used in most are those grown in the laboratory. The extent to which survival of these strains may differ from that of field isolates is unknown. Even with all these limitations, it is quite apparent from the data summarized that virtually all viruses, including those that are generally regarded as fragile, can retain their infectivity for at least a few hours in a dried state on a variety of inanimate materials.

Virus survival on animate surfaces

Table 8.2 is a summary of published information on the ability of human pathogenic viruses to survive on the skin of hands. As is the case with inanimate surfaces, non-enveloped viruses survive better on skin than do enveloped viruses. Also, in general, viruses retain their infectivity for much shorter periods on human skin than on inanimate materials. Data on the

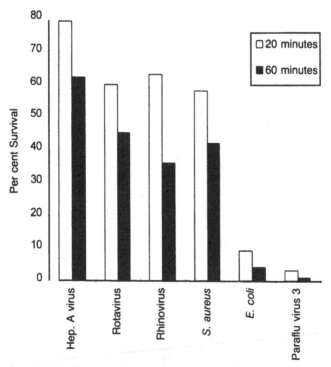

Figure 8.3 Comparative survival of selected viruses and bacteria on human hands. Fingertips of adult volunteers were contaminated with 10 μl of the test organism suspended in the appropriate organic load. The inoculum was allowed to dry for either 20 or 60 min under ambient conditions and eluted. The eluates and controls were then assayed for viable organisms. S. A. Sattar *et al.*, unpublished data.

Table 8.1. *Summary of information from published studies on the survival of human pathogenic viruses on porous and non-porous inanimate materials*

Virus	Suspending medium	Material(s) tested	Survival duration	Environment Relative humidity (%)	Temperature (°C)	Reference
Adenovirus 2	Culture medium	Glass, vinyl, asbestos, ceramic tile, stainless steel	3–8 weeks	3–96	25–37	Mahl & Sadler (1975)
Adenovirus 3	Liquid gum	Gum dried on paper	10 days	Ambient	Ambient	Selwyn (1965)
Adenovirus 5, 8, 19	Culture medium	Plastic and metal disks	35–49 days	Ambient	22	Gordon et al. (1993)
Adenovirus 40	Buffered saline, 20% fecal suspension	Paper, cloth, aluminum, china, glazed tile, latex and polystyrene	2–>15 days	50, 85, 90	4, 20	Abad et al. (1994)
Coxsackievirus B3	Culture medium	Glass, vinyl, asbestos, ceramic tile, stainless steel	2 weeks	3–96	25–37	Mahl & Sadler (1975)
Coxsackievirus B3	Culture medium	Plastic, glass and paper	24 h	Ambient	6–37	McGeady et al. (1979)
Echovirus 1	Liquid gum	Gum dried on paper	30 days	Ambient	Ambient	Selwyn (1965)
Enterovirus 70	Phosphate buffered saline	Stainless steel disks	Irrespective of temperature, virus survived >20 h at ultrahigh relative humidity (RH); it survived for 2–5 h at the lower RH levels	20, 33 or 35	20, 50, 80, or 95	Sattar et al. (1988)
Hepatitis A virus	Feces	Glass vials	30 days	42	25	McCaustland et al. (1982)
Hepatitis A virus	10% fecal suspension in saline	Stainless steel disks	Half-life ranged from >7 days at the low RH and 5 °C to 2 h at ultrahigh RH and 35 °C	25, 55, 80 or 95	5, 20 or 35	Mbithi et al. (1991)
Hepatitis A virus	Buffered saline, 20% fecal	Paper, cloth, aluminum, china, glazed tile, latex and	30–60 days	50, 85, 90	4, 20	Abad et al. (1994)

Virus	Suspending medium	Surface	Survival			Reference
Hepatitis B virus	Dried blood	Stainless steel and cotton	14 days	42	23	Favero et al. (1974)
Hepatitis B virus	Virus-positive plasma	Silanyzed tubes and stainless steel	>1 week	42	25 (in dark)	Bond et al. (1981)
Herpesvirus 1	Culture medium	Glass, vinyl, asbestos, ceramic tile, stainless steel	8 weeks	3–7	25–37	Mahl & Sadler (1975)
		Plastic	4–24 h	55–96	25–37	
Herpesvirus 1	Saliva	Cloth	>4 h >3 h	Ambient	Ambient	Turner et al. (1982)
Herpesvirus 1	Culture medium	Plastic doorknobs and tap handles with a chrome-plated surface	>2 h	Ambient	23–27	Bardell (1990)
Herpesvirus 1 and 2	Vesicular fluid from genital or oral lesions	Cotton gauze Plastic gloves Speculum Plastic	88 h 1 h 18 h 1.5 h	Ambient	Ambient	Larson & Bryson (1985)
Herpesvirus 2	Culture medium	Polystyrene Petri dishes (wet)	4.5 h	Humid	37–40	Nerurkar et al. (1983)
Human immuno-deficiency virus 1	Culture medium	Culture dishes	>7 days	Ambient	20–22	Barré-Sinoussi et al. (1985)
Human immuno-deficiency virus 1	Culture medium with 2.5% serum	Glass slides	Half-life about 32 h	Ambient	20–22	Tjøtta et al. (1991)
Human immuno-deficiency virus 1	Culture medium with 10% serum	Glass coverslips	D-values for cell-associated and cell-free virus were 17.5 and >70.5 h, respectively	Ambient	20–28	van Bueren et al. (1994)
Human immuno-deficiency virus 1	Culture medium with 50% human plasma	Plastic culture plates	D-value about 9 h for cell-free virus	Ambient	23–27	Resnick et al. (1986)
Influenzavirus A	Suspensions of in-fected mouse lung	Glass Cotton	6 weeks 2 weeks	Ambient	Ambient (in dark)	Edward (1941)
Influenzavirus A and B	Culture medium	Stainless steel, plastic, cotton handkerchief and flame-retardant fabric, paper facial tissue, non-glossy magazine paper	24–48 h < 8–12 h	35–56 35–56	26–29 26–29	Bean et al. (1982) Bean et al. (1982)

Table 8.1. *Cont.*

Virus	Suspending medium	Material(s) tested	Survival duration	Environment		Reference
				Relative humidity (%)	Temperature (°C)	
Parainfluenzavirus 3	5 mg/ml bovine mucin in saline	Stainless steel disks	>2 h	Ambient	Ambient	Ansari et al. (1991b)
Parainfluenzavirus 1, 2 and 3	Culture medium with 0.5% gelatin	Plastic Petri dishes	1–12 days	Inside	21	Parkinson et al. (1983)
		Plastic Petri dishes	1–17 days	Outside	−22 to +33	Parkinson et al. (1983)
Parainfluenzavirus 1, 2 and 3	Culture medium	Stainless steel, laminated plastic, hospital gown, facial tissue and lab coat	up to 10 h on nonabsorptive materials; up to 4 h on absorptive materials	Ambient	Ambient	Brady et al. (1990)
Poliovirus 1	10% fecal suspension in saline	Stainless steel disks	>4 h at low RH, 12 h at high RH	25 or 95	5, 20 or 35	Mbithi et al. (1991)
Poliovirus 2	Culture medium	Cotton	1–4 weeks	35	25	Dixon et al. (1966)
		Wool	20 weeks	35	25	
		Cotton	2–10 weeks	78	25	
		Wool	10–12 weeks	78	25	
Poliovirus 2	Culture medium	Glass, vinyl, asbestos, ceramic tile, stainless steel	1 day to 8 weeks	3–96	25–37	Mahl & Sadler (1975)
Poliovirus 1	Buffered saline, 20% fecal suspension	Paper, cloth, aluminum, china, glazed tile, latex and polystyrene	1–5 days	50, 85, 90	4, 20	Abad et al. (1994)
Respiratory syncytial virus	Culture medium and nasal secretions	Formica, rubber, plastic gloves, cloth, paper tissue	30 min–8 h	35–50	22–25	Hall et al. (1980)
Rhinovirus 2	Culture medium and nasal washings	Metal spoons, stainless steel, enamel, ball point pens and plastic	24 h–7 days	Ambient	23	Reed (1975)

Virus	Suspending medium	Surface/material	Survival time	RH (%)	Temperature (°C)	Reference
Rhinovirus 2 and 14	Saline or Tris buffer with and without serum albumin	Polyethylene vials (wet and dry)	>24 h	Ambient	6, 23 and 37	Reagan et al. (1981)
Rhinovirus 14	Tryptose phosphate broth (TPB), bovine mucin at 5 mg/ml in saline or human nasal discharge	Stainless steel disks	Half-life ranged from 0.09 h in nasal discharge at the low RH to 14 h in TPB and the high RH	20, 50 or 80	20	Sattar et al. (1987)
Rhinovirus 39	Culture medium and nasal secretions	Formica, stainless steel, wood, eight different fabrics	3 h	Ambient	23	Hendley et al. (1973)
		Paper tissue	1 h	Ambient	23	Hendley et al. (1973)
		Plastic Petri dish	72 h	Ambient	23	Hendley et al. (1973)
Rotavirus	Feces	Glass slides	12–>35 days	13–92	20	Moe & Shirley (1982)
		Cotton wool	12–45 days	92–94	4–37	
Rotavirus	5% non-fat dry milk	Plastic Petri dish	>6 days	<50	24–28	Ward et al. (1991)
Rotavirus	10% fecal suspension in saline	Metal, plastic or glass cotton–polyester, writing and currency paper or poster card	On nonporous materials, >10 days at low or medium RH and about 2 days at high RH and room temperature; highly variable on porous materials	25, 55 or 85	4 or 22	Sattar et al. (1986)
Rotavirus	Buffered saline, 20% fecal suspension	Paper, cloth, aluminum, china, glazed tile, latex and polystyrene	30–60 days	50, 85, 90	4, 20	Abad et al. (1994)
Vaccinia virus	Culture medium	Glass, vinyl, asbestos, ceramic tile, stainless steel	< 5 weeks	3–96	25–37	Mahl & Sadler (1975)
Vaccinia virus	Culture medium	Cotton	14 weeks	35	25	Sidwell et al. (1966)
		Wool	< 2–4 weeks	35	25	
		Cotton	< 1–2 weeks	78	25	
		Wool	< 2–4 weeks	78	25	

Table 8.2. *Virus survival on human hands*

Virus	Suspending medium	Survival duration	Reference
Hepatitis A virus	10% fecal suspension in saline	>7 h	Mbithi *et al.* (1992)
Herpesvirus 1	Saliva	2 h	Turner *et al.* (1982)
Herpesvirus 1	Culture medium 10% serum	>2 h	Bardell (1989)
Parainfluenzavirus 1, 2 and 3	Culture medium (percentage of serum not specified)	>1 h	Brady *et al.* (1990)
Parainfluenzavirus 3	Bovine mucin at 5 mg/ml in saline	>1 h	Ansari *et al.* (1991c)
Respiratory syncytial virus	Nasal secretions	< 1 h	Hall *et al.* (1980)
Rhinovirus 2	Culture medium with 2% serum	>3 h	Reed (1975)
Rhinovirus 14	Bovine mucin at 5 mg/ml in saline	>4 h	Ansari *et al.* (1991c)
Rotavirus	10% feces in saline	>4 h	Ansari *et al.* (1988)

comparative survival of selected viruses and bacteria of human origin on hands of adult volunteers are presented in Figure 8.3.

We are not aware of any studies that have examined virus survival on other parts of the body, such as the facial or genital areas. How long viruses can survive on mucous membranes is also unknown. Such studies are difficult to conduct, because of the confounding effect of the rapid virus-adsorbing and/or inhibitory properties of mucous surfaces. Whereas intact skin forms an effective barrier against viruses, this may not be the case with intact mucous membranes containing virus-susceptible cells. Such membranes in the eye as well as the respiratory, gastrointestinal and genital tracts represent portals of entry for a variety of viruses.

As far as we are aware, there is no information on what host or external factors affect virus survival on animate surfaces. The fact that viruses generally survive less well on skin when compared to non-porous inanimate surfaces may be due to higher skin temperature, higher levels of moisture, natural enzymatic or other degradative activity on the dermis and antagonism from skin microflora. We do not know if sebaceous skin secretions or the sweat composition or even age of the host have an effect on virus survival. Nor has the question been addressed of whether circulating antibodies against a given virus can influence its survival. Anecdotally, we noticed, during studies of the survival of a human *Rotavirus* (Ansari *et al.*, 1988), that reduced virus survival on the hands of one of the volunteers coincided with excessive perspiration. However, while safety and ethical considerations would exclude children from any such testing, dozens of adult volunteers representing various age groups would be required to properly answer these questions.

Virus transfer between vehicles

Apart from direct contamination, both animate and inanimate surfaces can become contaminated indirectly through transfer of virus-containing material from other vehicles. Of these, deposition by settling of virus-containing aerosols or contact with contaminated recreational waters are obvious. Less obvious are the transfers that may occur between different types of surfaces.

Several studies have shown that clean hands can readily become contaminated when objects or surfaces with infectious virus on them are touched or handled. The reverse has also been shown to be true. Transfer of *Rhinovirus* was observed in 15/16 trials in which a plastic surface, contaminated 1–3 h previously, was touched by a volunteer (Hendley *et al.*, 1973). Individuals with acute *Rhinovirus* colds were shown to deposit infectious *Rhinovirus* particles on objects they touched (Reed, 1975). Infectious *Rhinovirus* particles could be recovered from fingertips of volunteers who handled objects such as doorknobs previously touched by virus-contaminated (donor) hands (Pancic, Carpentier & Came, 1980). *Rhinovirus* transfer has also been shown to occur by direct hand-to-hand contact (Hendley *et al.*, 1973; Gwaltney, Moskalski & Hendley 1978; Pancic *et al.*, 1980).

Transfer of other viruses, such as influenzavirus (Bean *et al.*, 1982) and respiratory syncytial virus (Hall *et al.*, 1980), to hands from both porous and non-porous surfaces has been demonstrated. Polio- and vaccinia viruses could be acquired by sterile fabrics during contact with experimentally contaminated clothes in tumble dryers (Sidwell *et al.*, 1970). Further studies with rhinoviruses (Gwaltney, 1980) and rotaviruses (Ward *et al.*, 1991) also established in human volunteer studies that self-inoculation with fingers contaminated with infectious virus particles transferred from contaminated animate or inanimate surfaces could lead to infection.

The findings of more recent laboratory-based studies on virus transfer between animate–inanimate, animate–animate and inanimate–animate surfaces are summarized in Table 8.3, and the influence of pressure and friction on the amount of infectious virus transferred is shown in Figure 8.4. These data demonstrate that (1) virus transfer readily occurs irrespective of the nature of donor and recipient surfaces, (2) increase in the age of the inoculum reduces the relative amounts of virus transferred; this is most probably due to the increased loss of moisture, (3) the amount of virus transferred is directly proportional to the amount of pressure applied dur-

Table 8.3. Rotavirus *transfer through casual contact between surfaces*

Time between virus contamination and transfer (min)	Number of volunteers	% infectious virus transferred (SD)		
		Finger to metal disk	Metal disk to finger	Finger to finger
20	5	16.1 ± 5.43	16.8 ± 5.17	6.6 ± 2.61
60	6	1.8 ± 1.50	1.6 ± 1.81	2.8 ± 2.11

Three fingers of each volunteer were tested in each experiment. Ten µl of fecally suspended *Rotavirus* (2×10^4 to 8×10^4 plaque forming units) was placed on each donor surface. A 5-s contact with pressure ($1 \ kg/cm^2$) was made between the donor and target surface either 20 or 60 min after virus contamination. The extent of virus transfer was determined by recovery of infectious virus from the target surface. *Source:* Ansari *et al.* (1989).

ing contact, and (4) as would be expected, application of friction during contact substantially increases the quantity of infectious virus transferred.

The minimal infective dose

Inoculation can occur with either the primary surface on which virus was deposited and has survived, or a secondary one to which the virus was transferred. If the level of virus inoculated exceeds the minimal infective dose for that agent and portal of entry, then infection could result. The minimal dose of a virus required to initiate an infection is specific to the host and its portal of entry as well as to the virus. Host age and immune status also play a critical role.

Virus inoculation from contaminated animate or inanimate surfaces is to a large extent dependent on the habits and hygiene of individuals. When hands are involved, self-inoculation by touching of eyes and nose may occur unconsciously. Contaminated hands of caregivers also represent a significant risk to patient infection in a healthcare setting. Sucking of objects and surfaces is common in infants and younger children (Black *et al.*, 1981) and may be particularly important when many young children are housed together in settings such as day-care centers. Virus-contaminated toys could be a significant factor in transmission of viral infections in such settings (Hutto *et al.*, 1986).

Theoretically, the minimal infective dose (MID) is a single infectious virus particle. The MID for most viruses for which it has been measured is relatively small (Gerba & Rose, 1993) and this suggests that even in cases of very low levels of contamination, such as may occur after secondary or even tertiary transfer to other surfaces, there may be sufficient amounts of infectious virus to represent a potential threat to health. For animal diseases, obtaining information about the MID may be relatively easy. However, for human viral infections it is usually considered unethical to conduct such studies. Therefore, in modeling human viral disease certain assumptions may have to be made about this stage in the infectious process.

Virus specificity and levels of modeling

One thing to emphasize in the model shown in Figure 8.2 is that some of the steps are controlled more by the host, whereas others are more virus-specific and sometimes both host and virus influence the outcome. In particular, two steps can be cited for which virus characteristics are extremely important. The first of these is the survival stage, which, as mentioned above, is also influenced by host secretions/excretions and environmental variables. The second is the MID, in which host factors are clearly equally or more important in the disease process, if not in the initial infection. In the other steps shown in Figure 8.2, the nature of the virus may be less important than the nature of the material in which it is suspended or other factors discussed above.

Two distinct levels for modeling virus transmission can be conceived.

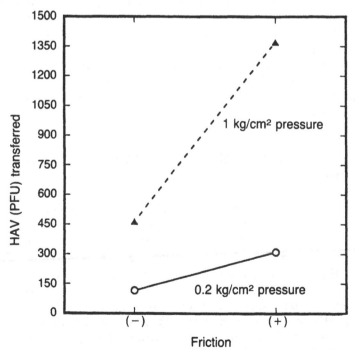

Figure 8.4   Influence of pressure and friction in the transfer of hepatitis A virus from contaminated to clean surfaces. From Mbithi *et al.* (1992).

The first of these is a simple model that can be elaborated in the laboratory under tightly controlled conditions and which can examine under a variety of appropriate environmental settings the transfer of the infectious virions from infected host to a susceptible host via a variety of contaminated vehicles. The information obtained from this first type of 'simulation' model can permit scientists to make 'best guesses' about the prevailing conditions which are likely to favor transmission of a particular virus, but gives no information of how the virus may spread naturally within a community. Simulation models rely on measured environmental parameters and known numbers of infectious viruses. While such models can be quite effective for animal diseases, the ethics of human experimentation usually limits their completeness to individual or groups of steps in the overall transmission process.

The second more complex 'natural community' model uses data obtained in simulation models in conjunction with observed or estimated behavior, in order to evaluate how viruses may spread within a community. In addition, the day-to-day lives of the population infected or exposed to the viral pathogen is of vital importance; a model that accurately describes community transmission in one setting may not be directly applicable to another environment of different socioeconomic and/or geographic climate. We are only just beginning to be able to tackle these more complex community models, and little relevant information has been obtained so far. Attempts to model interruption of transmission by germicide-impregnated tissues (Farr et al., 1988; Longini & Monto, 1988) met with little success, because placebo tissues seemed to be nearly as effective. It is likely that the investigators failed to consider the biocidal or barrier effects of regular tissue. We cite this to show that failure to understand the model system sufficiently may prevent consideration of all the relevant factors.

Inanimate and animate surfaces in the spread of viral infections

The information given above clearly shows that both animate and inanimate surfaces have potential involvement in the transmission of infections, but have such surfaces or their interactions been implicated on the basis of the evidence available in the scientific and medical literature? Unfortunately, the available evidence for the spread of viral infections through animate and inanimate surfaces is mostly indirect and circumstantial. As has been argued before, direct evidence in this regard is extremely difficult to generate, due to a large number of uncontrollable variables. However, a careful review of experimental and circumstantial evidence strongly suggests that animate and inanimate surfaces play a role as vehicles for the spread of a variety of viral infections.

Recovery of infectious viruses from naturally contaminated inanimate and animate surfaces

As can be seen from Table 8.4, recovery of infectious human pathogenic viruses from naturally contaminated surfaces has been reported in a limited number of studies. In most settings, such attempts at virus recovery are difficult to justify, because of the cost, questions of how, how often and what materials to sample, and the perceived lack of significance of the findings to preventing and controlling the spread of infections. Moreover, field strains of many viruses are refractory to culture and quantitation in the laboratory. However, with the mounting evidence for the role of environmental surfaces as vehicles for viruses, attempts are being made to apply sensitive and highly specific molecular biological techniques for detecting genomes of specific types of viruses in institutional settings. This is exemplified by the recent investigations of Butz *et al.* (1993) in which the polymerase chain reaction (PCR) technique was used to study the prevalence of rotaviral contamination of environmental surfaces in a day-care center. Not only did they find evidence for rotaviral contamination on a variety of surfaces, but the number of positive samples was higher during periods of outbreaks of gastroenteritis. Whereas the presence of rotaviral genome as detected by PCR cannot be taken to correspond with

Table 8.4. *Infectious human pathogenic viruses recovered from naturally contaminated surfaces*

Virus	Location and surfaces	Reference
Adenovirus	Drinking glasses in coffee shops and bars	Candeias *et al.* (1969)
Cytomegalovirus	Inanimate surfaces and hands in a day-care center	Hutto *et al.* (1986)
Echovirus	Drinking glasses in coffee shops and bars	Candeias *et al.* (1969)
Enterovirus 70	Fomites in a refugee camp	Arnow *et al.* (1977)
Herpesvirus	Drinking glasses in coffee shops and bars	Candeias *et al.* (1969)
Human immuno-deficiency virus	Contact lenses	Tervo *et al.* (1987)
Rhinovirus	Inanimate surfaces in homes of rhinovirus-positive patients	Gwaltney *et al.* (1978)
Rhinovirus	Hands of infected volunteers	Hendley *et al.* (1973)
Rhinovirus	Objects handled by infected volunteers	Reed (1975)
Rotavirus	Inanimate surfaces in a day-care center	Keswick *et al.* (1983)

the presence of infectious virus, it does make it possible to identify 'hot spots' for contamination in a given setting and perhaps the development of a more logical and selective regimen for routine environmental cleaning and decontamination, at least in institutional settings. Rotaviral antigens have also been recovered from the hands of caregivers (Samadi, Huq & Ahmed, 1983).

Is there evidence for the role of animate and/or inanimate surfaces as vehicles for viruses?

The answer to this question must be a qualified 'yes', because, as stated above, direct evidence in this regard is very difficult to generate. Nevertheless, the following types of indirect evidence strongly implicate hands and inanimate surfaces as vehicles for a variety of enteric as well as respiratory viruses:

(1) Hands and inanimate surfaces frequently become contaminated with viruses, and infectious viruses have been recovered from naturally contaminated hands as well as fomites (Table 8.4).
(2) Hands can readily acquire infectious virus particles through casual contact with other contaminated surfaces, and vice versa (Table 8.3).
(3) Many types of viruses can survive from several minutes to several hours on skin (Figure 8.3) and often for much longer periods on inanimate surfaces (Table 8.1)
(4) Hands frequently come in contact with portals of virus entry such as the eyes and the nose. Inanimate surfaces such as eating utensils are regularly placed in the mouth, and writing implements are often sucked or chewed absentmindedly. Infants and young children have been observed to place things in their mouths every 2–3 min (Black et al., 1981).
(5) In experimental settings, treatment of virus-contaminated hands with germicidal chemicals has been found to interrupt virus transmission through self-inoculation (Hayden et al., 1985a,b). In animal husbandry thorough cleaning and disinfection of pens has been shown to interrupt the rotaviral disease cycle in weanling pigs (Lecce, King & Dorsey, 1978).
(6) Frequent and thorough washing of hands by caregivers and food handlers in field settings has been found to help in preventing and controlling the spread of viral and other infections (Black et al., 1981; Cliver & Kostenbader, 1984).
(7) There has been an anecdotal suggestion that, in day-care facilities, using two parallel sets of toys that are alternated, with one set undergoing cleaning and disinfection while the other is in use, has

reduced the incidence of viral infections. This is supported by the recovery of viruses from toys (Hutto *et al.*, 1986).

Does transfer of infectious viruses occur between animate and inanimate surfaces in the field and what is the significance of such virus transfer for infection control?

Although molecular techniques are now available which would allow virus transfer to be modeled with a reasonable degree of certainty, we know of very little information in this regard. It is salient, however, that recovery of virus antigen, or even infectious virus, has been reported from surfaces that would not be expected to be directly contaminated, but are frequently touched by hands – refrigerator door handles, for example (Keswick *et al.*, 1983). Also, it has been reported (Samadi *et al.*, 1983) that virus antigen can be demonstrated on the hands of personnel who had no direct contact with cases of the infection for which the antigen was recovered. Although inconclusive, this does suggest possible acquisition from virus-contaminated surfaces and fomites in the immediate surroundings. Virus antigens have also been found on laboratory surfaces and equipment (Lauer *et al.*, 1979) and on the computer cards which accompanied specimen containers to the laboratory (Pattison *et al.*, 1974); since this came to light because of cases of hepatitis B amongst the personnel working with such materials, their hands may well have transferred the virus.

Although it was not technically a field study, it is worth drawing attention here to some work done to show that, during normal hand washing, infectious virus can be transferred from a contaminated to a clean hand as well as being spread over a larger area of the contaminated hand (S. A. Sattar *et al.*, unpublished data).

Infectious virus transfer between contaminated and clean surfaces, therefore, can increase the risk of virus transmission by spreading localized contamination over a wider area and to different surfaces. Even when only a small proportion of the virus is transferred (Table 8.3), the numbers of infectious virions originally present may be sufficient for a few virus types to ensure that transmission of the infection may be possible. This has direct implications for infection control. In institutions such as hospitals and day-care centers due emphasis is placed on hand washing, but in many instances properly washed hands may become immediately recontaminated by contact with contaminated inanimate surfaces and objects in the vicinity. This requires an increased degree of awareness on the part of caregivers and a realization that minimizing the contamination of frequently contacted inanimate surfaces in a given setting should be a necessary adjunct to emphasis on hand washing. If one accepts the model (Figure 8.2), it becomes clear that hand washing alone will not necessarily be adequate without a similar intervention measure applied to inanimate surfaces, especially in healthcare settings.

Interruption of virus transmission

In theory, at least, interruption of virus transmission by cleaning, disinfection or sterilization should be relatively easy to simulate in laboratory controlled models. This is indeed so, but the actual similarity of any of these simulations to practical in-use conditions is clearly the key to whether such models have relevance. Disinfection of virus-contaminated surfaces has been reviewed in detail (Springthorpe & Sattar, 1990) and it is clear that many factors impact on its efficacy.

Inanimate surfaces

Apart from the type of virus and the nature and concentration of chemical disinfectant, the key factor is the organic and inorganic matrix in which the virus is shed from the infected host. In fact, this last factor, as for virus survival, may be the most important of all, as (1) disinfectants may be unable to make contact with viruses that are fully embedded in an organic matrix, and (2) disinfectants are relatively non-specific in their chemical reactivity; they frequently react with the organic matrix and may, at least partially, be neutralized by it. In addition, the nature of the surface to be disinfected may play a role, because (1) its microtopography could occlude viruses and prevent disinfectant contact, (2) it may not have a readily wettable surface and therefore disinfectants may not make contact with viruses, or (3) it may be coated with a residue that interacts with and partially neutralizes the disinfectant. This last factor is important, because all environmental surfaces and objects tend to build up residues during their routine use and cleaning, and specific products used for cleaning surfaces prior to their disinfection may be incompatible with some disinfectants used for general disinfection, particularly certain phenolics or quaternary ammonium compounds.

The difficulties with organic material and surface build-up can be overcome through proper pre-cleaning of the surfaces. However, in addition to ensuring that the cleaner and disinfectant are compatible, this presents an operational dilemma, because pre-cleaning may not always be safe, and in many instances it is an inefficient use of time to treat surfaces twice. Moreover, pre-cleaning may range anywhere from a perfunctory to a thorough treatment. Therefore, if one is modeling disinfection of virus-contaminated surfaces from a prudent viewpoint, surfaces are not cleaned and the virus is present either in its natural matrix or a simulation thereof.

On the other hand, a more idealistic approach would use clean and often new surfaces and purified test virus. The real field situation probably lies somewhere in between these two extremes.

For the many laboratory-based studies in the literature on the disinfection of virus contaminated surfaces (Lloyd-Evans, 1985; Mbithi Springthorpe & Sattar, 1990), as is the case for publications on virus survival, the variations between studies in terms of the virus preparations and suspending media, the type of surface, the method of drying, the length of contact with the disinfectant, the method of recovering surviving viruses and the reporting of the data preclude a meaningful comparison between the studies and make the data they present almost useless for plugging into a model for a different setting and different conditions. Many such studies also fail to incorporate important controls, thus making interpretation of the data difficult. Therefore, a reader of work in this field should be prepared to examine the methodology used and the data obtained from a critical viewpoint (Springthorpe & Sattar, 1990; Sattar & Springthorpe, 1991). One further point worth mentioning, especially for disinfectant chemicals which have fixative properties, is that measured kill can be confounded with lack of recovery.

Animate surfaces

Whereas topical antiseptics are used for a variety of purposes, here we will focus on those formulations and chemicals that are meant primarily for 'hygienic hand washing', a term that describes the washing of hands, particularly by healthcare personnel and food handlers, in order to eliminate transient microflora from intact skin. In our view, this term is synonymous with 'healthcare personnel handwash' (Bruch, 1991).

In spite of the recognized importance of hands as vehicles for viruses and the role of hand washing in reducing such spread, there is a general lack of reliable information on the virus-eliminating ability of commonly used hand washing agents. In our view, formulations with antiseptic claims should be able to eliminate significantly more viruses than can be achieved mechanically by hand washing with water or with plain soap and water. Since soaps can vary considerably in pH and other factors, and tap water varies geographically, comparison with a standard hard water may be appropriate. Limited studies have shown that hand washing agents vary greatly in their ability to rid experimentally contaminated hands of viruses, and that, in general, they do not eliminate viruses as well as they remove bacteria (Ansari *et al.*, 1989). In fact, it has been shown that the use of an ineffective product could lead to the spread of localized viral contamination over a wider area of both hands during hand washing (S. A. Sattar *et al.*, unpublished data). Our studies have also shown that paper towels, cloth towels and warm air dryers vary in their ability to reduce the amount of residual infectious virus on washed hands (Ansari *et al.*, 1991*a*).

Requirements for effective interruption of virus transfer

Effective decontamination of surfaces, whether animate or inanimate, requires a proper procedure, diligently applied, with an effective product. It is often argued that all disinfectants are effective and that they are not applied with due diligence. However, we feel that even perfect procedures and diligence will not compensate for a product that is ineffective for the intended task. Indeed, properly designed studies on disinfection show distinct differences between different types and concentrations of disinfectants, particularly for viruses (Lloyd-Evans, Springthorpe & Sattar, 1986; Springthorpe *et al.*, 1986). A 'community' model, in which disinfection is incorporated as an intervention, cannot distinguish between the components of an effective inactivation procedure (e.g. pre-, and post-treatment rinsing in tap water, application of the hand-wash product itself and the drying of washed hands), unless prior and adequate simulation models had shown the effectiveness of the product for the disinfection required. If one extrapolates that to a field situation, then proper testing of disinfection in tightly controlled laboratory settings should be a prerequisite for any field trials of disinfection as an effective intervention measure.

Arguments could be advanced that tests which disinfectants and antiseptics are required to undergo prior to registration are equivalent to such simulation models. However, there are no generally accepted tests for products claiming virucidal efficacy on inanimate or animate surfaces, and those that are available generally examine virus disinfection under relatively ideal conditions (Sattar & Springthorpe, 1988; Springthorpe & Sattar, 1990). There are certain practical difficulties in designing a universally applicable organic challenge to use in simulating natural virus disinfection and a consensus on this issue has not yet been achieved (Sattar & Springthorpe, 1994).

For modeling purposes we can consider three types of surfaces: (1) inanimate non-porous surfaces, (2) inanimate porous surfaces, and (3) skin and mucous membranes. Inanimate non-porous surfaces can also be divided into three categories, namely, environmental surfaces that must be treated in place or are too large to soak, fomites that can be immersed for disinfection, and medical instruments that are heat-sensitive and must be subjected to high-level disinfection or chemisterilization. Practical considerations of how such surfaces are disinfected in the field must be built into disinfection models (Lloyd-Evans *et al.*, 1986; Mbithi *et al.*, 1993*b*; Best, Springthorpe & Sattar, 1994), and it is important to note that such considerations are not always compatible with the manufacturers' directions for use on the product label. This should be taken into account during disinfectant selection. Inanimate porous surfaces are mostly unsuitable for chemical disinfection. However, these surfaces have sometimes been impregnated with disinfectants to prevent their contamination with infectious agents, especially in healthcare settings. Virus-contaminated hands and mucous membranes present much more of a challenge for modeling. While some *in vivo* models are available for hands (Ansari *et al.*, 1988; Sattar & Springthorpe, 1994), for certain viruses that present hazards for *in vivo* testing on hands and for testing of the efficacy of products such as vaginal gels on mucous membranes, development of an *ex vivo* model would be useful.

Because of the limitations mentioned above, extrapolations from the testing of viruses in suspension have been made to both inanimate and animate surfaces (Sattar & Springthorpe, 1991). Such extrapolations are not valid scientifically, as there is evidence to show that viruses dried onto surfaces are more resistant than viruses suspended in an equivalent medium (Lloyd-Evans *et al.*, 1986; Springthorpe & Sattar, 1990). Similar extrapolation from inanimate to animate surface disinfection is also invalid (Schürrman & Eggers, 1983). Moreover, it must be emphasized that no laboratory test can be extrapolated to field efficacy, but what that test does measure is the *potential* of the product to be effective in the field.

From the available data on virucidal potency of general-purpose disinfectants and antiseptics it is clear that many of them have severe limitations (Sattar & Springthorpe, 1988). It is also clear that the kinetics of viral inactivation support the conclusion that virus inactivation may not occur at all if it is not relatively rapid (Lloyd-Evans *et al.*, 1986). Nor is there any convincing information to show that germicides in hand washing agents have residual activity against viruses. In view of this, germicidal chemicals added to vaginal gels, for example, will need to be fast-acting if they are to be effective in interrupting the venereal spread of viruses such as HIV.

Can chemical disinfectants interrupt the spread of viral infections?

Providing a direct answer to this question is difficult at this stage. Data from laboratory-based studies and a very limited number of field trials suggest a role for chemical germicides in this regard. The model (Figure 8.2) shows the stages that are amenable to interruption by cleaning and chemical disinfection. The findings of several studies show that virucidal chemicals can interrupt the spread of *Rhinovirus* colds (Hendley, Mika & Gwaltney, 1978; Hendley & Gwaltney, 1988). In the early experiments, many chemicals were found to inactivate rhinoviruses on human hands, and then Gwaltney (1980) demonstrated that aqueous solutions of iodine could successfully block the hand transmission of rhinoviral colds. Subsequent attempts at using chemically impregnated paper handkerchiefs to interrupt the spread of rhinoviral infections also showed promise in laboratory-based studies (Dick et al., 1986). However, thus far, field trials of such handkerchiefs have yielded equivocal results (Dick et al., 1980; Farr et al., 1988; Longini & Monto, 1988).

In a unique set of experiments, Ward et al. (1991) assessed the ability of a disinfectant spray to interrupt the spread of a rotavirus from experimentally contaminated inanimate surfaces to human volunteers. All the subjects (8/8) who licked the surface with the dried virus inoculum on it became infected, and 5/8 showed evidence of infection when they first touched the contaminated inanimate surface with a finger and then licked the finger. Prior treatment of the contaminated inanimate surface with the disinfectant spray completely blocked the spread of the virus, since none of the 14 subjects became infected.

Our own studies with rhino- and rotaviruses have demonstrated that virus transfer from experimentally contaminated inanimate surfaces to clean fingers can be interrupted through the use of suitable environmental surface disinfectants. Adult volunteers were required to touch the dried virus inoculum on stainless steel disks with the fingerpads for 10 s with a pressure of 1 kg/cm^2. The results of these experiments are summarized in Table 8.5. The disinfectant spray and bleach (800 parts per million available chlorine) could successfully interrupt virus transfer, but the amount of virus transferred after treatment of the disks with the quaternary ammonium-based and phenol-based compounds was higher than that from the contaminated but untreated disks. This points to the care needed in the selection of products for inanimate surface disinfection.

Table 8.5. *Interruption of virus transfer from metal disks to fingerpads*

	% Virus transferred		
Treatment	Human *Rotavirus*[a]	*Rhinovirus* 14[b]	*Poliovirus* 1 (Sabin)[c]
Top water	5.6 ± 1.1	Not done	3.5 ± 1.5
A spray product with 0.1% o-phenylphenol and 79% ethanol	0	0	0
Bleach (800 ppm)	0	0	3.6 ± 2.4
Quaternary ammonium (550 ppm)	7.6 ± 2.5	8.4 ± 3.6	5.3 ± 2.4
Phenolic (575 ppm)	0	3.3 ± 1.9	11.6 ± 6.6

Note: ppm, parts per million.
[a] Adapted from Sattar et al. (1994a)
[b] Adapted from Sattar et al. (1993).
[c] Sattar et al. (unpublished data).

Viral infections from animate and inanimate sources 249

Table 8.6. *Transfer of two types of viruses to metal disks from washed fingerpads*

Washing medium	Active ingredients	Hepatitis A	Poliovirus
		% PFU transferred (mean ± SD)	
Alcare (undiluted)	62% (v/v) emolliented ethanol foam	Undetectable	Undetectable
Aquaress (undiluted)	None	1.57 ± 0.50	0.96 ± 0.84
Bacti-Stat soap (undiluted)	0.3% Triclosan	0.63 ± 0.13	0.62 ± 0.62
Bioprep soap (undiluted)	0.1% chlorhexidine gluconate + didecyl dimethyl ammonium chloride + 5% isopropanol	0.34 ± 0.12	0.26 ± 0.06
Dettol (undiluted)	4.8% 4-chloro-3,5-xylenol + 9.4% isopropanol	0.5 ± 0.20	0.37 ± 0.09
Ethanol	70% (v/v) ethanol	Undetectable	Undetectable
Savlon (1:30 in 70% ethanol)	1.5% chlorhexidine gluconate, 15% cetrimide	Undetectable	Undetectable
Scrub Stat IV (undiluted)	4% chlorhexidine gluconate + 4% isopropanol	0.64 ± 0.02	0.23 ± 0.20
Septisol (1:2)	0.75% hexachlorophene	0.44 ± 0.09	0.41 ± 0.10
Tap water	Approximately 0.5 ppm free chlorine	3.88 ± 0.63	3.15 ± 1.32
Triclosan hand soap (undiluted)	0.5% Triclosan	1.75 ± 0.35	1.35 ± 0.02

Note: Virus transfer was determined by comparison with the amount of infectious virus left on a fingerpad after washing with the test agent. PFU, plaque-forming units.
Source: adapted from Mbithi et al. (1993a).

Hand-washing agents also differ in their ability to interrupt the transfer of viruses from experimentally contaminated hands to clean inanimate surfaces. Results of such experiments with hepatitis A virus and poliovirus type 1 are summarized in Table 8.6.

Even though the information summarized here suggests a role for disinfectants and antiseptics in interrupting virus transmission, and even if the role of inanimate surfaces as well as hands becomes more accepted, it is clear that the role of surfaces must be seen in the context of the overall transmission of a particular virus that may be primarily transmitted through other vehicles. Therefore, the impact of disinfectants in interrupting surface transmission may have little or no measurable impact on disease incidence. However, where it does become of importance is in special care facilities in hospitals where patients may be immunocompromised or otherwise debilitated. In these instances it may be particularly important to take every possible precaution for patient protection. The modeling required to provide this may include the potential of surfaces to transmit particular viruses and the potency of any particular chemicals to interrupt virus transfer.

Disinfectant use is, however, a double-edged sword; for use in a community setting the benefits should clearly outweigh the hazards due to human and environmental toxicity. With many of our present disinfectant formulations, the chemical hazards of keeping them in households with young children may be greater from accidental poisonings, etc., than can be justified from their ability to prevent infections in the home.

In view of this and other issues such as disinfectant cost, the more relevant question that should be asked from a modeling perspective is what other barriers can be put in place? Is cleaning just as good as chemical disinfection?

Summary

Viruses are important human pathogens and even those viruses that cause relatively mild diseases are responsible for a significant burden on the economy and the healthcare system. Human pathogenic viruses are not a part of the normal resident microflora of the human skin, but many types of pathogenic viruses, acquired as transient microflora, can survive for several hours on human hands and infectious virus particles have been recovered from naturally contaminated hands of caregivers. Transfer of infectious virus can readily occur to and from hands upon casual contact with objects or other animate surfaces. Hands can, therefore, spread many types of viruses, and proper and regular hand washing, particularly by caregivers and food handlers, has been found to be helpful in preventing and controlling such spread.

The question of the role of environmental surfaces in the spread of viral infections is only now beginning to be addressed. It should, however, be noted that evidence from experimental as well as epidemiological studies indicates a role for such surfaces in the spread of infections caused by hepatitis A virus, rotaviruses, rhinoviruses, adenoviruses and respiratory syncytial virus.

Considerable effort is still needed to develop safer, faster-acting and broad-spectrum chemical germicide formulations. Improvements and standardization is also needed in our methodology for testing the virucidal activity of such formulations and in the procedures for recovering infectious viruses from animate and inanimate surfaces. Since the detection of infectious viruses on animate and inanimate surfaces in the field will remain limited, due to cost and other constraints mentioned above, we need to

develop molecular biological techniques the results of which should indicate whether the signals detected are those of viable or non-viable viruses.
Eventually, we need to mount properly planned and long-term studies to
address the question of the role of chemical germicides in interrupting the
spread of viral and other infections.

Acknowledgments

We wish to thank our graduate students Shamim Ansari, Nellie Lloyd-
Evans and John Mbithi for generating much of our data summarized in
this chapter.

References

Abad, F. X., Pintò, R. M. & Bosch, A. (1994). Survival of enteric viruses on
environmental fomites. *Applied and Environmental Microbiology*, **60**, 3704–10.
Ansari, S. A., Sattar, S. A., Springthorpe, V. S., Wells, G. A. & Tostowaryk, W.
(1988). Rotavirus survival on human hands and transfer of infectious virus to
animate and non-porous inanimate surfaces. *Journal of Clinical Microbiology*,
26, 1513–18.
Ansari, S. A., Sattar, S. A., Springthorpe, V. S., Wells, G. A. & Tostowaryk, W.
(1989). *In vivo* protocol for testing efficacy of hand washing agents against viruses
and bacteria: experiments with rotavirus and *Escherichia coli*. *Applied and
Environmental Microbiology*, **55**, 3113–18.
Ansari, S. A., Springthorpe, V. S. & Sattar, S. A. (1991a). Comparison of cloth-,
paper- and warm air-drying in eliminating viruses and bacteria from washed
hands. *American Journal of Infection Control*, **19**, 243–9.
Ansari, S. A., Springthorpe, V. S. & Sattar, S. A. (1991b). Survival and vehicular
spread of human rotaviruses: possible relationship with seasonality of outbreaks.
Reviews of Infectious Diseases, **13**, 448–61.
Ansari, S. A., Springthorpe, V. S., Sattar, S. A., Rivard, S. & Rahman, M. (1991c).
Potential role of hands in the spread of respiratory viral infections: studies with
human parainfluenzavirus 3 and rhinovirus 14. *Journal of Clinical Microbiology*,
29, 2115–19.
Arnow, P. M., Hierholzer, J. C., Higbee, J. & Harris, D. H. (1977). Acute
hemorrhagic conjunctivitis: a mixed virus outbreak among Vietnamese refugees
on Guam. *American Journal of Epidemiology*, **105**, 68–74.
Aston, J. W. (1992). Post-polio syndrome. An emerging threat to polio survivors.
Postgraduate Medicine, **92**, 249–60.
Bardell, D. (1989). Survival of herpes simplex virus type 1 on some frequently
touched objects in the home and public buildings. *Microbios*, **59**, 93–100.
Bardell, D. (1990). Hand-to-hand transmission of herpes simplex virus. *Microbios*,
63, 145–50.
Barré-Sinoussi, F., Nugeyre, M. T. & Chermann, J. C. (1985). Resistance of AIDS
virus at room temperature. *Lancet*, **ii**, 721–2.

Bean, B., Moore, B. M., Sterner, B., Peterson, L. R., Gerding, D. N. & Balfour, H. H. (1982). Survival of influenza viruses on environmental surfaces. *Journal of Infectious Diseases*, **146**, 47–51.

Best, M., Springthorpe, V. S. & Sattar, S. A. (1994). Feasibility of a combined carrier test for disinfectants: studies with a mixture of five types of microorganisms. *American Journal of Infection Control*, **22**, 152–62.

Black, R. E., Dykes, A. C., Anderson, K. E., Wells, J. G., Sinclair, S. P., Gary, G. W., Hatch, M. W. & Gangarosa, E. J. (1981). Handwashing to prevent diarrhea in day-care centers. *American Journal of Epidemiology*, **113**, 445–51.

Bond, W. W., Favero, M. S., Petersen, N. J., Gravelle, C. R., Ebert, J. W. & Maynard, J. E. (1981). Survival of hepatitis B virus after drying and storage for one week. *Lancet*, **i**, 550–1.

Brady, M. T., Evans, J. & Cuartas, J. (1990). Survival and disinfection of parainfluenza viruses on environmental surfaces. *American Journal of Infection Control*, **18**, 18–23.

Bruch, M. (1991). Methods of testing antiseptics: antimicrobials used topically in humans and procedures for hand scrubs. In *Disinfection, Sterilization and Preservation*, 4th edn, ed. S. S. Block, pp. 1028–46. Philadelphia: Lea Febiger.

Buckland, F. E. & Tyrrell, D. A. J. (1962). Loss of infectivity on drying various viruses. *Nature*, **195**, 1062–4.

Butz, A. M., Fosarelli, P., Dick, J., Cusack, T. & Yolken, R. (1993). Prevalence of rotavirus on high-risk fomites in day-care facilities. *Pediatrics*, **92**, 202–5.

Candeias, J. A. N., De Almeida Christovao, D. & Cotillo, Z. L. G. (1969). Isolation of virus from drinking glasses in coffee shops and bars in the city of Sao Paulo. *Abstracts of Hygiene*, **44**, 930.

Centers for Disease Control and Prevention (1992). *Proceedings of the International Conference on Child Day Care Health: Science, Prevention and Practice*. Atlanta, Georgia: CDC.

Centers for Disease Control and Prevention (1993). Recommendations of the International Task Force for Disease Eradication. *Morbidity Mortality Weekly Reports*, **42**, 1–38.

Centers for Disease Control and Prevention (1994). Addressing infectious disease threats: a prevention strategy for the United States. *Morbidity Mortality Weekly Reports*, **43**, 1–18.

Cliver, D. O. & Kostenbader, K. D. (1984). Disinfection of virus contaminated hands for prevention of foodborne disease. *Journal of Food Microbiology*, **1**, 75–87.

Dick, E. C., Hossain, S. U., Mink, K. A., Meschievitz, C. K., Schultz, S. B., Raynor, W. J. & Inhorn, S. L. (1986). Interruption of transmission of rhinovirus colds among human volunteers using virucidal paper handkerchiefs. *Journal of Infectious Diseases*, **153**, 352–6.

Dick, E. C., Jennings, L. C., Meschievitz, C. K., MacMillan, D. & Goodrum, J. (1980). Possible modification of normal winter fly-in respiratory disease outbreaks at McMurdo Station. *Antarctic Journal of the United States*, **15**, 173–4.

Dixon, G. J., Sidwell, R. W. & McNeil, E. (1966). Quantitative studies on fabrics as disseminators of viruses. II. Persistence of poliomyelitis virus on cotton and wool fabrics. *Applied Microbiology*, **14**, 183–8.

Edward, D. G. (1941). Resistance of influenza virus to drying and its demonstration on dust. *Lancet*, **241**, 664–6.

England, B. L. (1981). Detection of viruses on fomites. In *Methods in Environmental Virology*, ed. C. P. Gerba & S. M. Goyal, pp. 179–220. New York: Marcel Dekker.

Farr, B. M., Hendley, J. O., Kaiser, D. L. & Gwaltney, J. M. (1988). Two randomized trials of virucidal nasal tissues in the prevention of natural upper respiratory infections. *American Journal of Epidemiology*, **128**, 1162–72.

Favero, M. S., Bond, W. W., Petersen, N. J., Berquist, R. K. & Maynard, E. J. (1974). Detection methods for study of the stability of hepatitis B antigen on surfaces. *Journal of Infectious Diseases*, **129**, 210–12.

Fox, J. P., Cooney, M. K., Hall, C. & Foy, H. M. (1985). Rhinoviruses in Seattle families, 1975–1979. *American Journal of Epidemiology*, **122**, 830–46.

Friman, G. & Fohlman, J. (1993). The epidemiology of viral heart disease. *Scandinavian Journal of Infectious Diseases*, **88** (Suppl.), 7–10.

Gerba, C. P. & Rose, J. B. (1993). Estimating viral disease risk from drinking water. In *Comparative Environmental Risk Assessment*, ed. C. R. Cothern. Ann Arbor, Michigan: Lewis Publishers.

Gordon, J. Y., Gordon, R. Y., Romanowski, E. & Araullo-Cruz, T. P. (1993). Prolonged recovery of desiccated adenoviral serotypes 5, 8 and 19 from plastic and metal surfaces *in vitro*. *Ophthalmology*, **100**, 1835–40.

Gwaltney, J. M. (1980). Epidemiology of the common cold. *Annals of the New York Academy of Sciences*, **353**, 54–60.

Gwaltney, J. M., Moskalski, P. B. & Hendley, J. O. (1978). Hand-to-hand transmission of rhinovirus colds. *Annals of Internal Medicine*, **88**, 463–7.

Hadler, S. C. & McFarland, L. (1986). Hepatitis in day-care centers: epidemiology and prevention. *Reviews of Infectious Diseases*, **8**, 548–57.

Hall, C. B., Douglas, R. G. & Geiman, J. M. (1980). Possible transmission by fomites of respiratory syncytial virus. *Journal of Infectious Diseases*, **141**, 98–102.

Hayden, G. F., Gwaltney, J. M., Thacker, D. F. & Hendley, J. O. (1985a). Rhinovirus inactivation by nasal tissues treated with virucide. *Antiviral Research*, **5**, 103–9.

Hayden, G. F., Hendley, J. O. & Gwaltney, J. M. (1985b). The effect of placebo and virucidal paper handkerchiefs on viral contamination of the hand and transmission of experimental rhinoviral infection. *Journal of Infectious Diseases*, **152**, 403–7.

Hendley, J. O. & Gwaltney, J. M. Jr (1988). Mechanisms of transmission of rhinovirus infections. *Epidemiological Reviews*, **10**, 242–58.

Hendley, J. O., Mika, L. A. & Gwaltney, J. M. (1978). Evaluation of virucidal compounds for inactivation of rhinovirus on hands. *Antimicrobial Agents and Chemotherapy*, **14**, 690–4.

Hendley, J. O., Wenzel, R. P. & Gwaltney, J. M. (1973). Transmission of rhinovirus colds by self-inoculation. *New England Journal of Medicine*, **288**, 1361–4.

Ho, M. S., Glass, R. I., Pinsky, P. F., Young-Okoh, N., Sappenfield, W. M., Buehler, J. W., Gunter, N. & Anderson, L. J. (1988). Diarrheal deaths in American children, are they preventable? *Journal of the American Medical Association*, **260**, 3281–5.

Holdaway, M. D., Kalmakoff, J., Schroeder, B. A., Wright, G. C. & Todd, B. A. (1982). Rotavirus infection in Otago: a serological study. *New Zealand Medical Journal*, **95**, 110–112.

Hoshino, Y. & Kapikian, A. A. (1994). Rotavirus vaccine development for the prevention of severe diarrhea in infants and young children. *Trends in Microbiology*, **2**, 242–9.

Hutto, C., Little, A., Ricks, R., Lee, J. D. & Pass, R. F. (1986). Isolation of cytomegalovirus from toys and hands in a day-care center. *Journal of Infectious Diseases*, **154**, 527–30.

Keswick, B. H., Pickering, L. K., DuPont, H. L. & Woodward, W. E. (1983). Survival and detection of rotaviruses on environmental surfaces in day-care centers. *Applied and Environmental Microbiology*, **46**, 813–16.

Larson, T. & Bryson, Y. J. (1985). Fomites and herpes simplex virus. *Journal of Infectious Diseases*, **151**, 746–7.

Lauer, J. L., VanDrunen, N. A., Washburn, J. W. & Balfour, H. H. (1979). Transmission of hepatitis B virus in clinical laboratory areas. *Journal of Infectious Diseases*, **140**, 513–16.

Lecce, J. G., King, M. W. & Dorsey, W. E. (1978). Rearing regimen producing piglet diarrhea (rotavirus) and its relevance to acute infantile diarrhea. *Science*, **199**, 776–8.

Lederberg, J., Shope, R. E. & Oaks, S. C. (1992). *Emerging Infections: Microbial Threats in the United States*. Washington, DC: Institute of Medicine/National Academy of Sciences.

Lloyd-Evans, N. (1985). Studies on the survival and chemical disinfection of human rotavirus (Wa) on inanimate surfaces, MSc Thesis, Department of Microbiology and Immunology, University of Ottawa, Ottawa, Ontario, Canada.

Lloyd-Evans, N., Springthorpe, V. S. & Sattar, S. A. (1986). Chemical disinfection of human rotavirus-contaminated inanimate surfaces. *Journal of Hygiene*, **97**, 163–73.

Longini, I. M. & Monto, A. S. (1988). Efficacy of virucidal nasal tissues in interrupting familial transmission of respiratory agents: a field trial in Tecumseh, Michigan. *American Journal of Epidemiology*, **128**, 639–44.

Mahl, M. C. & Sadler, C. (1975). Virus survival on inanimate surfaces. *Canadian Journal of Microbiology*, **21**, 819–23.

Mbithi, J. N., Springthorpe, V. S., Boulet, J. R. & Sattar, S. A. (1992). Survival of hepatitis A virus on human hands and its transfer on contact with animate and inanimate surfaces. *Journal of Clinical Microbiology*, **30**, 757–63.

Mbithi, J. N., Springthorpe, V. S. & Sattar, S. A. (1990). Chemical disinfection of hepatitis A virus on environmental surfaces. *Applied and Environmental Microbiology*, **56**, 3601–4.

Mbithi, J. N., Springthorpe, V. S. & Sattar, S. A. (1991). Effect of relative humidity and air temperature on survival of hepatitis A virus on environmental surfaces. *Applied and Environmental Microbiology*, **57**, 1394–9.

Mbithi, J. N., Springthorpe, V. S. & Sattar, S. A. (1993a). Comparative *in vivo* efficiency of hand washing agents against hepatitis A virus (HM-175) and poliovirus type 1 (Sabin). *Applied and Environmental Microbiology*, **59**, 3463–9.

Mbithi, J. N., Springthorpe, V. S., Sattar, S. A. & Paquette, M. (1993b). Bactericidal, virucidal, and mycobactericidal activities of reused alkaline glutaraldehyde in an endoscopy unit. *Journal of Clinical Microbiology*, **31**, 2988–95.

McCaustland, K. A., Bond, W. W., Bradley, D. W., Ebert, J. W. & Maynard, J. E.

(1982). Survival of hepatitis A virus in feces after drying and storage for 1 month. *Journal of Clinical Microbiology*, **16**, 957–8.

McGeady, M. L., Siak, J. S. & Crowell, R. L. (1979). Survival of coxsackievirus B3 under diverse environmental conditions. *Applied and Environmental Microbiology*, **37**, 972–7.

McGinnis, J. M. & Foege, W. H. (1993). Actual causes of death in the United States. *Journal of the American Medical Association*, **270**, 2207–12.

Moe, K. & Shirley, J. A. (1982). The effects of relative humidity and temperature on the survival of human rotavirus in feces. *Archives of Virology*, **72**, 179–86.

Naides, S. J. (1994). Viral infection including HIV and AIDS. *Current Opinion in Rheumatology*, **6**, 423–8.

National Health Survey (1991). *Vital and Health Statistics* Hyattsville, Maryland: *1990*. US Department of Health and Human Services.

Nerurkar, L. S., West, F., May, M., Madden, D. L. & Sever, J. L. (1983). Survival of herpes simplex virus in water specimens collected from hot tub spa facilities and on plastic surfaces. *Journal of the American Medical Association*, **250**, 3081–3.

Pancic, F., Carpentier, D. C. & Came, P. E. (1980). Role of infectious secretions in the transmission of rhinovirus. *Journal of Clinical Microbiology*, **12**, 567–71.

Parkinson, A. J., Muchmore, H. G., Scott, E. N. & Scott, L. V. (1983). Survival of human parainfluenza viruses in the south polar environment. *Applied and Environmental Microbiology*, **46**, 901–5.

Pattison, C. P., Boyer, K. M., Maynard, J. E. & Kelly, P. C. (1974). Epidemic hepatitis in a clinical laboratory. Possible association with computer card handling. *Journal of the American Medical Association*, **230**, 854–7.

Reagan, K. L., McGeady, M. L. & Crowell, R. L. (1981). Persistence of human rhinovirus infectivity under diverse environmental conditions. *Applied and Environmental Microbiology*, **41**, 618–20.

Reed, S. (1975). An investigation of the possible transmission of rhinovirus colds through direct contact. *Journal of Hygiene*, **75**, 249–58.

Resnick, L., Veren, K., Salahuddin, S. Z., Tondreau, S. & Markham, P. D. (1986). Stability and inactivation of HTLV-III/LAV under clinical and laboratory environments. *Journal of the American Medical Association*, **255**, 1887–91.

Samadi, A. R., Huq, M. I. & Ahmed, Q. S. (1983). Detection of rotavirus in hand washings of attendants of children with rotavirus diarrhoea. *British Medical Journal*, **286**, 188.

Sattar, S. A., Dimock, K. D., Ansari, S. A. & Springthorpe, V. S. (1988). Spread of acute hemorrhagic conjunctivitis due to enterovirus-70: effect of air temperature and relative humidity on virus survival on fomites. *Journal of Medical Virology*, **25**, 289–96.

Sattar, S. A. & Ijaz, M. K. (1987). Spread of viral infections by aerosols. *CRC Critical Reviews in Environmental Control*, **17**, 89–131.

Sattar, S. A. Jacobsen, H., Rahman, H., Cusack, T. & Rubino, R. (1994a). Interruption of rotavirus spread through chemical disinfection. *Infection Control and Hospital Epidemiology*, **15**, 751–6.

Sattar, S. A. Jacobsen, H., Springthorpe, V. S., Cusack, T. & Rubino, R. (1993). Chemical disinfection to interrupt transfer of rhinovirus type 14 from environmental surfaces to hands. *Applied and Environmental Microbiology*, **59**, 1579–85.

Sattar, S. A., Karim, Y. G., Springthorpe, V. S. & Johnson-Lussenburg, C. M. (1987). Survival of human rhinovirus type 14 dried onto non-porous inanimate surfaces: effect of relative humidity and suspending medium. *Canadian Journal of Microbiology*, **33**, 802–6.

Sattar, S. A., Lloyd-Evans, N., Springthorpe, V. S. & Nair, R. C. (1986). Institutional outbreaks of rotavirus diarrhea: potential role of fomites and environmental surfaces as vehicles for virus transmission. *Journal of Hygiene*, **96**, 277–89.

Sattar, S. A. & Springthorpe, V. S. (1988). Test procedures for virucides: need for improvements. *Proceedings of the 4th International Conference on Progress in Chemical Disinfection*, State University of New York, Binghamton, NY, April, pp. 369–86.

Sattar, S. A. & Springthorpe, V. S. (1991). Survival and disinfectant inactivation of the human immunodeficiency virus: a critical review. *Reviews of Infectious Diseases*, **13**, 430–47.

Sattar, S. A. & Springthorpe, V. S. (1994). Methods under development for evaluating the antimicrobial activity of chemical germicides. *Proceedings of the International Symposium on Chemical Germicides in Health Care*, May 26–27, Cincinnati, Ohio.

Sattar, S. A., Springthorpe, V. S., Conway, B. & Xu, Y. (1994*b*). Inactivation of the human immunodeficiency virus: an update. *Reviews in Medical Microbiology*, **5**, 139–50.

Schürmann, W. & Eggers, H. J. (1983). Antiviral activity of an alcoholic hand disinfectant: comparison of the *in vitro* suspension test with *in vivo* experiments on hands, and on individual fingertips. *Antiviral Research*, **3**, 25–41.

Selwyn, S. (1965). The transmission of bacteria and viruses on gummed paper. *Journal of Hygiene*, **63**, 411–16.

Sidwell, R. W., Dixon, G. J. & McNeil, E. (1966). Quantitative studies on fabrics as disseminators of viruses. I. Persistence of vaccinia virus on cotton and wool fabrics. *Applied Microbiology*, **14**, 55–9.

Sidwell, R. W., Dixon, G. J., Westbrook, L. & Forziati, F. H. (1970). Quantitative studies on fabrics as disseminators of viruses. *Applied and Environmental Microbiology*, **19**, 950–4.

Springthorpe, V. S., Grenier, J. L., Lloyd-Evans, N. & Sattar, S. A. (1986). Chemical disinfection of human rotaviruses: efficacy of commercially available products in suspension test. *Journal of Hygiene*, **97**, 139–61.

Springthorpe, V. S. & Sattar, S. A. (1990). Chemical disinfection of virus-contaminated surfaces. *CRC Critical Reviews in Environmental Control*, **20**, 169–229.

Tervo, T., Laatikaninen, L., Valle, S.-L., Tervo, K., Vaheri, A. & Suni, J. (1987). Updating of methods for prevention of HIV transmission during ophthalmological procedures. *Acta Ophthalmologica*, **65**, 13–18.

Tjøtta, E., Hungnes, O. & Grinde, B. (1991). Survival of HIV-1 activity after disinfection, temperature and pH changes, or drying. *Journal of Medical Virology*, **35**, 223–7.

Turner, R., Shehab, Z., Osborne, K. & Hendley, J. O. (1982). Shedding and survival of herpes simplex virus from 'fever blisters'. *Pediatrics*, **70**, 547–9.

van Bueren, J., Simpson, R. A., Jacobs, P. & Cookson, B. D. (1994). Survival of human immunodeficiency virus in suspension and dried onto surfaces. *Journal of Clinical Microbiology*, **32**, 571–4.

Wadia, N. H., Katrak, S. M., Misra, V. P., Wadia, P. N., Miyamura, K., Hashimoto, K., Ogino, T., Hikiji, T. & Kono, R. (1983). Polio-like motor paralysis associated with acute hemorrhagic conjunctivitis in an outbreak in 1981. *Journal of Infectious Diseases*, **147**, 660–8.

Ward., R. L., Bernstein, D. I., Knowlton, D. R., Sherwood, J. R., Young, E. C., Cusack, T. M., Rubino, J. R. & Schiff, G. M. (1991). Prevention of surface-to-human transmission of rotavirus by treatment with disinfectant spray. *Journal of Clinical Microbiology*, **29**, 1991–6.

Ward, R. L., Knowlton, D. R. & Pierce, M. J. (1984). Efficiency of human rotavirus propagation in cell culture. *Journal of Clinical Microbiology*, **19**, 748–53.

World Health Organization (1992). Communicable disease epidemiology and control. *World Health Statistical Quarterly*, **45**, 166–7.

Chapter 9

The role of chemical disinfectants in controlling bacterial contaminants on environmental surfaces

DONNA J. GABER, TIMOTHY M. CUSACK and
ELIZABETH SCOTT

Introduction

The concept of acquiring an infection from contact with contaminated surfaces is age old, and in fact is cited in the Old Testament of the Bible in Leviticus. Formal, scientific studies which conclusively demonstrate the acquisition of infection from direct or indirect contact with contaminated surfaces have not been extensively conducted. However, we believe that there is a sufficient body of evidence to support this form of disease transmission.

The purpose of this chapter is to take a closer look at contaminated environmental sites, surfaces and fomites to assess their role as vehicles of bacterial disease transmission. From the perspective of general levels of risk, three different settings will be examined: the home (low risk), day care (intermediate risk) and healthcare (high risk). It should be understood that within each setting all three levels of risk may be present.

In determining the role that surfaces play in the transmission of disease, it is important first to develop a working definition of surface contamination. The chain of events leading to the occurrence and spread of infection will be discussed. Finally, with the recognition that chemical disinfection of environmental surfaces may not be necessary in all situations, the prudent use of chemical disinfectants as a possible means of reducing the incidence of disease will be discussed.

When surface contamination is defined, several factors must be considered. These factors include the presence of potentially pathogenic contaminants in the environment, their minimal infectious dose, their ability to survive on environmental surfaces and their likelihood of being transferred. We must also consider the range and types of surfaces in homes, day care and healthcare settings that are most likely to become contaminated.

Occurrence and spread of infection via surfaces

Four factors must be present in order for an infection to develop from contact with contaminated surfaces. They are the presence of an infectious agent, a reservoir for growth, a means or mode of transmission and a susceptible host. These factors, linked together, increase the probability that disease transmission will occur (Figure 9.1). In order to interrupt disease transmission via surfaces, one of the factors must be eliminated. The following is a discussion of each factor.

Infectious agents

Many types of infectious agent are capable of causing disease in humans. As the focus of this chapter is on bacterial contaminants, our discussion will be limited to some of the more significant pathogens. Bacterial pathogens are found in persons who have active or incubating disease, or in those who are colonized by an agent but have no apparent disease (carriers).

Reservoirs or sources of bacterial contaminants

Potentially infectious sources include infected persons, inanimate objects or fomites, food, pets or animals, and equipment which may have become contaminated. In order to control or interrupt the spread of disease from sources in the home and day-care centers, one must have an acute awareness of the possible presence of potential pathogens. In clinical settings, rapid and accurate identification of pathogenic organisms is required. Appropriate control measures may then be taken to eliminate or reduce their presence from the environment, thereby reducing the risk of exposure

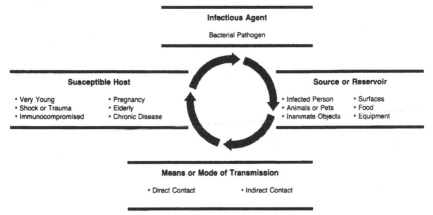

Figure 9.1   Factors necessary for the transmission of bacterial infection.

to susceptible persons. Control measures to break the chain of transmission via environmental surfaces include the development of protocols for hand washing, ongoing infection control education, proper cleaning and chemical disinfection, and isolation as necessary.

Means or mode of disease transmission via environmental surfaces

Microorganisms are transmitted by two means or modes of contact: direct and indirect. As the focus of this chapter is on the role of disinfection of contaminated surfaces, our discussion will be limited to the indirect contact mode. Indirect contact involves personal contact with contaminated inanimate objects or environmental surfaces by a susceptible host. In order to control the spread of disease via the indirect mode of transmission, the following actions are necessary: (1) objects and surfaces must be routinely and correctly processed in a manner that renders them appropriately sterile, disinfected or sanitized; and (2) hands must be frequently washed.

Susceptible host

The fourth factor needed in the spread of infection is a susceptible host. Individual susceptibility depends on a person's resistance to microorganisms. Age, pregnancy, immunosuppression, shock, trauma or chronic disease, such as diabetes, may influence susceptibility. Care must be taken to ensure that underlying or chronic high-risk conditions are recognized as a factor that may place a person at greater risk.

In the three sections of this chapter, we attempt to support the following hypotheses:

Humans are colonized or infected with human pathogenic bacteria.

Interaction with environmental surfaces may be intimate, resulting in contamination *of* the environment with organisms, and acquisition of these contaminants *from* the environment.

Organisms have the ability to survive in the environment in quantities capable of initiating infection.

Disinfectants can reduce populations of certain pathogenic contaminants on environmental surfaces.

Therefore, it is reasonable to suggest that:

(1) Contact with contaminated surfaces plays a role in the transmission of certain pathogens; and
(2) The appropriate use of chemical disinfectants applied to surfaces could reduce the incidence of infection caused by these pathogens.

Assessing the risk of disease transmission from environmental surfaces

In the home

Over the last decade, we have observed an increase in the incidence of foodborne disease in most countries of the developed world. While some of the increase can be attributed to improvements in data collection systems and increased surveillance, there is no doubt that much of the increase is genuine (Sockett, 1993*a*).

Until recently, it was generally assumed that foodborne disease was primarily associated with commercial catering establishments such as hotels, restaurants, cafeterias, fast-food chains, etc., and not with the kitchen at home. However, many countries with reliable data collection systems are noting the high proportion of outbreaks affecting single households. For example, data published in the *CDR Review* (Sockett *et al.*, 1993) for England and Wales, covering the three-year period 1989–1991, indicates that there were a total of 2766 *Salmonella* outbreaks, 86% of which were classified as family outbreaks. In addition, there were 1097 *Campylobacter* outbreaks, 97% of which were classified as family outbreaks. It should be noted here that although the classification 'family outbreak' is defined in this context as 'two or more cases confined to the same household', it is possible that there may be other related, but unknown, cases within the community. Therefore, it is possible that not all family outbreaks are confined to the home. Detailed sentinel studies (involving a group of physicians acting as 'sentinels' and reporting on their defined populations of patients) and community studies in the Netherlands suggest that 80% of foodborne illness arises within the home (Hoogenboom-Verdegaal, 1989). In Spain, detailed investigations show that private homes account for nearly 50% of all outbreaks of food poisoning (World Health Organization, 1992). The exact extent of the foodborne disease involvement in the home in the USA is not well known. A survey published in 1970 by Aserkoff, Schroeder & Breckman suggested that the largest proportion of food poisoning outbreaks originate in the home. Recent estimates for foodborne *Campylobacter* and *Salmonella* disease suggest that more than 4 million cases occur in the USA annually (Anon., 1993). Various estimates have put the total incidence of foodborne disease in the USA in the range of 6.3 million to 81 million cases (Todd, 1991), most probably around 20 million.

Recent estimates of the national costs of foodborne disease have shown that the economic burden to national agencies, industry and the individual is great. Sockett (1993*b*) estimated that the real cost of just salmonellosis in England and Wales is approximately £200 million to £300 million annually, and the total annual cost of all foodborne disease in England and Wales alone may be as high as £1 billion. Half of these costs are due to lost production resulting from absence from work due to sickness, or time off from work to care for relatives as a result of foodborne illness (Sockett & Roberts, 1991). Recent cost estimates for all foodborne illness in the USA, as reported by Todd (1991) range from US $7.7 billion to 23 billion. As yet, there is no indication as to what proportion of the total national costs are attributable to the origination of foodborne disease from the home. Although the actual per episode cost resulting from breakdowns in hygienic practices in the home are likely to be less than for similar break-downs in the food industry, total costs related to illness associated with the home are likely to be substantial, due to the high number of episodes.

Data from various countries indicate that up to 80% of foodborne disease outbreaks occur within the home, but there are no data at this time to indicate what proportion is attributable to poor hygiene practice. Commenting upon the high proportion of outbreaks affecting single households in England and Wales, Sockett (1993a) suggested that, although some outbreaks are likely to have resulted from food prepared elsewhere and brought into the home, data indicate that inappropriate food practices in the domestic kitchen may also be a significant factor. In investigation of the causes of food poisoning outbreaks, the occurrence of cross contamination involving inanimate surfaces or objects, or the hands of food handlers, is difficult to determine. Many experts believe, however, that cross contamination plays a significant role (Van Schothorst, Huisman & Van Os, 1978; De Wit, Broekhuizen & Kempelmacher, 1979; Gilbert, 1986). Various workers have reported the potential role of environmental sites and surfaces, including those found in the home, as vehicles for cross contamination. Finch, Prince & Hawksworth (1978) commented on the large number of Enterobacteriaceae isolated from domestic sinks, and the potential for cross contamination as seen in a survey of 21 homes. A survey of over 200 homes (Scott, Bloomfield & Barlow, 1982) indicated that, in the kitchen, moist areas such as sinks, wastetraps and surrounding areas act as reservoirs which harbor and encourage proliferation of enterobacteria. In addition, small numbers of potentially pathogenic bacteria were isolated from 49% of all food contact surfaces and 28% of all other hand contact surfaces in the survey. Subsequent laboratory experiments showed that various species of bacteria can survive on soiled (but clean looking) kitchen surfaces, and both clean and soiled dishcloths, for 4 h or up to 24 h (Scott & Bloomfield, 1990). The same experiments also indicated the ease of cross contamination of surfaces such as food preparation surfaces, dish-cloths, utensils and the hands.

The introduction of contaminated food or the presence of an ill family member may also influence the likelihood of cross contamination. The potential for cross contamination in kitchens resulting from the preparation of frozen chickens was examined by De Wit *et al.* (1979). The results showed that cross contamination occurred in a high proportion of the kitchens. Contaminants were found on surfaces such as chopping boards, dishcloths and sinks, even after rinsing and washing-up actions had taken place. It has been suggested that *Campylobacter* infections are usually acquired through the consumption of undercooked meat, especially poultry, or through cross contamination of kitchen surfaces and utensils (Blaser, 1992). Van Schothorst *et al.* (1978) sampled homes where a case of infant salmonellosis had occurred. The survey found that both human contacts and domestic animals were frequently found to be contaminated with the same serotype as the infant. In addition, kitchen sinks and other kitchen surfaces were also contaminated. The authors commented that surfaces in the kitchen may be contaminated by the hands of the mothers or of the nurses, and that this may result in cyclic infection.

Despite the data on foodborne disease in the home, and studies demonstrating the role that surfaces and utensils may play in the chain of cross contamination, little attention has been paid to the theory and practice of disinfection in the home. In order to formulate an effective disinfection/cleaning policy in any environment, an assessment of risk of pathogen acquisition is required. The risks associated with any particular site depend upon the frequency with which the site may be contaminated and the probability of transfer of this contamination to a more sensitive surface or site (e.g. food). Increased risk would occur, for example, during an outbreak of enteric infection, in which levels of contamination tend to increase. In addition to these factors, the likelihood of susceptible individuals coming in contact with the contamination should also be taken into consideration.

A risk assessment for the home

Using data on the bacterial contamination of the home (Scott *et al.*, 1982), sites were grouped together as either reservoir sites (e.g. toilets, sinks and draining boards, sink wastetraps), reservoir–disseminators (e.g. wet cleaning utensils, dishcloths), or contact sites (e.g food preparation surfaces, handles and faucets). The frequency of occurrence of potentially pathogenic organisms together with a quantitation of levels of contamination at these various sites was assessed. These data were then used to develop the contamination risks shown in Table 9.1.

There is sufficient evidence available to make an assessment of the likelihood of contamination transfer, either directly or indirectly, from a site of lower risk to a more sensitive site or surface. Much of this evidence is reviewed in research on the survival and transfer of pathogenic bacteria from environmental sites and surfaces (Scott, 1990). For reservoir sites such as toilets and sink wastetraps in the home, it would seem that, under

Table 9.1. *Frequency of occurrence of contamination at reservoir, reservoir and disseminator, and contact sites in the domestic environment*

Site classification	Frequency of occurrence of enterobacteria (%)	Frequency of occurrence of high counts (%)
Reservoir	30–64	10–80
Reservoir and disseminator	24–30	28–80
Contact	4–20	4–40

Source: Scott *et al.* (1982).

normal circumstances, the risks of infection transfer are relatively low (Scott, 1990). In contrast, there is considerable evidence that wet cleaning utensils, and dishcloths in particular, can act not only as reservoirs of free-living organisms, but also as disseminators of contamination into the environment. In laboratory experiments, $\geq 100\,000$ bacteria (i.e. *Staphylococcus aureus*, *Escherichia coli* or *Klebsiella aerogenes*) were inoculated onto a dishcloth and then put into mock use. In each case, significant numbers of both gram-positive (*S. aureus*) and gram-negative (*E. coli* and *K. aerogenes*) organisms were deposited onto the fingers and also transferred to a previously clean surface (Scott 1990). In another evaluation, colony counts from dishcloths that were returned to the laboratory following three days of use in the home indicated that these cloths were more heavily contaminated (up to 10^6 organisms/cm^2 of cloth) than were those that were artificially contaminated and maintained in the laboratory. Therefore, the risk of contamination transfer from dishcloths is considerable, and such items must be considered a constant risk. In-use studies have shown that the use of an effective self-disinfecting cloth was associated with a significant improvement in hygiene on food contact surfaces (Scott & Bloomfield, 1992).

Experiments with contact surfaces, such as food preparation areas, have indicated that a substantial number of both gram-positive and gram-negative bacteria could be transferred via hands or a utensil. These studies demonstrated that bacteria from an apparently dry work surface, which had been contaminated with relatively small numbers (i.e. several hundreds per contact area) of potential pathogens, could be transferred to a more crucial site or surface (e.g. food preparation areas, directly to the mouth; Scott, 1990). Therefore, as there is no immediate way of determining whether a particular surface is contaminated or not, contact surfaces must also be regarded as a constant risk. In assessing the potential hazard associated with the transfer of contamination via hands, dishcloths, utensils, etc., it must be remembered that the risk of infection depends not only on the type, but also on the numbers of organisms transferred. For example, the minimal infectious dose of *Salmonella* may be ≤ 100 organisms (Gill *et al.*, 1983; Greenwood & Hooper, 1983; D'Aoust, 1985), and infection with *Campylobacter* or verotoxigenic *E. coli* may be caused by ingestion of relatively few organisms (Council for Agricultural Science and Technology, 1994). As a result, surface transfer of even a small number of these highly virulent pathogens to a site of higher risk (e.g. food preparation area, utensils, food) represents a potentially significant hazard.

A disinfection model for the home

With the use of the criteria developed above, it is possible to construct a model for disinfection in the home, as shown in Table 9.2. The model allows flexibility in response to changing circumstances. For other general sites and surfaces not included in the model, where the risks of contamination transfer and infection are usually very low, a flexible response is also required. For example, for floors and soft furnishings, which would normally be classified as low risks in the home and would not routinely require disinfection, the risk would increase in the event of contamination with feces, blood or vomitus.

Table 9.2. *An example of a risk evaluation for environmental sites and surfaces as evaluated for the domestic environment (from Scott, 1990)*

Site(s)	Frequency of occurrence of 'significant' contamination	Risk of transfer of contamination	Assessment of need for disinfection	Approach to disinfection
Reservoirs, (sinks, wastetraps, etc.)	High	Relatively low, occasional	Relatively little, except where known risk	Use continuous-release or sustained-action disinfectants where required
Reservoir and disseminators (wet cloths and cleaning utensils)	High	Constant (i.e. all of the time)	Always	Use disposable cloths or disinfect immediately before use
Contact sites (e.g. food preparation areas and hand contact surfaces)	Sometimes	Constant	Always	Disinfect critical surfaces immediately before and between uses

Summary

One of the common contributory factors for foodborne disease is the occurrence of cross contamination involving food-contact surfaces, hand-contact surfaces, cloths and kitchen utensils. Cross contamination may be prevented by identifying those sites and surfaces most likely to be involved, and targeting them for disinfection. The principle upon which this approach is based is that of recognizing hazards and blocking routes of contamination transfer.

Finally, whenever and wherever we suggest that the use of a disinfection process could improve hygiene, it is important to offer information to the consumer, who must put the process into practice. This information should include appropriate preparation instructions for the product, as well as suggestions for the frequency and site of application. It is important that consumers understand that the basic infection control principles apply to all settings, including the home.

In day-care settings

As women in the United States have entered the workforce in increasing numbers over the last several decades, there has been a growing necessity for child day-care. During the past 30 years, an increasing percentage of this care has taken place outside of the home, typically in day-care centers. These centers can be responsible for the care of up to several hundred children. At present, about 90% of families with preschool children use some form of child day-care (Thacker et al., 1992).

Infants and children under the age of six represent the most infection-susceptible segment of our population, since they lack prior exposure or antibodies to most pathogens. The increased levels of disease experienced by children in day-care is viewed by some as insignificant, since it is believed that children are only developing typical childhood diseases at a younger age than children from the 'pre-day-care era'. This may be true for some diseases, but conditions present in the day-care environment actually promote the transmission of certain types of disease (e.g. enteric). Close contact, coupled with poor hygienic practices, facilitate the transmission of infectious agents. This problem is amplified by the fact that many of the illnesses commonly spread in day-care are most infectious *prior* to the appearance of symptoms. In addition, exposure to some of the pathogens (e.g. *Rotavirus*) found in this environment provide only limited or transient immunity, resulting in the potential for multiple infections with the same agent. Once infected, these children serve as a source of infection for their family contacts, who in turn may transmit these pathogens throughout the community (Polis et al., Reves & Pickering, 1990).

There is substantial evidence that children in out-of-home child care are at increased risk for a variety of illnesses (reviewed by Osterholm *et al.*, 1992; Pickering & Morrow, 1991). Children who attend day-care have a 1.6–3.5 times higher risk of acquiring diarrheal illness than children who are cared for at home (Thacker *et al.*, 1992). Similarly, children under two years of age in day-care have an incidence of acute respiratory illness that is 1.6 times higher than age-matched children not in day-care. Day-care centers have been identified as the setting for outbreaks of enteric disease caused by such bacteria as *Shigella* (Pickering *et al.*, 1981; Briley, Teel & Fowler, 1994), *Salmonella* (Lieb, Gunn & Taylor, 1982), *Clostridium difficile* (Kim, DuPont & Pickering, 1983), *Campylobacter* (Itoh, Saito & Maruyama, 1980; Bartlett *et al.*, 1985) and *Escherichia coli* (Paulozzi, Johnson & Komahele, 1986; Spika *et al.*, 1986; Fornasini *et al.*, 1992). The acquisition of *E. coli* 0157:H7 in a day-care center has already been linked to significant morbidity and at least one death (Centers for Disease Control, 1993).

A major problem associated with this increased incidence of infection is the economic burden that it places on families of children in day-care. The average family in the United States spends 10% of its gross income on child-care services (LaMaure & Thompson, 1984). This expenditure can be greatly impacted by illness. Most day-care centers have exclusion policies that limit the participation of obviously ill children. When a family is faced with this situation, they must either find alternative care or stay home from work. Both of these options result in costs to the family. It has been estimated that over 60% of employee absenteeism in the workplace, at an approximate annual cost of $12.7 billion, is related to unmet child-care needs, especially those of sick children (Thacker *et al.*, 1992). These costs may be reduced by improvements in disease prevention strategies.

The emphasis of this chapter is on the role of chemical disinfectants in limiting the spread of bacterial disease. We will, therefore, concentrate on transmission involving contact with contaminated surfaces and objects. Bacterial diseases transmitted via the fecal–oral route will be stressed, since they frequently involve surfaces as intermediates in the transmission chain. The overall concepts presented here are also applicable to the transmission of many viruses (e.g. *Rotavirus*, hepatitis A) and parasites (e.g. *Giardia*, *Cryptosporidium*). It also bears mentioning that many of the sanitation problems that exist in day-care settings for children would also apply to day-care and residential-care facilities for the elderly and disabled. Some of these residents are kept in diapers, personal habits may be poor, and there are shared bathing and toilet facilities.

Transmission

In order for a pathogen to be effectively transmitted, a number of factors must be in place. As described in the introduction to this chapter, these factors include an infectious agent, survival in a reservoir, a means of transmission and a susceptible host. In the day-care environment each of these factors is readily available.

By their very nature, young children tend to have little control of their secretions and excretions and therefore often shed infectious organisms into the environment. For example, fecal material is not always adequately contained within diapers, and nasal secretions are often wiped with childrens' hands and then indiscriminately spread. This common lack of personal hygiene provides a ready source of infectious organisms.

Once these organisms have left their host, they must then be able to survive. Numerous studies have documented the contamination of common objects and surfaces (e.g. communal toys, diaper-changing areas and floors) in day-care facilities. Most of these studies have looked for the presence of coliforms as a marker for fecal contamination. Although rates of recovery of coliforms at specific sites have varied among studies (Table 9.3), widespread fecal contamination was commonly noted. Several studies (Ekanem et al., 1983; Van et al., 1991a,b) have also demonstrated a significant increase in the rates of environmental contamination during outbreaks of diarrhea. The general trend noted in all studies was that coliforms were found most often in areas in which diapered children were present. These studies have pointed out the potential role of environmental surfaces as intermediates or reservoirs in the transmission of enteric agents.

In the day-care environment, the mode of transmission of organisms is

Table 9.3. *Percentage of cultures positive for fecal coliforms in the day-care environment during routine sampling*

Study	Sites			
	Inanimate objects and surfaces[a]	Toys	Floors	Diaper-changing areas
Weniger et al. (1983)	3	0	10	7
Ekanem et al. (1983)	0–35[b]	10	5–20[b]	30
Holaday et al. (1990)	12	2	16	0
Van et al. (1991a)	15	26	15	15

[a] Includes doorknobs, faucet handles, kitchen counter, table tops, furniture (e.g. chair, crib), diaper pail handle.
[b] Rates varied depending upon specific site. No raw data given to calculate cumulative mean.

Table 9.4. *Assessment of transfer risk from various surfaces in day-care centers and the resulting need for disinfection*

Site	Frequency of occurrence of contamination	Risk of transfer	Need for disinfection
Diaper-changing area	High	High	Disinfect after every use
Toys	Moderate	High	Remove from circulation when mouthed or soiled. Clean and disinfect daily
Floors	High	Moderate	At least daily or when obviously soiled
Tabletops and other inanimate surfaces	Moderate–high	Moderate	At least weekly or when obviously soiled

closely related to the actions of the susceptible hosts. As stated previously, young children are generally susceptible to most infectious agents, due to a lack of prior exposure and inadequately developed immune systems. Children of this age, especially toddlers, are very active in exploring the environment, often placing potentially contaminated items into their mouths. Schumann (1983) observed that toddlers placed their fingers or an object into their mouth once every 2–3 min. As children explore the environment, and share their findings with other children, they can very easily pass on microorganisms. The feasible routes of transmission involving surface contact include surface-to-hand and surface-to-mouth. In addition, caregivers are often responsible for all aspects of child care, including diaper changing, food preparation and coordination of play. Any breakdowns in proper hygienic practices by the caregivers could very easily result in the microbial contamination of food, objects or surfaces.

Risk of transmission from surfaces and objects

The risk of disease transmission associated with many surfaces and objects in the day-care environment is relatively high, especially when children in diapers are present. We have attempted to delineate the risk associated with specific sites and surfaces through an assessment of the frequency of contamination, and the probability of transfer from these surfaces. The risk associated with each of these areas is probably increased during outbreaks of disease, due to increased levels of contamination (Ekanem *et al.*, 1983). The need for chemical disinfection of these areas, based on this risk, has also been addressed (Table 9.4).

Diaper-changing areas pose a particularly high risk as a reservoir for pathogens, due to the high probability of fecal soiling. These areas are typically used quite frequently, resulting in a high risk of transfer. In order to reduce this risk, this area should be disinfected after every use.

Bacterial contamination of toys is also common in this setting. Since communal toys are generally passed from child to child, and frequently mouthed, they can rapidly expose large numbers of children to significant levels of infectious organisms. This exposure can be lessened if obviously soiled or mouthed toys are taken out of circulation, cleaned and disinfected. Toys, and other frequently mouthed items, should be cleaned and disinfected on a daily basis, especially in day-care centers that enrol diapered children (American Academy of Pediatrics/American Public Health Association, 1992).

Areas like floors and tabletops, which may pose minimal risk for transfer of infectious agents in other settings, have also been shown to be potential reservoirs in day-care centers (Ekanem et al., 1983; Van et al., 1991a) Young children, especially toddlers, contact many of these surfaces and frequently place their hands into their mouths. The nature of their interaction with these surfaces demonstrates the potential for transfer and the resulting need for disinfection.

Infection control in day-care centers

In order to address the risks associated with these and other sites in day-care centers, strict attention to basic infection control practices is necessary. Practices such as hand washing, environmental disinfection, containment of body fluids and isolation of ill children are necessary. Frequent hand washing is considered to be the cornerstone of infection control, and is extremely important in this setting. One study has described the role of increased compliance with established hand washing practices as a means of successfully reducing the transmission of diarrhea (Black et al., 1981). Another group of researchers investigated the role of education, combined with the implementation of specific barrier precautions (i.e. gloves, diaper-changing pads, alcohol hand rinses). Their findings also demonstrated significant decreases in symptoms associated with enteric disease (i.e. diarrhea, vomiting) (Butz et al., 1990). In an effort to elucidate the effect of containment of body fluids, Van et al. (1991b) examined the role of diaper type and overclothing. They found that disposable paper diapers maximized the containment of stool, and thereby minimized the spread of fecal contamination in day-care facilities. All of these studies illustrate the need for infection control in the day-care environment.

The role of surface disinfection

The close and direct physical contact of children and caregivers results in numerous interactions involving hands and surfaces or objects. Frequent contact by soiled hands can deposit pathogens onto these surfaces. Similarly, frequent handling of contaminated surfaces or objects can act as a source of pathogen acquisition, even for recently washed hands. The roles of surfaces and hands appear to be complementary, and possibly even synergistic, in the transfer of enteric pathogens (Ansari *et al.*, 1988). As a result, hard-surface disinfection, as an adjunct to frequent hand washing, is essential.

It has been suggested that priority should be given to the cleaning and disinfection of the day-care environment, to help reduce the spread of enteric pathogens during outbreaks (Pickering *et al.*, 1986). Investigators who have looked at the impact of cleaning and disinfection in child care have described specific decreases in the microbial load of surfaces. Petersen & Bressler (1986) noted that toilet and diapering areas contained lower colony forming units of fecal coliforms than did other areas. They attributed this reduction to greater use of disinfectants on these surfaces. In addition, they suggested that the ability to clean surfaces should be given strong consideration in the design and construction of day-care centers. Sullivan *et al.* (1984) also observed that surfaces such as diaper-changing areas, bathrooms, floors and play areas were generally cleaned and disinfected on a daily basis, but toys were not. They advised that more attention should be given to the toys, since they are often contaminated with coliforms and are frequently mouthed.

A joint panel of experts of the American Academy of Pediatrics and the American Public Health Association (1992) has recently created national standards for infection control in out-of-home child-care programs. These standards have recognized the role that commonly contacted environmental surfaces and objects play in the transmission of disease, and cite the necessity of cleaning and disinfection of these areas (e.g. toys, diaper-changing areas, floors, table tops, play areas). This effort represents an important first step in this country to develop practices that will help to promote the health of children and their family contacts.

Summary

All of the elements for the transmission of disease via contaminated fomites have been shown to be present in the day-care environment. In addition, the high incidence of disease in this setting suggests that any reasonable means of reducing transmission should be instituted. While quantitative proof of the impact of disinfectant usage has not been adequately studied, a great deal of evidence suggests that the inanimate environment plays a role in the transmission of pathogens in day-care. Therefore, the judicious use of disinfectants to control the spread of disease is warranted and recommended.

In healthcare settings

Prior to 1981, infection control programs in hospitals were driven by three main factors: state or federal regulations, surveillance for nosocomial infections and the results obtained from routine environmental cultures. Infection control programs in nursing home facilities were practically non-existent, except as required to satisfy operating regulations. With few exceptions, the finding of surface contamination could rarely be correlated with the occurrence or prevalence of nosocomial infection (Maki, 1982). Furthermore, 'acceptable' levels of microbial contamination could not be established, and still do not exist. In 1981, the Center For Disease Control, as it was then called, issued guidelines that discouraged the practice of *routine* environmental culturing of surfaces in hospitals. This recommendation was quickly echoed and endorsed by other regulating agencies such as the Joint Commission For Accreditation of Hospitals, as it was then called (Joint Commission for the Accreditation of Hospitals, 1976). Gone were the days of culturing ice machines, floors, walls, meat slicers and other surfaces on a monthly basis. Routine sampling programs were, and still are, considered to be expensive, time-consuming efforts, yielding culture results that may be difficult to interpret. In recent years, however, multi-drug-resistant pathogens have emerged, one of which has caused the Center for Disease Control to call for a return to limited environmental sampling (Simmons, 1981). These 'super bugs' have produced increased morbidity and mortality accompanied by skyrocketing healthcare costs. The following is an overview of three of the more significant pathogens associated with healthcare: *Clostridium difficile*, methicillin-resistant *Staphylococcus aureus* (MRSA) and antibiotic-resistant strains of *Enterococcus*. During outbreaks, studies have demonstrated that these pathogens have been found on multiple environmental surfaces and equipment frequently contacted by patients and healthcare workers. The risk of disease transmission from these organisms when found on surfaces in healthcare settings will now be discussed.

Clostridium difficile

Clostridium difficile is a spore-forming, anaerobic bacillus. It is capable of producing severe diarrhea and colitis in hospitalized patients and those with extended care, especially those receiving antimicrobial therapy. Outbreaks have been well documented (Kofsky *et al.*, 1991; Stratton, 1990). Some of these outbreaks have been associated with the use of shared equipment (Table 9.5), such as commodes and electronic thermometers (Smith *et al.*, 1981; McFarland *et al.*, 1989; Brooks *et al.*, 1992; Skoutelis *et al.*, 1994). While the vegetative state of *C. difficile* is oxygen-sensitive and easily killed, its spores are extremely hardy, and are capable of being recovered from environmental surfaces after long periods of time (Fekety *et al.*, 1981). In a recent publication, floor cultures were positive for *C. difficile* spores up to eight weeks after initial, deliberate contamination (Skoutelis *et al.*, 1994). The risk of acquiring a nosocomial infection by this organism is dependent upon several factors. These include the level of environmental contamination in a facility, the number of symptomatic and asymptomatic infectious patients, the degree of patient susceptibility, and the frequency and thoroughness of cleaning and disinfection practices. McFarland *et al.* (1989) showed that colonized patients constituted an important reservoir within a facility. It was also reported that person-to-person spread and environmental reservoirs played the major roles in disease transmission. Due to the viability of this organism in its spore state, environmental contamination may be quite persistent, thus raising the risk of nosocomial infection. McFarland *et al.* further reported that surface cleaning and disinfection, proper isolation, the use of gloves and an antiseptic hand washing preparation seemed the most likely measures to reduce nosocomial transmission of *C. difficile*. It is important to note that while products registered as 'hospital disinfectants' are not effective against spores, frequent environmental cleaning and disinfection will help reduce environmental bioburden and vegetative bacteria. Only products registered as 'high-level disinfectants' or 'sterilants' are capable of killing spores.

Multi-drug-resistant *Staphylococcus* and *Enterococcus*

In recent years there has been much in the literature about the frightening emergence of two infectious, multi-drug-resistant pathogens. Specifically, these new 'super bugs' include the methicillin-resistant strains of *Staphylococcus aureus* (MRSA) and, more recently, certain strains of *Enterococcus* that have become resistant to vancomycin (VRE) and vancomycin and ampicillin (VAREC) (Haley, Hightower & Khabbaz, 1982; Reboli *et al.*, 1990, Layton *et al.*, 1993; Boyce *et al.*, 1994).

Table 9.5. *Environmental surfaces in healthcare settings contaminated with* Clostridium difficile

Investigator	Study date	Contaminated sites
Brooks *et al.*	1992	Electronic thermometer
McFarland *et al.*	1989	Bed rail, floor, commode, call button, toilet, sink, slipper sole, dialysis machine, nasogastric alimentation solution, windowsill
Skoutelis *et al.*	1994	Hospital carpeting
Smith *et al.*	1981	Electronic thermometer

Methicillin-resistant *Staphylococcus aureus* (MRSA)

Since the mid-1970s, the incidence of MRSA has been increasing in healthcare facilities in the United States (Boyce, 1989). While the prevalence of MRSA varies from facility to facility, sooner or later most will experience an outbreak. MRSA is problematic in hospitals of all sizes, and may account for 30–50% of all *S. aureus* isolates recovered from patients. It has been shown that once MRSA strains become endemic in a facility, eradication of the pathogen is difficult (Boyce, 1989). MRSA is also of considerable concern for nursing facilities. Due to a lack of understanding about controlling this pathogen, hospitalized patients colonized with MRSA have been denied admission into some nursing facilities.

While MRSA is no more virulent than antibiotic-sensitive strains of *S. aureus*, it is of special concern, because treatment options are limited. Infected patients are generally treated with vancomycin. This drug requires intravenous administration, is expensive and may be more toxic than agents used to treat methicillin-sensitive strains of *S. aureus* (Boyce *et al.*, 1994). Methods to control MRSA were recently reviewed by the Methicillin-Resistant *Staphylococcus aureus* Task force of The American Hospital Association (Boyce *et al.*, 1994). Their work concluded that (1) colonized and infected patients represented the major reservoir of MRSA in acute care and nursing facilities, (2) MRSA was most frequently spread from patient to patient via the hands of healthcare workers, and (3) the efficacy of measures commonly used to control MRSA had not been established in controlled trials. They also suggested, as have others, that contaminated environmental surfaces were not a significant reservoir in most outbreaks. This last conclusion is puzzling. Investigators have repeatedly shown that MRSA strains are capable of surviving in a dry environment, and contamination of surfaces has been documented (Table 9.6) during several MRSA outbreaks (Crossley, Landesman & Zaske, 1979; Thompson, Cabezudo & Wenzel, 1982; Rutala *et al.*, 1983; Bartzokas *et al.*, 1984; Bradley, Terpen-

Table 9.6. *Environmental surfaces in healthcare settings contaminated with methicillin-resistant* Staphylococcus aureus *(MRSA)*

Investigator	Study date	Contaminated sites
Bartzokas *et al.*	1984	Surgical unit
Bradley *et al.*	1991	Bedside table, floor
Crossley *et al.*	1979	Burn unit
Layton *et al.*	1993	Blood pressure cuff, shower
Moore & Williams	1991	Bidet spout, hand basin, toilet seat, bath tub, bath mat, mattresses
Rutala *et al.*	1983	Burn unit
Thompson *et al.*	1982	Burn unit

ning & Ramsey, 1991; Moore & Williams, 1991; Layton *et al.*, 1993). MRSA can spread rapidly through a hospital or nursing facility and is capable of causing considerable morbidity and mortality, especially in high-risk populations (Peacock, Marsik & Wenzel, 1980). In support of our original hypothesis, we submit that while the role of environmental surfaces as a means of transmission of this bacterial pathogen has not been conclusively studied, until more research is conducted it should not be ignored.

Vancomycin-resistant *Enterococcus faecalis* (VRE) and vancomycin- and ampicillin-resistant *Enterococcus faecium* (VAREC)

The emergence of VRE and VAREC has recently been described as a 'national crisis' (D. G. Maki, unpublished data). Maki warned that VRE and VAREC may place infection control back in the pre-antibiotic era of untreatable infections, and that VAREC may be a virtual 'Andromeda strain' that leaves few therapeutic options for clinicians treating infections. Attributable mortality for VRE is probably in the range of 40–50% (Maki, 1994). What is most frightening about VRE is that it appears to be capable of transferring its vancomycin resistance genes to other organisms outside of its own species. In laboratory experiments the *vanA* gene of *Enterococcus* has been transferred to isolates of *S. aureus*, but clinical isolates of vancomycin-resistant *S. aureus* have not yet been reported (Noble, Virani & Cree, 1992). This sets the stage for the development of resistance to vancomycin in organisms such as MRSA or *E. coli*.

Data are now pointing to the possibility of VRE or VAREC transmission via contaminated patient-care equipment and environmental surfaces (Table 9.7). Multi-drug-resistant strains appear to be able to survive for prolonged periods of time. Storsor *et al.* (1994) reported that VRE survived on countertops for ⩾ 7 days, on bedrails for ⩾ 3 days, and on the diaphragmatic surface of a stethoscope and telephone receiver for at least one hour. In a recent study by Yamaguchi *et al.* (1994), VAREC was recovered from cultures of bedside stand tables, over-bed tables, used linen and bedside rails. In still another publication, VAREC was recovered from similar environmental sites as well as from a stethoscope and a pulse oximeter (Boyce, Opal & Chow, 1993).

Drug-resistant strains of *Enterococcus* appear to be extremely hardy. Strains of *Enterococcus faecium* were found before and after decontamination of a fluidized microsphere bed. Microsphere beds are a popular method of caring for compromised patients. Due to the fluid's high pH, the bed has been described as 'self sterilizing'. Following use by a patient with a draining leg wound positive for *E. faecium*, cultures of the bed continued to grow the pathogen even after repeated decontamination by

Table 9.7. *Environmental surfaces in health care settings contaminated with vancomycin-resistant enterococci (VRE) or vancomycin and ampicillin-resistant enterococci (VAREC)*

Investigator	Study date	Contaminated sites
Gould & Freeman	1993	Microsphere bed
Karanfil *et al.*	1992	Intensive care unit
Livornese *et al.*	1992	Electronic thermometer
Storsor *et al.*	1994	Countertop, bedrails, telephone receiver, stethoscope diaphragm
Yamaguchi *et al.*	1994	Bedside table, overbed table, used linen, bedrails

the bed's manufacturer. The next patient to use this bed subsequently developed a draining rectal fistula with *E. faecium*, after several previous culture specimens from the wound site had been negative. Isolates from both patients and the bed were shown to be identical. It appeared that even with the pH, dryness and raised temperature (56 °C) of this type of bed, *Enterococcus* were still present (Gould & Freeman, 1993).

Nosocomial transmissions of multi-drug-resistant *Enterococcus* are being reported with increasing frequency (Karanfil *et al.*, 1992; Livornese *et al.*, 1992). According to the Centers for Disease Control and Prevention (1994), for the period 1989–1993 there has been a 34-fold increase (0.4–3.6%) in VRE occurring in intensive care units. This alarming trend has prompted them to issue guidelines for controlling VRE (Centers for Disease Control and Prevention, 1995). These guidelines are unlike any others issued since 1981. They include a recommendation for a return to limited environmental culturing to verify that procedures are adequate for cleaning and disinfecting environmental surfaces such as bed rails, charts, carts, doorknobs, faucet handles and bedside commodes. To date, there is no evidence to support the hypothesis that VRE is any more resistant to hospital-grade chemical disinfectants (i.e. quaternary ammonium compounds or phenolics), than non-VRE (Lundstrom & Lerner, 1994). Therefore, routine cleaning and chemical decontamination of environmental surfaces in healthcare facilities is essential, and should be included in infection control protocols aimed at effectively controlling VRE.

Summary

Clinical reports linking infections with objects or environmental surfaces are frequently based on circumstantial evidence. The time lapse between a person's exposure to a contaminated surface and overt infection may be great, making the association between contact and infection difficult. Some organisms survive in the environment for shorter periods of time than others; therefore, culturing surfaces may prove inconclusive. With the advent of more sensitive molecular biological techniques (e.g. the polymerase chain reaction), researchers are now able to detect the presence of nucleic acid from pathogenic microorganisms that are no longer viable or able to grow in culture (Fang *et al.*, 1993). Additional research using these new methods may provide a more sensitive and specific means of demonstrating the role of the inanimate environment as either a contributor to, or predictor of, an impending outbreak (Back *et al.*, 1993). For now, we need to acknowledge all potential reservoirs of contamination, including environmental surfaces, and take the necessary steps required for infection control in healthcare settings.

Conclusion

Research studies clearly show that potentially pathogenic bacteria can and do originate from all parts of the environment including food, people, equipment and surfaces in healthcare, day-care and domestic settings. Pathogenic contamination has been associated with many inanimate environmental surfaces and objects. What has been less clear, is the frequency with which various surfaces may be directly or indirectly responsible for producing infection.

Within the three settings discussed in this chapter it is apparent that unique conditions exist which afford opportunities for disease transmission. The healthcare setting has entered an era characterized by the emergence of bacteria resistant to multiple antibiotics and, in some cases, all known antibiotic therapies have been shown to be ineffective. Adding to this problem, extended care facilities are struggling to accommodate the ever-increasing number of elderly patients, and hospitals are faced with the presence of immunocompromised individuals. In day-care settings greater numbers of children are being exposed to a wider range of infectious agents at an earlier age. In the short term, as more families are using preschool day-care services, the economic impact of children who must stay home due to illness is substantial. The long-term health and economic effects of illness suffered by children in day-care, and their parents, has yet to be determined. In the home, where more meals are prepared and consumed than anywhere else, there appears to be an increasing risk of acquiring foodborne disease.

Throughout this chapter the theme has been one of risk evaluation. Certain settings must be considered as higher risk than others on the basis of the health and vulnerability of the occupants. We have assessed that, in general, hospitals and healthcare settings for the elderly present the highest risk, followed by an intermediate risk in day-care for young children, with the home presenting the lowest risk. However, it has been shown that certain environmental sites and surfaces within each setting may present all three levels of risk.

As discussed in the chapter, there are three basic points to be considered in order to develop an effective chemical disinfection model for environmental sites and surfaces. First, it is necessary to know the probability of significant bacterial contamination at the site or surface under consideration. In this context it is important to know the likely levels of contamination (high, low or occasional) as well as the type of contamination. Second, it is important to consider the likelihood of contamination transfer, either directly or indirectly, from a low-risk site to a higher-risk site. Third, as discussed, the possibility of increased risk, as would happen in any setting with an outbreak of infection, must be taken into account. A successful disinfection model requires a continuous assessment or awareness of environmental and other factors, together with a flexible response capability.

In addition to developing a disinfection model, it is also important to consider the disinfection process itself. Reliable and effective disinfectants are available, and assuming that the user understands the constraints of chemical disinfectants, there are many situations where a policy of effective disinfection should be continuously and rigorously enforced. Considering the available evidence, it is time to acknowledge the role of chemical disinfectants in controlling bacterial contaminants on environmental surfaces.

References

American Academy of Pediatrics/American Public Health Association (1992). Caring for our Children – *National Health and Safety Performance Standards: Guidelines for Out of Home Child Care Programs*. Elk Grove Village, IL/Washington, DC: American Academy of Pediatrics/American Public Health Association.

Anon. (1993). Mandatory safe handling statements on labeling of raw meat and poultry. *Federal Register*, **58**, 58922–34.

Ansari, S. A., Sattar, S. A., Springthorpe, V. S., Wells, G. A. & Tostowaryk, W. (1988). Rotavirus survival on human hands and transfer of infectious virus to animate and nonporous inanimate surfaces. *Journal of Clinical Microbiology*. **26**, 1513–18.

Aserkoff, B., Schroeder, S. A. & Breckman, P. S. (1970). Salmonellosis in the United States – a five year review. *American Journal of Epidemiology*, **92**, 13.

Back, N. A., Linnemann, C. C., Morthland, B. S., Pfaller, M. A. & Staneck, J. L. (1993). Recurrent epidemics caused by a single strain of erythromycin-resistant *Staphylococcus aureus*. *Journal of the American Medical Association*, **270**, 1329–33.

Bartlett, A. V., Moore, M., Gary, G. W., Starko, K. M., Erben, J. J. & Meredith, B. A. (1985). Diarrheal disease among infants and toddlers in day-care centers. I. Epidemiology and pathogens. *Journal of Pediatrics*, **107**, 495–502.

Bartzokas, C. A., Croton, R. S., Gibson, M. F., Graham, R., McLaughlin, G. A. & Paton, J. H. (1984). Control and eradication of methicillin-resistant *Staphylococcus aureus* in a surgical unit. *New England Journal of Medicine*, **311**, 1422–25.

Black, R. E., Dykes, A. C., Anderson, K. E., Wells, J. G., Sinclair, S. P., Gary, G. W. & Hatch, M. H. (1981). Handwashing to prevent diarrhea in day-care centers. *American Journal of Epidemiology*, **113**, 445–51.

Blaser, M. J. (1992). Infections due to *Campylobacter* and *Helicobacter* species. In *Infectious Diseases*, ed. S. H. Gorbach, J. G. Bartlett & N. R. Blacklow, pp. 596–601. Philadelphia: W. B. Saunders.

Boyce, J. M. (1989). Methicillin-resistant *Staphylococcus aureus*: detection, epidemiology and control measures. *Infectious Disease Clinics of North America*, **3**, 901–13.

Boyce, J. M., Jackson, M. M., Pugliese, R. N., Batt, M. D., Fleming, D., Garner, J. S., Hartstein, A. I., Kauffman, C. A., Simmons, M., Weinstein, R., O'Boyle-Williams, C. & the American Hospital Association Technical Panel on Infections Within Hospitals (1994). Methicillin-resistant *Staphylococcus aureus* (MRSA): a briefing for acute care hospitals and nursing facilities. *Infection Control and Hospital Epidemiology* **15**, 105–13.

Boyce, J. M., Opal, S. M. & Chow, J. W. (1993). Epidemiology of vancomycin-resistant *Enterococcus faecium* (VRE) (abstract 26). Presented at the *Third Annual Meeting of the Society for Hospital Epidemiology of America*, Chicago, IL.

Bradley, S. F., Terpenning, M. S. & Ramsey, M. A. (1991). Methicillin-resistant *Staphylococcus aureus*: colonization and infection in a long term care facility. *Annals of Internal Medicine*, **115**, 417–22.

Briley, R. T., Teel, J. H. & Fowler, J. P. (1994) Investigation and control of a *Shigella sonnei* outbreak in a day-care center. *Journal of Environmental Health*, **56**, 23–5.

Brooks, S. E., Veal, R. O., Kramer, M., Dore, L., Schupf, N. & Adachi, M. (1992). Reduction in the incidence of *Clostridium difficile* – associated diarrhea in an acute care hospital and a skilled nursing facility following replacement of electronic thermometers with single-use disposables. *Infection Control and Hospital Epidemiology*, **13**, 98–103.

Butz, A. M., Larson, E., Fosarelli, P. & Yolken, R. (1990). The occurrence of infectious symptoms in children in day-care homes. *American Journal of Infection Control*, **18**, 347–53.

Centers for Disease Control and Prevention (1993). Update: multistate outbreak of *Escherichia coli* 0157:H7 infection from hamburgers – Western United States 1992–1993. *Morbidity and Mortality Weekly Reports*, **42**, 258–63.

Centers for Disease Control and Prevention (1994). Nosocomial enterococci resistant to vancomycin – United States, 1989–1993 *Morbidity and Mortality Weekly Reports*, **264**, 375–82.

Centers for Disease Control and Prevention (1995). Recommendations for preventing the spread of vancomycin resistance. Hospital Infection Control

Practices Advisory Committee. *Infection Control and Hospital Epidemiology*, **16**, 105–13.

Council for Agricultural Science and Technology (1994). *Foodborne Pathogens: Risks and Consequences*, Task Force Report 122, p. 11–13. Ames, Iowa: Council for Agricultural Science and Technology.

Crossley, K., Landesman, B. & Zaske, D. (1979). An outbreak of infections caused by strains of *Staphylococcus aureus* resistant to methicillin and aminoglycosides. II. Epidemiologic studies. *Journal of Infectious Disease*, **139**, 280–7.

D'Aoust, J. Y. (1985). Infective dose of *Salmonella typhimurium* in cheddar cheese – brief report. *American Journal of Epidemiology*, **122**, 717–20.

De Wit, J. C., Broekhuizen, G. & Kempelmacher, E. H. (1979). Cross contamination during the preparation of frozen chickens. *Journal of Hygiene, Cambridge*, **82**, 27–32.

Ekanem, E. E., DuPont, H. L., Pickering, L. K., Selwyn, B. J. & Hawkins, C. M. (1983). Transmission dynamics of enteric bacteria in day-care centers. *American Journal of Epidemiology*, **118**, 562–72.

Fang, F. J., McClelland, M., Guiney, D. G., Jackson, M. M., Hartstein, A. I., Morthland, V. H., Davis, C. E., McPherson, D. C. & Welsh, J. (1993). Value of molecular epidemiologic analysis in a nosocomial methicillin-resistant *Staphylococcus aureus* outbreak. *Journal of the American Medical Association*, **270**, 1323–8.

Fekety, R., Kim, K. H., Brown, D., Batts, D. H., Cudmore, M. & Silva, J. Jr. (1981). Epidemiology of antibiotic-associated colitis: isolation of *Clostridium difficile* from the hospital environment. *American Journal of Medicine*, **70**, 906–8.

Finch, J. E., Prince, J. & Hawksworth, M. (1978). A bacteriological survey of the domestic environment. *Journal of Applied Bacteriology*, **45**, 357–64.

Fornasini, M., Reves, R. R., Murray, B. E., Morrow, A. L. & Pickering, L. K. (1992). Trimethoprim-resistant *Escherichia coli* in households of children attending day-care centers. *Journal of Infectious Diseases*, **166**, 326–30.

Gilbert, R. J. (1986). Food poisoning – the latest developments. In *Food Poisoning – Reversing the Trend*, Proceedings of a seminar organized by the Institution of Environmental Health Officers, London, October 29th.

Gill, O. N., Bartlett, C. L. R., Sockett, P. N. & Vaile, M. S. B. (1983). Outbreak of *Salmonella napoli* infection caused by contaminated chocolate bars. *Lancet*, **i**, 574–5.

Gould, F. K. & Freeman, R. (1993). Nosocomial infections with microsphere beds. *Lancet*, **342**, 241–2.

Greenwood, M. W. & Hooper, W. L. (1983). Chocolate bars contaminated with *Salmonella napoli*. *British Medical Journal*, **286**, 1394.

Haley, R. W., Hightower, A. W. & Khabbaz, R. F. (1982). The emergence of methicillin-resistant *Staphylococcus aureus* infections in United States hospitals. Possible role of the house staff–patient transfer circuit. *Annals of Internal Medicine*, **97**, 297–308.

Holaday, B., Pantell, R. & Lewis, C. L. (1990). Patterns of fecal coliform contamination in day-care centers. *Public Health Nursing*, **7**, 224–8.

Hoogenboom-Verdegaal, A. M. M. (1989). *Epidemiollogisch en mikrobiologisch onderzoek m.b.t. de mens in de regio's Amsterdam en Helmond, in 1987.* Bilthoven: *RIVM Rapport.*

Itoh, T., Saito, K. & Maruyama, T. (1980). An outbreak of acute enteritis due to *Campylobacter fetus* subspecies *jejuni* at a nursery school in Tokyo. *Microbiology and Immunology*, **24**, 371–9.

Joint Commission for the Accreditation of Hospitals (1976). *Manual for the Accreditation of Hospitals*, pp. 49–56. Chicago: Joint Commission for the Accreditation of Hospitals.

Karanfil, L. V., Murphy, M. & Josephson, A. (1992). A cluster of vancomycin-resistant *Enterococcus faecium* in an intensive care unit. *Infection Control and Hospital Epidemiology*, **13**, 195–200.

Kim, K., DuPont, H. L. & Pickering, L. K. (1983). Outbreaks of diarrhea associated with *Clostridium difficile* and its toxin in day-care centers: evidence of person-to-person spread. *Journal of Pediatrics*, **102**, 376–82.

Kofsky, P., Reed, J., Rosen, L., Tolmie, M. & Ufberg, D. (1991). *Clostridium difficile* – a common and costly colitis. *Diseases of the Colon and Rectum*, **34**, 244–8.

LaMaure, S. E. & Thompson, K. (1984). Industry sponsored day-care. *Personnel Administrator*, **29**, 53–65.

Layton, M. C., Heald, P., Patterson, J. E. & Perez, M. (1993). An outbreak of mupirocin-resistant *Staphylococcus aureus* on a dermatology ward associated with an environmental reservoir. *Infection Control and Hospital Epidemiology*, **14**, 369–75.

Lieb, S., Gunn, R. A. & Taylor, D. N. (1982). Salmonellosis in a day-care center. *Journal of Pediatrics*, **100**, 1004–5.

Livornese, L. L. Jr, Dias, S., Samel, C., Romanowski, B., Taylor, S., May, P., Pitsakis, P., Woods, G., Kaye, D., Levison, M. E. & Johnson, C. C. (1992). Hospital-acquired infection with vancomycin-resistant *Enterococcus faecium* transmitted by electronic thermometers. *Annals of Internal Medicine*, **117**, 112–6.

Lundstrom, T. S. & Lerner, S. A. (1994). Vancomycin-resistant enterococci – an overview. *Association for Professionals in Infection Control and Epidemiology, Inc. News*, **13**, 10–11.

Maki, D. G. (1982). Relation of inanimate hospital environment to endemic nosocomial infections. *New England Journal of Medicine*, **307**, 1562–6.

Maki, D. G. (1994). VRE and other resistant pathogens threaten return to pre-antibiotic era. *Hospital Infection Control*, **21**, 89–93.

McFarland, L. V., Mulligan, M. E., Kwok, R. Y. Y. & Stamm, W. E. (1989). Nosocomial acquisition of *Clostridium difficile* infection. *New England Journal of Medicine*, **320**, 204–10.

Moore, E. P. & Williams, E. W. (1991). A maternity hospital outbreak of methicillin-resistant *Staphylococcus aureus*. *Journal of Hospital Infections*, **19**, 15–16.

Noble, W. E., Virani, Z. & Cree, R. G. A. (1992). Co-transfer of Vancomycin and other resistance genes from *Enterococcus faecalis* NCTC12201 to *Staphylococcus aureus*. *Federation of European Microbiological Societies Microbiology Letters*, **93**, 195–8.

Osterholm, M. T., Reves, R. R., Murph, J. R. & Pickering, L. K. (1992). Infectious diseases and child day-care. *Pediatric Infectious Disease Journal*, **11**, S31–S41.

Paulozzi, L. J., Johnson, K. E. & Komahele, L. M. (1986). Diarrhea associated with adherent enteropathogenic *Escherichia coli* in an infant and toddler center. *Pediatrics*, **77**, 296–300.

282   D. J. GABER, T. M. CUSACK AND E. SCOTT

Peacock, J. E. Jr, Marsik, F. J. & Wenzel, R. P. (1980). Methicillin-resistant *Staphylococcus aureus*: introduction and spread within a hospital. *Annals of Internal Medicine*, **93**, 526–32.

Petersen, N. J. & Bressler, G. K. (1986). Design and modification of the day-care environment. *Reviews of Infectious Disease*, **8**, 618–21.

Pickering, L. K., Bartlett, A. V. & Woodward, W. E. (1986). Acute infectious diarrhea among children in day-care: epidemiology and control. *Reviews of Infectious Disease*, **8**, 539–47.

Pickering, L. K., Evans, D. G., DuPont, H. L., Vollet, J. J. & Evans, D. J. (1981). Diarrhea caused by *Shigella*, rotavirus, and *Giardia* in day-care centers: prospective study. *Journal of Pediatrics*, **99**, 51–6.

Pickering, L. K. & Morrow, A. L. (1991). Contagious diseases of child day-care. *Infection*, **19**, 61–3.

Polis, M. A., Tuazon, C. U., Alling, D. W. & Talmanis, E. (1986). Transmission of *Giardia lamblia* from a day-care center to the community. *American Journal of Public Health*, **76**, 1142–4.

Reboli, A. C., Canty, J. R., John, J. F. Jr & Platt, C. G. (1990). Methicillin-resistant *Staphylococcus aureus* outbreak at a veterans' affairs medical center: importance of carriage of the organism by hospital personnel. *Infection Control and Hospital Epidemiology*, **11**, 291–6.

Reves, R. R. & Pickering, L. K. (1990). Infections in child day-care centers as they relate to internal medicine. *Annual Reviews of Medicine*, **41**, 383–91.

Rutala, W. A., Katz, E. B. S., Sherertz, R. J. & Sarubbi, F. A. Jr (1983). Environmental study of a methicillin-resistant *Staphylococcus aureus* epidemic in a burn unit. *Journal of Clinical Microbiology*, **18**, 683–8.

Schumann, S. H. (1983). Day-care associated infections: more than meets the eye (editorial). *Journal of the American Medical Association*, **249**, 76.

Scott, E. (1990). The survival and transfer of potentially pathogenic bacteria from environmental sites and surfaces. PhD thesis, University of London.

Scott, E. & Bloomfield, S. F. (1990). The survival and transfer of microbial contamination via cloths, hands and utensils. *Journal of Applied Bacteriology*, **68**, 271–8.

Scott, E. & Bloomfield, S. F. (1992). An in-use study of the relationship between bacterial contamination of food preparation surfaces and cleaning cloths. *Letters in Applied Microbiology*, **16**, 173–7.

Scott, E., Bloomfield, S. F. & Barlow, C. G. (1982). An investigation of microbial contamination in the home. *Journal of Hygiene, Cambridge*, **89**, 279–93.

Simmons, B. P. in consultation with Hooten, T. M. & Mallison, G. F. (1981). *Guidelines for the Prevention and Control of Nosocomial Infection: Guidelines for Hospital Environmental Control, Cleaning, Disinfection and Sterilization of Hospital Equipment*. Center for Disease Control, US Department of Health and Human Services

Skoutelis, A. T., Westenfelder, G. O., Beckerdite, M. & Phair, J. P. (1994). Hospital carpeting and epidemiology of *Clostridium difficile*. *American Journal of Infection Control*, **22**, 212–17.

Smith, L., Prince, H. N. & Johnson, E. (1981). Bacteriologic studies on electronic hospital thermometers. *Infection Control*, **2**, 315–16.

Sockett, P. N. (1993*a*). Foodborne statistics: Europe and North America. In

Encyclopaedia of Food Science, Food Technology and Nutrition, pp. 2023–31. London: Academic Press.

Sockett, P. N. (1993b). Social and economic aspects of foodborne disease. *Food Policy* **April**, 110–19.

Sockett, P. N., Cowden, J. M., Baigue, S. L., Ross, D., Adak, G. K. & Evans, H. (1993). Foodborne disease surveillance in England and Wales: 1989–1991. In *Communicable Disease Review 3*, pp. R159–173. London: Public Health Laboratory Service.

Sockett, P. N. & Roberts, J. A. (1991). The social and economic impact of salmonellosis: a report of a national survey in England and Wales. *Epidemiology and Infection*, **107**, 335–47.

Spika, J. S., Parson, J. E., Nordenburg, D., Wells, J. G., Gunn, R. A. & Blake, P. A. (1986). Hemolytic uremic syndrome and diarrhea associated with *Escherichia coli* 0157:H7 in a day-care center. *Journal of Pediatrics*, **109**, 287–91.

Storsor, V., Frauchiger, W., Cooper, I., Peterson, L. & Noskin, G. (1994). Survival of enterococci on environmental surfaces (abstract L-42). Presented at the *American Society of Microbiology General Meeting*, Las Vegas, May.

Stratton, C. W. (1990). *Clostridium difficile* colitis in the hospital setting: a potentially explosive problem. *Infection Control and Hospital Epidemiology*, **11**, 281–2.

Sullivan, P., Woodward, W. E., Pickering, L. K. & DuPont, H. L. (1984). Longitudinal study of diarrheal disease in day-care centers. *American Journal of Public Health*, **74**, 987–91.

Thacker, S. R., Addiss, D. G., Goodman, R. A., Holloway, B. R & Spencer, H. C. (1992). Infectious diseases and injuries in child day-care: opportunities for healthier children. *Journal of the American Medical Association*, **268**, 1720–6.

Thompson, R. L., Cabezudo, I. & Wenzel, R. P. (1982). Epidemiology of nosocomial infections caused by methicillin-resistant *Staphylococcus aureus*. *Annals of Internal Medicine*, **97**, 309–17.

Todd, E. (1991). Epidemiology of foodborne illness: North America. In *Foodborne Illness, A Lancet Review*, editorial advisors: W. M. Waites & J. P. Arbuthnott, pp. 9–15. London: Edward Arnold.

Van, R., Morrow, A. L., Reves, R. R. & Pickering, L. K. (1991a). Environmental contamination in child day-care centers. *American Journal of Epidemiology*, **133**, 460–9.

Van, R., Wun, C. C., Morrow, A. L. & Pickering, L. K. (1991b). The effect of diaper type and overclothing on fecal contamination in day-care centers. *Journal of the American Medical Association*, **265**, 1840–4.

Van Schothorst, M., Huisman, J. & Van Os, M. (1978). *Salmonella-ondetzoek* in huishoudens met Salmonellose bij zuigelingen. *Nederlands Tijdschrift voor Geneeskunde*, **122**, 1121–5.

Weniger, B. G., Ruttenber, A. J., Goodman, R. A., Juranek, D. D., Wahlquist, S. P. & Smith, J. D. (1983). Fecal coliforms on environmental surfaces in two day-care centers. *Applied and Environmental Microbiology*, **45**, 733–5.

World Health Organization (1992). *Surveillance Programme for Control of Foodborne Infections and Intoxications in Europe, Fifth Report 1985–1989*. Berlin: Robert von Ostertag-Institute, FAO/WHO Collaborating Centre.

Yamaguchi, E., Valena, F., Smith, S. M., Simmons, A. & Eng, R. H. K. (1994). Colonization pattern of vancomycin-resistant *Enterococcus faecium*. *American Journal of Infection Control*, **22**, 202–6.

Chapter 10

Sterilization and disinfection of medical devices

AARON B. MARGOLIN and
VIRGINIA C. CHAMBERLAIN

Introduction

An important aspect in the prevention of nosocomial infections is the use of sterile drugs and devices in patient treatment. The increasing number of immunocompromised persons, including those with cancer and AIDS, transplant patients and the elderly, has led to observations of infections caused not only by known pathogenic microorganisms but also by opportunistic microorganisms. Sterilization and disinfection of medical devices can take place industrially, in healthcare facilities or, for certain devices, in the home.

According to the Centers for Disease Control and Prevention (CDC), about 5% of all hospital patients acquire an infection while hospitalized (General Accounting Office/Human Resources Division, 1993). Hospital-acquired infections prolong hospital stays, increase the costs of patient care and, in some cases, can lead to death. Research has linked some hospital-acquired infections to the use of medical devices, such as endoscopes, that have been inadequately sterilized or disinfected. The inadequacy of proper sterilization or disinfection may have several foundations. Medical instruments today are much more complicated than those used years ago. The complexity of such instruments presents new challenges for hospital disinfection policies. Use of fiberoptics in scopes and the advent of noninvasive surgery require cleansing and proper sterilization or disinfection of equipment that may be the thickness of only a human hair, yet may still be able to harbor highly infectious organisms. Compounding the problem is the economic factor, which must be considered. The complexity of certain instruments may require long labor-intensive processes for adequate instrument decontamination. The costs of such instruments and the expected life of the instrument are all too often the major considerations when sterilization and disinfection policies are designed. The problem is compounded when the claims about liquid sterilants and disinfectants are not valid. The lack of proper microbicidal activity is a problem that arises from both antiquated testing procedures of the Association of Official Analytical Chemists (AOAC) and false marketing claims by manufacturers of liquid germicides.

Within the United States the regulation of hospital sterilants and disinfectants falls under several different federal agencies. The United States Environmental Protection Agency (EPA) regulates them as pesticides. When it is used on medical devices, the sterilant/disinfectant itself is considered a medical device, and is therefore also regulated by the Food and Drug Administration (FDA). The Federal Trade Commission (FTC) is the government agency responsible for preventing false or deceptive product advertising of hospital sterilants and disinfectants (General Accounting Office/Human Resources Division, 1993). Recently, the EPA, FDA and FTC have taken regulatory action against several germicide manufacturers, addressing problems such as test failures in microbicidal testing, manufacturing problems and false and deceptive advertising.

This chapter includes a discussion of the role of healthcare facilities in prevention of infection from medical devices. Also included are discussions on sterilizing and disinfecting agents, a description of the impact of the FDA's regulation of standards for medical device sterilization or disinfection, and information on national and international sterilization and disinfection standards for industry and healthcare facilities.

The infection control team and its role

The range of nosocomial infections is not limited to the hospital. The spread of infectious organisms by inadequately cleaned and disinfected or sterilized instruments has been documented for private clinicians as well. Establishment of an infection control policy should not be limited to large hospital institutes and, in one form or another, needs to be adopted by all medical personnel responsible for the care and preparation of medical instruments. All types of infection control policies begin with a basic understanding of how organisms are transmitted, where they come from and how they can be controlled.

Nosocomial pathogens may arise, especially in hospitals or institutions, *de novo*, by such means as the transfer of antibiotic resistant genes, or they may be brought in by staff or patients admitted to the unit (S. Mehtar, personal communication). There are two main routes of transmission for nosocomial infections: airborne spread and direct contact. Direct contact can be further classified into that of body parts or clothing, or medical instruments. Medical instrumentation is then further subdivided into clinical equipment, e.g. viewing devices (termed scopes), surgical metals, ventilators, etc. and non-clinical equipment, e.g. bedpans, bowls, etc. Both of these categories can be implicated in the spread of infectious material if they are not properly sterilized or disinfected. In most cases, the more invasive of the body the medical device is, the greater the chance that serious infection or fatality can arise. This is true also because invasive medical instrumentation is complex and more difficult to clean and properly decontaminate than are simple devices.

Infection control teams, whether they be a single nurse or clinician, or a number of staff members, have certain responsibilities that can help limit the spread of nosocomial infections by medical devices. One such responsibility is the establishment of a decontamination, sterilization and disinfection policy. Many hospitals and physician's offices have no clear policy, and practices are often based more on tradition and habit than on logic. This is often not the fault of the individuals. Complexities in medical instruments, false marketing claims and lack of knowledge often contribute to improper instrument preparation and care. Certain basic rules, such as cleansing of the instrument, will always apply and should always be considered when medical instruments are prepared. It is no surprise that the efficiency of all microbicidal procedures is greatly reduced, to the point at times of not being effective at all, by the amount of organic matter present on the instrument. Further practices concerning the microbicidal preparation of medical instruments will be discussed later in this chapter. It is the role of the infection control team to ensure that individuals preparing medical instruments are knowledgeable about their actions and the implications of not doing their job properly or taking it seriously. Many times individuals responsible for washing instruments or preparing cold sterilizing solutions are not aware of the ramifications of improper quality control. It is one of the responsibilities of the infection control team to see that quality control measures are made available to participating individuals. These measures should range from education of personnel to microbiological and chemical testing of solutions and medical instruments. In one study by Kaczmarek *et al.* (1992), an examination of 22 hospitals and four ambulatory care centers in three states revealed that 78% of the facilities failed to sterilize all biopsy forceps and that in 23.9% of 71 gastrointestinal endoscopes more than 100 000 colony forming units (CFU) of bacteria were isolated from the internal channels of the scopes.

The infection control team can begin to develop sterilization and disinfection policies by assessing the current practices of the institute. One of several points that needs to be emphasized is that disinfection refers to the destruction of those microorganisms that are known to be pathogenic, and sterilization carries this precaution further by destroying all microorganisms, regardless of whether or not they have been identified as pathogens. Other questions that need to be addressed include:

(1) Would cleaning be sufficient?
(2) If disinfection is needed, then is the required level low, intermediate or high?

(3) Is sterilization required?
(4) Could disposables be used economically instead?
(5) What are the germicide requirements given by the manufacturer of the device?

Coates & Hutchinson (1994) developed several guidelines for choosing the type of cleaning, disinfecting and sterilizing required by the different items usually encountered in medical facilities. Their guidelines are not very different from those offered by the CDC.

In summary, the infection control team, whether it is composed of one or more than one individual, needs to be actively involved in the care and preparation of medical equipment. This process begins with education of personnel and continues with the development of monitoring practices. The infection control team, like medical instruments, must be able to change to meet the needs of the medical community. This is not a small task, and underestimating the role of the infection control team in establishing sterilization and disinfection policies will increase the possibility of the spread of nosocomial infections by medical instruments.

Definition of terms

To fully understand the decision-making process that is required in choosing the type of process appropriate for a particular device, it is important to understand the terms sterilization, disinfection, decontamination and antisepsis. Often these terms are used interchangeably and mistakenly.

Sterilization is an absolute term and refers to the absence of all life and subcellular infectious material. This last category includes viruses and a host of infectious particles that have still not been fully elucidated, such as prions and satellite RNAs. Most important is the concept that sterilization is an absolute term, that is, things cannot be 'almost sterilized' or 'partially sterilized'; for something to be sterile, all infectious agents must be destroyed. Since bacterial endospores usually present the most resistance to inactivation, they are commonly used as indicators or barometers to measure the effectiveness of the sterilization procedure.

In hospitals and large institutions, sterilization by either moist or dry heat is the most common way to sterilize items. When items are heat labile, the second most used method is a process known as ethylene oxide gas sterilization. Unfortunately, ethylene oxide processes require long recovery periods for exhaustion of the gas, making this method less desirable for items that are required quickly.

Disinfection is generally less rigorous than sterilization. Ideally, disinfection would remove any known and potential pathogens, but not necessarily all forms of microbial life. It is for this reason that disinfection practices may not be sufficient for instruments that are going to be used on immunocompromised hosts, since their weakened immunological defenses can be overcome by nonpathogens.

The effectiveness of disinfection is influenced by many factors, each able profoundly to influence the end results. Major factors that influence the outcome of disinfection include the type of germicide and the killing activity it is directed towards, the microbial load, the extent of contaminating organic material, the shape and complexity of the medical instrument, the type of material the instrument is made from, the temperature of the germicide and disinfection conditions, the germicide exposure time and the concentration of the germicide.

Confusion between 'sterilization' and 'disinfection' usually does not occur with heat or ethylene oxide, since neither method is used as a form of disinfection. However, confusion can arise when liquid germicides are used as the method of microbial inactivation. Quite often the difference between sterilization and disinfection with use of these compounds is due to the concentration of the germicide and the length of time it is used. Time and concentration have an inverse relationship; as one decreases the other needs to increase for the same microcidal activity to occur. The problem is increased by the reuse of germicides for more than one medical device ('germ' is used here as a term that refers to pathogenic or potentially pathogenic microorganisms). When a germicide is formulated at a concentration used for sterilization and then is used repetitively, organic material, water from wash solutions and other material can dilute the germicide concentration sufficiently that it no longer acts as a sterilant, but as a disinfectant.

'Decontamination' can be a very misleading term. In most instances, decontamination makes the instrument safe to handle, but does not adequately prepare a medical instrument for reuse. Confusion arises when this term is used with personnel who are not acutely aware of the differences between disinfection practices, which prepare instruments for patient use, and decontamination, which prepares an instrument for safe handling by medical care workers. In most cases, 'decontamination' should be used to describe a 'used' or 'dirty' medical device and disinfection for a 'clean' or 'to be used' device.

'Antisepsis' is defined as a germicide that is used on skin or living tissue for the purpose of inhibiting or destroying microorganisms (Favero & Bond, 1991). A favorable distinction between disinfection and antisepsis is based upon the material on which they are used. Although many antiseptics contain the same active ingredient as that found in a disinfectant, they are usually found at much lower concentrations. Disinfection compounds are usually meant to be used on inanimate objects, and the antiseptic is meant for living tissue. Use of a disinfectant on living tissue can cause great harm and may result in permanent scarring of the skin. Use of an antiseptic on inanimate surfaces will not destroy the microbes, especially spores, which must be killed before the instrument can be reused on a subsequent patient. Antiseptics are regulated solely by the FDA.

Disinfection and reprocessing recommendations

For medical devices, the level of disinfection is dependent on the type of instrument and how it is used. In 1972, Dr E. H. Spaulding devised a system of ranking sterilization into three categories, based on the nature and use of the instrument and its ability to transmit infectious organisms. The categories are: (1) critical; (2) semicritical; and (3) non-critical. The types of disinfection used for each category have been classified as sterilization and high-level, intermediate-level and low-level disinfection. This scheme was adapted by the Centers for Disease Control (1985).

Critical instruments are devices normally required to be sterile prior to use. This category contains devices that are introduced directly into the bloodstream or into areas of the body that are normally sterile. Examples include needles, scalpels, transfer forceps, cardiac catheters, implants and also the inner surface components of extracorporeal blood-flow devices such as the heart–lung oxygenator and the blood-side of artificial kidneys (hemodialysers) (Favero & Bond, 1991).

Semicritical instruments are devices that usually do not penetrate the surface of the body. They do come in contact with mucous membranes; however, in most cases they do not penetrate through the mucous membrane barrier. Examples of this category include endoscopes, endotracheal and aspirator tubes, bronchoscopes, respiratory therapy equipment, cystoscopes, vaginal specula and urinary catheters. Although sterility is desired for this category, it is often impractical, since these objects must be used repetitively during the day. Their multi-use limits sterility by such means as ethylene oxide. Also, many of these instruments are not heat tolerant, and therefore cannot be sterilized by steam or dry heat. The infection control team and healthcare individual must use caution when making a decision on how to decontaminate these items. Additionally, the infection control team and healthcare provider must be assured that such instruments are cleaned properly. Most instruments that carry flexible fiberoptic channels and/or biopsy channels can easily become clogged with organic matter. This further compounds the problem of nosocomial infections, since the efficiency of any germicidal practices will be greatly reduced by the presence of even small amounts of contaminating organic material (Gelinas & Goulet, 1982).

Semicritical instruments should at a minimum be subjected to a powerful, broad-spectrum procedure that can be expected to destroy a few bacterial spores, most fungal spores, all ordinary vegetative bacteria, tubercle bacilli and small or nonlipid viruses, and medium-sized or lipid viruses. However, it cannot be emphasized enough that cleaning of the instrument is crucial for successful liquid sterilization or disinfection.

For many 'noncritical' reusable objects, such as blood-pressure cuffs, bedside utensils, wheelchairs, crutches, etc., cleansing with a mild disinfecting detergent will be adequate. These instruments rarely transmit disease and only sometimes come in contact with human skin.

Commonly used methods

Heat sterilization is currently the best form of sterilization, but as mentioned previously, it is not always practical. The same is true for ethylene oxide, due to the long time it takes before all the residual gas has dissipated and the type of instrumentation that is required for the safe handling of the compound. The use of liquid germicides offers an alternative method that, when used properly, can provide a relatively quick turn-around time and ensure the level of disinfection or sterility required, and will not harm the instrument being sterilized.

Since it is neither necessary nor practical to sterilize all environmental objects, hospitals and healthcare facilities should have policies that explain the type of care medical instruments should require. Device manufacturers' instructions should be included in the policy. Which process is indicated for an object depends on the object's intended use and, sometimes, the type of contamination. Cleaning is the physical removal of organic material or soil from objects, and it is usually done by using water with or without detergents. Generally cleaning is not designed to kill microorganisms but to remove them. Sterilization, on the other hand, has as its goal the complete removal or destruction of all forms of microbial life. Disinfection achieves the intermediate measure between physical cleaning and sterilization, and is carried out by several different methods, one of which is the use of liquid chemical germicides. The degree of disinfection accomplished depends upon several factors, but principally upon the strength of the agent and the nature of the contamination. Some high-level disinfection procedures are capable of producing sterility, if they are continued long enough. The length of time a germicide is used, the concentration of germicide used, the material it is used upon, the contaminating organic load and the type of microbe will determine whether high-level, intermediate-level or low-level disinfection is achieved (US Department of Health and Human Services, 1982).

Choosing the correct disinfectant is not an easy task, and therefore should be part of the infection control team's recommendations. If everything could be steam sterilized, it would not be necessary to have these discussions, but steam sterilization is not always practical, and therefore alternative forms of sterilization and disinfection, such as the use of liquid germicides, must be explored. Currently, there are approximately 14 000 germicidal products that have been registered with the EPA, and, on the labels of these products, approximately 300 active ingredients have been listed. Fourteen of these 300 active ingredients are listed in 92% of the registered products (Favero & Bond, 1991). Due to the competitiveness of this field, the infection control practitioner should be aware that claims of sterilization and disinfection can be exaggerated.

The following is a brief overview of the commonly found active ingredients in liquid germicides. It is by no means an exhaustive list and the reader is encouraged to seek out more in-depth information by consulting the FDA in the United States or other appropriate national and international organizations prior to making their choice of germicide and disinfection procedure for their medical devices. At times, although this is not always cost-effective, use of disposable instruments will help solve this problem.

One of the most common active ingredients in germicides, and the one that usually has the greatest amount of microbicidal activity against a wide range of organisms, including bacterial spores, is glutaraldehyde (Boucher, 1972). Glutaraldehyde is a saturated dialdehyde that is chemically related to formaldehyde and has been shown to be 2–8 times more sporicidal than formaldehyde (Borick, 1968). The mode of action of glutaraldehyde on microorganisms is by alkylation reactions with the amino and sulfhydryl groups of proteins and ring nitrogen atoms of purine bases (Favero & Bond, 1991). Although glutaraldehyde is an excellent microbicidal compound, its activity decreases over repeated uses and storage. In one study, the concentration of alkaline glutaraldehyde (2–2.5%) was shown to fall to < 1% after 20 endoscopes were processed in a manual bath (Hanson et al., 1989). Factors that glutaraldehyde users should be aware of that influence its microbicidal activity include dilution, contact time, pH, contaminating organic load, aging and the temperature at which it is used.

Glutaraldehyde-based germicides are some of the most widely used sterilants and disinfectants of instruments for endoscopic and respiratory therapy and anesthesia. There are some inherent dangers to the patient and the user from glutaraldehyde-based germicides. These range from noxious and irritating effects to a variety of toxic reactions in the patient and especially the healthcare worker. Individuals working with glutaraldehyde solutions directly should always have proper ventilation in their work areas and use personal barriers (e.g. gloves, face protection) to avoid adverse reactions (Favero & Bond, 1991).

Formaldehyde, depending on its concentration, is classified as a high-level (8% formaldehyde plus 70% alcohol) or intermediate- to high-level (4–8% formaldehyde in water) disinfectant. The activity of formaldehyde is much like that of glutaraldehyde, producing alkylation of amino and sulfhydryl groups of proteins and ring nitrogen atoms of purine bases such as guanine. Even though formaldehyde is classified as a high-level disinfectant, there are several drawbacks to its use that make it undesirable as a disinfectant. These include contact times, which can range up to 18 h or longer, irritating fumes, potential carcinogenicity and its toxicity for tissue which requires that disinfected material be thoroughly rinsed before use.

Alcohol has been widely used as an antiseptic for cleansing skin, for example prior to the use of a needle for injecting solutions into the body or for removal of blood. However, the low microbicidal activity of alcohol and its rapid evaporation when exposed to air makes alcohol-based liquid germicides a poor choice for disinfecting medical instruments. Further, rubber articles absorb alcohol, and irritation of the skin or mucous membranes may follow from exposure of patients to those alcohol-treated articles.

Phenol, or carbolic acid, is one of the oldest germicidal agents used in hospitals and in the home for disinfecting environmental surfaces. Solutions containing phenol are considered to be intermediate- to low-level disinfectants and have only moderate bactericidal activity. One problem with using phenol-based compounds on medical instruments, but which makes them attractive for use on environmental surfaces, is their residual effect. If all the phenol is not removed during rinsing, and this is difficult, then subsequent application of water to the phenol residual on a dry surface previously treated with a phenolic can redissolve the residual chemical so that it again becomes bactericidal. This is helpful for environmental surfaces, but it can be detrimental for critical or semicritical medical instruments that come in contact with the patient. One other advantage to using phenolic compounds on noncritical devices and environmental surfaces is that phenolic compounds in the concentration range of 1–2% remain active when in contact with organic soil. This makes phenolic compounds the ideal candidates for the general clean-up of soiled noncritical surfaces commonly found in the laboratory or the home.

Chlorine-based solutions or germicides have a wide microbicidal activity, depending upon the concentration of free chlorine. They have been reported to be effective against all forms of vegetative bacteria, bacterial spores, fungi and viruses at very low concentrations. Additionally, chlorine-based solutions are usually among the least expensive germicides and working solutions can be formulated with the use of household bleach. Unfortunately, due to the extremely reactive nature of chlorine, it is very corrosive to most medical devices. Chlorine-based disinfectants are ideally used on areas that are impervious to chlorine corrosion. Additionally, due to the low cost and readily available supply, chlorine-based solutions can be used for decontamination of objects when procedures such as autoclaving cannot immediately be performed. One caution about using chlorine as a decontaminate should be noted: it could lead to a false sense of security. Chlorine-based solutions, when left exposed to the atmosphere, lose their potential biocidal activity over time through evaporation of the chlorine. Additionally, chlorine reacts with contaminating organic material, thereby reducing the amount of free chlorine that is available to react with infectious agents. This can be rectified by covering the chlorine bath and periodically adding more chlorine. The amount of chlorine and frequency of replenishment must be evaluated by the infection control team or healthcare provider.

Methods for evaluating liquid chemical sterilants and disinfectants

Currently, the AOAC Sporicidal Test is used for evaluating liquid sterilants and high-level disinfectants to be used on medical instruments. Generally, it is felt that resistance to inactivation by a germicide follows the flow chart in Figure 10.1. Hence, bacterial spores become the ideal candidate, based upon resistance, for evaluating germicide efficiency. Unfortunately, the nature of the organism and the test method present problems that can often yield extremely variable results when certain chemical sterilants and disinfectants are evaluated.

In brief, spores of *Clostridium sporogenes* (ATCC No. 3584) and spores of *Bacillus subtilis* (ATCC No. 19659, or now also used is *B. subtilits* var. *niger* ATCC No. 9372) are grown and harvested. The spores are enumer-

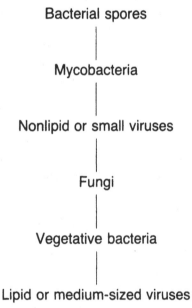

Bacterial spores
|
Mycobacteria
|
Nonlipid or small viruses
|
Fungi
|
Vegetative bacteria
|
Lipid or medium-sized viruses

Figure 10.1   Descending order of resistance in germicides.

ated by the CFU method and checked for 'hardiness' by evaluation of their resistance to hydrochloric acid. One problem is the lack of a defined growth medium and an understanding of the growth medium's effect on spore production. Growth media for both spore producing test organisms require the use of 'common' soil. Danielson (1993) demonstrated variability in spore resistance to hydrochloric acid when different batches of soil from different areas were used.

Once the spores have been enumerated and checked for resistance to hydrochloric acid, they are applied to either of two types of solid test surfaces, termed 'carriers': a suture loop (size 3 surgical silk) or a pennycylinder (porcelain and stainless steel). The AOAC Sporicidal Test does not require enumeration of the spores on carriers.

Application of the spores to the carriers is accomplished by dipping each carrier into a spore suspension. This part of the test is also very problematic. Although the spores may be enumerated by the CFU method, the ratio of live to dead spores is not determined and, consequently, the amount of non-viable organic matter placed on the carriers can vary. Additionally, enumeration of the amount of spores applied to each carrier has always been by subsequent removal of spores and enumeration by cultivation of organisms on control cylinders. Work by Gangi, Margolin & Brennen (1994) showed that this really only enumerated the amount of spores that could be removed, not the amount that could be applied. Their work demonstrated that many more removal steps (usually by sonication and washing) than are required by the procedure are necessary for complete removal of spores. Carrier material also played a critical role in the ability to remove and enumerate spores. Although it was possible to remove all spores from most of the pennycylinders, it was not possible to do the same from the suture loops. Cultivation of loops demonstrated spore retention, even after extensive sonication and washing.

Danielson (1993) has also shown that the adsorptive properties of the spores that have been applied to the carriers is partially dependent on how the spores were grown and processed. These factors combined to yield sporicide tests that were often very variable and inconsistent. Danielson's work demonstrated that different percentages of glutaraldehyde (0.05, 1.0 and 2.0%) were effective for destroying spores loaded onto porcelain carriers at a concentration of 10^4 spores per carrier, but were not effective when carriers were inoculated with 10^6 spores. The current AOAC test method does not dictate a consistent spore load on carriers which further increases the variability of test results.

Finally, the last problem associated with the current test is the material from which carriers are made. While porcelain and steel are relatively resistant to germicides, they do not satisfactorily represent the material from which the common flexible fiberoptic scopes are made. Today the materials used in manufacturing endoscopes consist more of plastics and

Teflon than of porcelain or steel. Work by Gangi *et al.* (1994) demonstrated that different materials have different protective properties and some can enhance spore survival when they are exposed to a chemical germicide. While spores are generally considered to be the most resistant form of microbe, this may not be true when they are thought of in combination with the material comprising the medical instrument. For instance, owing to size and electrostatic interactions, certain nonenveloped small viruses may be better protected from germicide exposure than are larger spores. So, although the AOAC Sporicidal Test may indicate that a specific germicide is efficient at destroying spores on a carrier, that same germicide may not be able to penetrate the minute cracks in the outer plastic covering of a sigmoidoscope, which are able to harbor viruses with hydrophobic capsid properties.

FDA regulation of devices and chemical sterilants

Except for the authority to ban harmful devices, the FDA's regulatory authority over medical devices began with the passage in 1976 of the medical device amendments to the Food, Drug and Cosmetic Act. Other refinements regarding devices have followed. The current Act defines medical devices as follows:

> The term 'device' ... means an instrument, apparatus, implement, machine, contrivance, implant, *in vitro* reagent, or other similar or related article, including any component, part, or accessory, which is –
> (1) recognized in the official *National Formulary*, or the *United States Pharmacopeia*, or any supplement to them,
> (2) intended for use in the diagnosis of disease or other conditions, or in the cure, mitigation, treatment, or prevention of disease, in man or other animals, or
> (3) intended to affect the structure or any function of the body of man or other animals, and which does not achieve its primary intended purposes through chemical action within or on the body of man or other animals and which is not dependent upon being metabolized for the achievement of its primary intended purposes.
> [Section 201(h), F, D & C Act]

Devices to be used in a sterile condition such as adhesive bandages, implants, intravenous sets, urinary catheters, sterilizers, washer disinfectors and sterilizers, and liquid chemical sterilants and disinfectants all meet the legal definition of a device and are regulated by the FDA. The term 'sterile' is an absolute term defined in the dictionary as 'free from living microorganisms'; 'sterilize' is defined as 'to free from living microorganisms'. It would be impractical to demonstrate that all product units are rendered sterile during a sterilization process, since sterility testing of all units would render them unusable. In practice, processes are designed to allow only a small probability of nonsterility. FDA policy has defined the appropriate probability of nonsterility for devices, depending on their intended use, as follows:

> The desired sterility assurance level (SAL), or probability of a unit being nonsterile after exposure to a valid sterilization process, varies according to the intended use of the device. Sterilized articles not intended to contact compromised tissues [i.e. tissue that has lost natural barrier integrity] are generally thought to be safe for use with an SAL of $10-3$; that is, a probability of one nonsterile unit in a thousand. Invasive and implantable devices should have an SAL of $10-6$, that is no more than one nonsterile unit in a million.
>
> Center for Devices and Radiological Health (1985)

Determining the extent of processing necessary to achieve the appropriate SAL involves development of a microbial lethality curve, followed by extrapolation to the intended SAL. The microbial lethality curve can be based on killing of the natural bioburden on the device or a known population of indicator microorganisms with high resistance to the chosen sterilization agent. An example of this type of extrapolation is shown in Figure 10.2. This study is part of the validation of the sterilization process for a

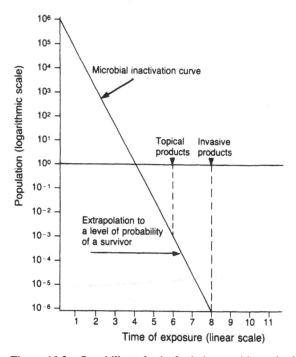

Figure 10.2 Overkill method of ethylene oxide cycle development. Reprinted from Association for the Advancement of Instrumentation (1988), with permission.

device. In addition to determining the microbial lethality, validation includes studies to assure that the sterilization process will not compromise the performance or safety of the device. The state of the art for industrial validation of sterilization processes is contained in national and international sterilization standards. More information on these standards and recommended practices for hospital sterilization of devices is provided later in this chapter.

Devices intended to be used in a sterile condition may be sterile when sold or may be nonsterile when sold, to be sterilized by the user facility. Reusable devices need to be reprocessed between uses, to prevent transmission of infectious agents between patients; the level of reprocessing necessary depends on the intended use. The FDA uses its premarket and postmarket regulatory authority to assure that devices sold in a sterile condition are processed to the appropriate SAL and that manufacturers provide, in labeling, adequate validated methods for devices to be sterilized or disinfected by the users.

As mentioned previously, Centers for Disease Control published guidelines in 1985 regarding the level of reprocessing necessary for categories of reusable medical devices, depending on their intended use. Those guidelines were the basis for portions of the FDA/EPA *Memorandum of Understanding on Chemical Germicides* discussed more fully later in this chapter. This stated the following:

> Critical devices are those that are intended to be introduced directly into the human body, either into or in contact with the bloodstream or normally sterile areas of the body. Critical devices should be sterilized between patients. Semicritical devices are those which are intended to contact intact mucous membranes but which do not ordinarily penetrate the blood barrier or otherwise enter normally sterile areas of the body. For these devices, sterilization is desirable, but not mandatory. These devices must be subjected at least to a high level disinfection process using a sterilant, but for a shorter time than that required for sterilization ... Noncritical devices [intended for contact with intact skin] and medical equipment surfaces must be subjected to intermediate or low level disinfection.

The FDA recommends that critical and semicritical devices be sterilized by processes that can be biologically monitored, such as steam, ethylene oxide and dry heat. Devices that cannot be subjected to a biological monitoring can be reprocessed by means of a liquid chemical sterilant.

Sterile devices

In 1978, the FDA published the *Regulation, Good Manufacturing Practice for Medical Devices*, establishing quality assurance requirements for device manufacturing under the United States Code of Federal Regulations (21CFR800). The FDA reviews information regarding industrial sterilization of devices either submitted to the FDA in premarket applications or obtained by the FDA through good manufacturing practice (GMP) inspections of device manufacturing operations. Most devices marketed prior to the 1976 amendments to the Act are exempt from premarket submissions to the FDA. Devices entering the market after that date or before the amendments with significant changes in the device, its intended use, or its manufacturing process are subject to FDA review prior to marketing, unless exemptions have been granted by regulation. There are two premarket review processes. The premarket notification, known as a 510(k) after the section in the Act, is required for devices that are similar to other existing devices on the market prior to the amendments. The manufacturer must submit information to demonstrate that the new device is substantially equivalent to the pre-amendment device. Under current agency policy, the manufacturer is required to submit information concerning the type of sterilization process to be used and the SAL for the device, but is not usually required to submit complete data on the sterilization process and its validation in a 510(k) (510(k) Sterility Review Guidance, February 12, 1990). Validation and control of the sterilization process is required to meet GMP requirements, which are discussed more fully below. Investigational devices, for which the use would pose a significant risk to the patient, require a Premarket Approval Application, including clinical studies and manufacturing information. Manufacturers are required to submit detailed information on the manufacturing processes, including sterilization and process validation studies. Of course, these devices are also subject to GMP requirements.

Manufacturers can petition the agency for an exemption from GMP requirements for a particular device; no exemptions have been granted for sterile devices, because of the importance of GMPs in assuring the sterility of these devices. The agency is required by the Act to conduct GMP inspections of device manufacturers biennially. The GMP regulation requires that a process such as sterilization be validated, controlled and monitored to assure that the sterile device meets its specifications, which would include sterility. Monitoring of sterilization processes involves continuous recording of physical and chemical parameters and biological monitoring with biological indicators when appropriate. Biological indicators are carriers impregnated with a known population of microorganisms, usually bacterial spores, having a defined resistance to the specified sterilization process. Assuring adequate sterilization involves controlling all parts of device manufacturing prior to sterilization, to ensure that the bioburden resulting prior to sterilization is controlled and its resistance will not exceed that used during validation of the process. GMP requirements regarding component acceptance and environmental monitoring and control during manufacturing address that issue. Additional requirements for manufacturing, equipment maintenance and calibration also impact on the adequacy of sterilization. In addition to the sterilization process itself, the packaging process must also be validated and controlled, to ensure that the packaging materials and seals can withstand the sterilization process.

The FDA has a written inspection program for sterile devices covering the most commonly used industrial sterilization methods such as steam ethylene oxide gas, radiation and aseptic processing as well as the packaging process. If the FDA finds during an inspection that the sterilization process cannot adequately ensure the device's sterility, then regulatory action is considered. That action can include sending a Warning Letter, seizure of the devices, mandatory recall of the devices, enjoining the firm from further manufacturing until corrections are made, or prosecution and civil penalties.

Currently, the FDA is in the process of revising the GMP regulation as mandated in the Safe Medical Device Act of 1990. This law directs the FDA to add design validation testing to the GMP requirements to bring it into line with the International Organization for Standardization (ISO) quality standards being used worldwide. This revision will give the FDA more authority to determine if devices are adequately designed for sterilization or reprocessing, if the devices are intended to be reused.

Single-use devices

The FDA has a Compliance Policy Guide (CPG) entitled *Reuse of Medical Disposable Devices*. This CPG states the following:

> The FDA ... finds that there is a lack of data to support the general reuse of disposable medical devices ... The fact that devices are labeled disposable is indicative of this lack of data. In order for a device to be considered 'reusable', it must be capable of withstanding necessary cleaning, and resterilization techniques and methods, and continue to be safe and reliable for its intended use. The FDA has concluded, therefore, that the institution or practitioner who reuses a disposable medical device should be able to demonstrate: (1) that the device can be adequately cleaned and sterilized, (2) that the physical characteristics or quality of the device will not be adversely affected, and (3) that the device remains safe and effective for its intended use. Moreover, since disposable devices are not intended by the manufacturer or distributor for reuse, any institution or practitioner who resterilizes or reuses a disposable medical device must bear full responsibility for its safety and effectiveness.

If in practice the use of a device changes from single use to multiple use, the FDA may consider it to be a reusable device subject to the requirements explained below for reusable devices.

User-sterilized or disinfected devices

This section includes single-use devices sold in a nonsteril condition to be sterilized by the user facility and reusable devices that must be sterilized or disinfected between uses on different patients. The FDA's labeling regulation for medical devices (21 CFR 801) covers general labeling requirements as well as requirements for prescription devices. General labeling requirements in 21 CFR 801, Subpart A provide for adequate directions for use, including the following:

> Statements of all ... uses for which it is prescribed, recommended or suggested in its oral, written, printed, or graphic advertising ... [801.5(a)].
> Preparation for use, i.e., ... manipulation or process [801.5(g).]

Detailed instructions for processing of a device prior to intended use are an essential part of adequate directions for use. In subpart D, 'Exemptions from Adequate Directions for Use', labeling requirements for prescription devices include the following:

> any relevant hazards, contraindications ... and precautions under which practitioners licensed by law to administer the device can use the device safely and for the purpose for which it is intended ...
> [801.809(c).]

In order to be used safely, a device must be labeled with essential precautions, including the specifications necessary for processing of the reused device to prevent transmission of infection between patients. Although this section also provides for the omission of the information 'commonly known' to the practitioner, it is apparent from the inquiries received by the FDA that knowledge of adequate processing methods for reused devices is incomplete in the healthcare community. The increasing complexity of devices and seriousness of health hazards necessitate more detailed labeling in this regard.

In the premarket application for these devices, the manufacturer must submit detailed information on at least one validated sterilization or disinfection process for the product. The labeling submitted with the application must contain adequate directions for the user to follow the validated process. These directions should include the specified parameters of the process. For example, if an ethylene oxide sterilization process is recommended, the temperature, vacuum level, pressure of ethylene oxide gas, humidity and time should be specified. The packaging to be used should also be specified, if not provided with the device. For an ethylene oxide process, aeration conditions such as temperature and time should be specified.

It is extremely important that reusable devices be designed so that they can be adequately cleaned prior to sterilization or disinfection. Otherwise, the microbicidal agent may not reach the surface of the device and may not be able sufficiently to kill the microbial contaminants. For reusable devices, reprocessing information in labeling should include cleaning methods as well as sterilization or disinfection methods.

Section 820.160, Finished Product Testing, of the GMP Regulation states the following:

> There shall be written procedures for finished device inspection to assure that device specifications are met. Prior to release for distribution, each production run, lot or batch shall be checked and, where necessary, tested for conformance with device specifications. Where practical, a device shall be selected from a production run, lot or batch and tested under simulated use conditions.
>
> [21 CFR 820.160]

Since the use of these devices includes processing or reprocessing by the user, the FDA requires the manufacturer to perform simulated use tests which assure that the device can be sterilized or disinfected to the appropriate level without compromising its performance specifications. This testing is similar to process validation conducted for industrial sterilization. For reusable devices, this would involve repeated reprocessing to simulate actual use; simulated testing should include cleaning as well as sterilization/disinfection. The use of a microbial and organic load may be appropriate to simulate in-use conditions.

FDA postmarket surveillance

The current FDA Medical Device Reporting regulation [21 CFR 803] requires that manufacturers report to the FDA any device-related patient injuries or deaths and any device malfunctions that could cause injury or death. The Safe Medical Device Act of 1990 required that, beginning in 1991, user facilities report device-related deaths to the FDA and to the manufacturer, if known. User facilities are also required to report device-related serious injuries and serious illnesses to the manufacturer or to the FDA if the manufacturer is not known. Information on this mandatory reporting program can be obtained by telephoning 1-800-638-2041. Health professionals can submit voluntary reports by telephoning 1-800-FDA-1088; the reporter's identity is not publicly available and will not be reported to the manufacturer if the reporter requests confidentiality. Mandatory and voluntary reporting are part of MedWatch, the FDA Medical Products Reporting Program.

Washer sterilizers and disinfectors

Washer sterilizers and disinfectors which are sold to healthcare facilities are medical devices which must undergo premarket review by the FDA unless they were marketed prior to the 1976 amendments to the Act. The FDA has issued draft guidance for 510(k)s for these devices: Guidance on Premarket Notification [510(k)] Submissions for Automated Endoscope Washers, Washer/Disinfectors, and Disinfectors Intended for Use in Health Care Facilities, August 1993; Guidance on Premarket Notification [510(k)] Submissions for Sterilizers Intended for Use in Health Care Facilities, March, 1993. The guidance indicates that a 510(k) submission for a sterilizer should provide a general description of the sterilizer, documentation concerning protocols and data analysis, labeling, test packs, physical and chemical performance tests, biological performance tests, toxicity of the sterilant and its by-products, elimination of toxic residues, device and material qualification and final process qualification. A 510(k) submission for a washer disinfector should provide a general description of the endoscope washer disinfector, labeling performance data including process parameter tests, simulated-use tests and in-use tests, software documentation and toxicological evaluation of residues. As for all 510(k)s, the submission must include a description of the comparison to a legally marketed device. Copies of the FDA guidance documents mentioned here and below can be obtained from the Division of Small Manufacturers Assistance by telephoning 1–800-638–2041 in the United States.

Liquid chemical sterilants

Until recently, liquid chemical germicides for use on medical devices were regulated by the EPA as pesticides and by the FDA as accessories to devices. In June of 1993, the EPA and FDA signed a Memorandum of Understanding which, when fully implemented, will result in the regulation of liquid chemical sterilants by the FDA and general purpose disinfectants by the EPA. Until the expected FDA and EPA regulations are promulgated exempting sterilants from EPA regulation and general purpose disinfectants from FDA regulation, the agencies are accepting each other's reviews of these products as an interim measure. The Memorandum of Understanding was amended in June of 1994 to revise the disclaimer statement required on the labels of all chemical germicides other than sterilants. As mentioned above, reprocessing of critical and semicritical devices, when a biologically monitored sterilization process cannot be used, requires the use of final processing with a liquid chemical sterilant for sterilization or high-level disinfection. Cleaners with general purpose disinfectant claims can be used on critical and semicritical devices prior to final processing with a sterilant; the disclaimer required on these general purpose disinfectants indicates that further processing with a sterilant or high-level disinfectant is required.

The FDA has issued 510(k) guidance for germicides: Guidance on the

Content and Format of Premarket Notification [510(k)] Submissions for Liquid Chemical Germicides, January, 1992; and Guidance on the Content and Format of Premarket Submissions for General Purpose Disinfectants, October, 1993. This guidance covers physical and chemical properties, labeling, efficacy data, residue data, toxicity of residues, device–material compatibility qualification, and chemical indicators for liquid chemical germicides. Basic recommendations in this guidance are summarized below, but manufacturers and others should refer to the complete guidance for a full explanation.

Microbiological tests must validate the effectiveness of the entire range of the specifications for the germicide. Microbicidal testing by manufacturers should be done under worst case conditions of chemical composition, pH, temperature, hard water and organic load. The product tested should be at the end of the expiration period; reusable germicides must be tested at the maximum reuse time. Currently the sterilization time for a germicide is based on the demonstration of a comparable time to that of a predicate sterilant for passage of the AOAC Sporicidal Test including end-point analysis. The time for high-level disinfection is based on the longest time for a six log reduction of *Mycobacterium tuberculosis* var *bovis* (TB) or other representative mycobacterium or the time required for 100% TB kill as based upon EPA labeling claims. Alternative tests may be used to demonstrate a challenge at least as rigorous as the AOAC test methods and EPA guidance. For sterilants and high-level disinfectants with lesser microbicidal claims, these claims must be based on tests with the microorganisms specified. In addition to the above potency tests, simulated microbicidal testing with devices is necessary.

The 510(k) application submitted to the FDA must include data confirming the compatibility of the germicide with the medical devices and component materials that are indicated in the germicide label. The studies from which data are submitted must also demonstrate that residues of the sterilant can be removed to safe levels for the patient.

Sterilizing and disinfecting medical devices

Affects of shape, material and instrument complexity

The evolution of medical instruments continues to present new challenges for effective rapid and reliable methods of sterilizing or disinfecting them. In particular, endoscopes present one of the greatest challenges. The use of flexible endoscopes to diagnose and treat gastrointestinal and respiratory diseases has increased considerably over the last decade. In recent years an estimated 8.7 million gastrointestinal and 580 000 pulmonary flexible endoscopic procedures have been performed in healthcare facilities in the United States alone (Spach, Silverstein & Stamm, 1993).

One major problem associated with the use of flexible endoscopes is the shortcomings associated with the manual procedures traditionally used to reprocess them. Most of these scopes are not heat resistant and therefore cannot be processed by steam sterilization. Additionally, endoscopes are usually used several times a day by healthcare facilities, and sterilization or disinfection by ethylene oxide is impractical, since that type of sterilization procedure requires extensive periods of time to allow for diffusion of the gas both during sterilization and after sterilization. In some cases disposable endoscopes can be used, but increased use of disposable items ultimately may add to the rising costs of medical care and will further potentiate the medical waste problem already facing society. The current alternative is the use of liquid germicides as a means for sterilizing or disinfecting these instruments.

The most effective method of ensuring the success of a disinfection or sterilization procedure is consistent adherence to an established, written protocol for the cleaning and disinfecting or sterilizing of these instruments. Before such a protocol or 'institutional policy' is established, a number of basic variables must be considered. These include (1) the relative infection risks inherent to the instrument, the medical procedure or the patient; (2) the relative resistance levels of a variety of microorganisms; (3) the comparative powers of a variety of germicidal chemicals; and (4) a number of procedural factors that can influence the effectiveness of germicidal procedures (Bond *et al.*, 1991).

Adherence to protocol for care and cleaning of these scopes may be best accomplished by the purchase of automatic reprocessors to achieve the following three objectives (Anon., 1994):

(1) Ensuring that the reprocessing procedure is performed consistently with the use of a recommended protocol to reduce the likelihood of using a contaminated endoscope;
(2) Reducing personnel exposure to vapors produced by some liquid chemical germicides (e.g. activated 2% glutaraldehyde) that have been reported to irritate the skin, eyes and nasal membranes;
(3) Decreasing the time, tedium and labor costs associated with endoscope reprocessing.

Unfortunately, due to the complexity of the instrument, the tortuous pathways, internal channels and complex valve systems, achievement of consistent and reliable high-level disinfection (Fraser, Jones & Murray, 1992) or sterilization (Cotton & Williams, 1990; Vesley, Norlien & Nelson, 1992) has been questioned. In addition to the reprocessing limitation associated with endoscope design, reprocessors themselves can pose risks. On several occasions, reprocessors were found to support the growth of pathogenic organisms in their internal components, resulting in flexible endoscope contamination during reprocessing.

Regardless of the method of sterilization or disinfection, cleansing of the endoscope and removal of as much of the contaminating organic matter as possible is the key to efficient processing. Areas of the scope that remain inaccessible to the germicide, due to clogging or blockage from organic material, will not be properly treated. In every known case in which infection outbreaks resulted following an endoscopic procedure, inadequate cleaning, damage to the instrument channel, improper disinfection (e.g. the liquid chemical germicide was not used in accordance with the instructions on its concentration or temperature), or contaminated rinse water was at fault (Fraser et al., 1992; Vesley et al., 1992; Spach et al., 1993).

The incidence of postendoscopic infections has been estimated to be approximately one in every 1.8 million gastrointestinal procedures (American Society for Gastrointestinal Endoscopy, 1988). This number seems relatively low, but it is probably an underestimation, since gastrointestinal procedures are generally not a focus of routine infection control surveillance once the patient leaves the healthcare facility (Gorse & Messner, 1991) and many endoscopic procedures are performed on an outpatient basis (Fraser et al., 1992). Additionally, the actual number of infections caused by endoscopes may be artificially low, because individual postendoscopic infections may be subclinical or may not be detected for several days, weeks or months.

Other studies have indicated a higher rate of infection. From October 1986 to June 1988, at a hospital in Wisconsin, Pseudomonas aeruginosa colonization or infection of the biliary tract, respiratory tract, or bloodstream occurred in 16 (6.7%) of 240 patients undergoing endoscopic retrograde cholangiopancreatography and in 99 (8.9%) of 1109 patients undergoing other upper gastrointestinal endoscopic procedures. Endoscopes used in these procedures were reprocessed with an automated reprocessing machine that flushed with a detergent solution, disinfected with one of two liquid chemical germicides (2% glutaraldehyde; or 2% glutaraldehyde, 7.0% phenol, 1.2% sodium phenate diluted 1:16 in tap water), and rinsed with tap water. An investigation carried out by the hospital in June 1988 found that a thick biofilm of P. aeruginosa had formed in the detergent holding tank, inlet water hose and air vents of the automated machine. Attempts to disinfect the machine by the manufacturer's instructions to use commercial preparations of glutaraldehyde were unsuccessful (Centers for Disease Control, 1991).

In August 1990, infection-control personnel in a hospital in Missouri noted an increase in the number of Mycobacterium chelonae isolates (20 isolates during January–August 1990, compared with a median of six isolates per year during 1984–1989). A phenotypically unique strain of M. chelonae subsp. absicessus, highly resistant to cefoxitin, was recovered from all 14 patients with pseudoinfections related to bronchoscopes or endo-scopes and from the rinse water from the automated reprocessing machine (Centers for Disease Control, 1991). The true incidence of postendoscopic infection is difficult to identify (Rutala *et al.*, 1990; Spach *et al.*, 1993) and is probably much higher than is currently reported.

National and international standards

Aside from the FDA, several national and international organizations have published standards, technical reports and guidelines on the subject of industrial sterilization, hospital sterilization, reprocessing of reusable devices and related test methodology. In 1990, the ISO Technical Commit-tee 198 (ISO/TC 198, Sterilization of Health Care Products) began developing international sterilization standards with the emphasis on indus-trial standards. Standards for moist heat and ethylene oxide industrial sterilization, biological indicators (for general requirements and for ethyl-ene oxide) have been published and adopted as United States Standards. These standards can be obtained from the Association for the Advance-ment of Medical Instrumentation (AAMI) in Arlington, Virginia, USA. Standards for radiation sterilization, moist heat biological indicators and chemical indicators are in the process of publication.

Other standards under development in the ISO committee include those covering microbiological methods, sterile device packaging, aseptic pro-cessing, ethylene oxide residues and the processing of devices of animal origin with liquid chemical sterilants. Many of these ISO standards draw heavily from sterilization standards published previously in the USA by AAMI and by other nations and drafts of European standards under devel-opment. In addition to the ISO standards, the International Electrotechn-ical Commission is developing safety standards for sterilizers. The European Committee for Normalization (CEN) has also published regional sterilization standards. In most cases, these have been harmonized with the ISO standards. One major difference is that CEN requires all devices labeled as sterile to have a 10–6 SAL, regardless of the intended use; exceptions to this may be entertained.

In addition to documents mentioned above, the AAMI Sterilization Standards Committee develops other national sterilization standards, tech-nical reports and recommendations of sterilization practices for healthcare facilities. These include standards for biological indicator–evaluator resistometer (BIER)/ethylene oxide and BIER/steam vessels for testing of biological indicators, hospital steam sterilizers, ethylene oxide sterilizers for healthcare facilities, hospital steam sterilization and sterility assurance, hospital ethylene oxide sterilization and sterility assurance, table-top dry heat sterilization and sterility assurance in dental and medical facilities, selection and use of reusable rigid sterilization container systems, handling and biological decontamination of reusable medical devices, ethylene oxide gas ventilation recommendations for safe use and flash sterilization.

AAMI Technical Information Reports have been published on the following subjects: microbiological methods for gamma irradiation, chemical sterilants and sterilization methods, safe handling of biologically contaminated medical devices in nonclinical and clinical settings, and designing, testing and labeling of reusable medical devices for reprocessing in healthcare facilities: a guide for device manufacturers. Other documents are under preparation including a technical information report on safe handling of glutaraldehyde. The United States Pharmacopeia *Sterility Test* monograph has been a longstanding guide to describing sterility testing of drugs and devices. Of course, due to the test's statistical limitations regarding the number of samples, as stated by the United States Pharmacopeia and others, the sterility of a product batch can only be assured by using an adequately validated and controlled process.

Other organizations such as the EPA, AOAC, Association of Operating Room Nursing, American Society of Testing and Materials, Association for Professional Infection Control and Epidemiology, and the Society of Gastroenterology Nurses and Associates have standards or guidelines to address issues concerning liquid chemical germicides and/or reprocessing of reusable devices such as endoscopes.

References

American Society for Gastrointestinal Endoscopy (ASGE) (1988). Infection control during gastrointestinal endoscopy – guidelines for clinical application. *Gastrointestinal Endoscopy*, **34** (Suppl.), 37S–40S.

Anon. (1994). Reducing endoscopic contamination levels: are liquid disinfecting and sterilizing reprocessors the solution? *Health Devices, Focus on Endoscopy*, **23**, 212.

Association for the Advancement of Instrumentation (1988). *Guideline for Industrial Ethylene Oxide Sterilization of Medical Devices*. Association for the Advancement of Medical Instrumentation.

Bond, W. W., Ott, B. J., Franke K. A. & McCracken, J. E. (1991). Effective use of liquid chemical germicides on medical devices: instrument design problems. In *Disinfection, Sterilization, and Preservation*, 4th edn, ed. S. S. Block, p. 1097. Philadelphia: Lea and Febiger.

Borick, P. M. (1968). Chemical sterilizers (chemo sterilizers). *Advances in Applied Microbiology*, **10**, 291–312.

Boucher, R. M. G. (1972). Advances in sterilization techniques: state of the art and recent breakthrough. *American Journal of Hospital Pharmacy*, **29**, 661–72.

Center for Devices and Radiological Health (1985). *Sterilization Questions and Answers*. Food and Drug Administration.

Centers for Disease Control (1985). *Guidelines for the Prevention and Control of*

Nosocomial Infections, Guidelines for Hand Washing and Hospital Environmental Control, Section 2: Cleaning, Disinfecting, and Sterilizing Patient-Care Equipment. Centers for Disease Control.

Centers for Disease Control and Prevention (1991). Nosocomial infection and pseudoinfection from contaminated endoscopes and bronchoscopes – Wisconsin and Missouri. *Morbidity and Mortality Weekly Reports*, **40**, 675–7.

Coates, D. & Hutchinson, D. N. (1994). Infection control in practice: how to produce a hospital disinfection policy. *Journal of Hospital Infections*, **26**, 57–68.

Cotton, P. B. & Williams, C. B. (1990). In *Practical Gastrointestinal Endoscopy*, 3rd edn, pp. 249–60. Oxford: Blackwell Scientific Publications.

Danielson, J. W. (1993). Evaluation of microbial loads of *Bacillus subtilis* spores on pennycylinders. *Journal of the Association of Official Analytical Chemists International*, **76**, 355–60.

Favero, M. S. & Bond, W. W. (1991). Chemical disinfection of medical and surgical materials. In *Disinfection, Sterilization, and Preservation*, 4th edn, ed. S. S. Block, p. 617. Philadelphia: Lea and Febiger.

Fraser, V. J., Jones, M. & Murray, P. R. (1992). Contamination of flexible fiberoptic bronchoscope with *Mycobacterium chelonae* linked to an automated bronchoscope disinfection machine. *American Reviews of Respiratory Diseases*, **145**, 853–5.

Gangi, V., Margolin, A. B. & Brennen, R. (1994). Effect of carrier material on survivability of bacterial endospores. *Food and Drug Administration Germicide Conference*, Rockville, Maryland, March.

Gelinas, P. & Goulet, J. (1982). Neutralization of the activity of eight disinfectants by organic matter. *Journal of Applied Bacteriology*, **54**, 243–7.

General Accounting Office/Human Resources Division (1993). Hospital sterilants: insufficient FDA regulation may pose a public health risk. In *FDA Regulation of Hospital Sterilants: A Report to the Ranking Minority Member Committee on Government Operations, House of Representatives*, pp. 2–5.

Gorse, G. J. & Messner, R. I. (1991). Infection control practices in gastrointestinal endoscopy in the United States: a national survey. *Infection Control and Hospital Epidemiology*, **12**, 1245–9.

Hanson, P. J. V., Clarke, J. R., Nicholson, G., Gazzard, B., Faya, H., Gore, D., Chadwick, M. V., Shah, N., Jeffries, D. J. & Collins, J. V. (1989). Contamination of endoscopes used in AIDS patients. *Lancet*, **ii**, 86–8.

Kaczmarek, R. G., Moore, R. M. Jr, McMrohan, J., Goldmann, D. A., Reynolds, C., Caquelin, C. & Israel, E. (1992). Multi-state investigation of the actual disinfection/sterilization of endoscopes in healthcare facilities. *American Journal of Medicine*, **92**, 257–61.

Rutala, W. A. (1990). Association of Practitioners in Infection Control (APIC) Guidelines Committee. APIC guidelines for selection and use of disinfectants. *American Journal of Infection Control*, **18**, 99–117.

Spach, D. H., Silverstein, F. E. & Stamm, W. E. (1993). Transmission of infection by gastrointestinal endoscopy and bronchoscopy. *Annals of Internal Medicine*, **15**, 117–28.

Spaulding, E. H. (1972). Chemical disinfection and antisepsis in the hospital. *Journal of Hospital Research*, **9**, 5–31.

US Department of Health and Human Services (1982). *Guideline for Hospital*

Environmental Control and Guideline Ranking Scheme, NTIS 66R469218323, pp. 2–3. Atlanta: US Department of Health and Human Services.
Vesley, D., Norlien, K. G. & Nelson, B. (1992). Significant factors in the disinfection and sterilization of flexible endoscopes. *American Journal of Infection Control*, **20**, 291–300.

PART 4 GENERAL MECHANISMS OF DISINFECTION

Chapter 11

Ultraviolet light disinfection of water and wastewater

PETER F. ROESSLER and
BLAINE F. SEVERIN

Introduction

The popularity of ultraviolet light (UV) disinfection of water and wastewater is being driven by current realizations that the use of halogenated disinfectants such as chlorine, chloramine and chlorine dioxide potentially create by-products detrimental to the environment, biological systems and human health (Brungs, 1973; Martens & Servizi, 1974; Rook, 1974; Jolley, 1975; Murphy, Zaloum & Fulford, 1975; Ward & DeGrave, 1978). These findings have prompted the United States Environmental Protection Agency to begin formulating a Disinfection By-Product (DBP) Rule to regulate the levels of these by-products in drinking water. The DBP Rule would by law limit the levels of contaminants such as trihalomethanes, haloacetic acids, bromate and chlorite allowed in drinking water, due to their carcinogenic potential. Although UV has played a lesser role in the disinfection of water, this may change in the future, due to the elevated public awareness and interest in the by-products of halogen disinfection. These concerns may dictate a more serious dependency on UV as a 'primary' disinfectant, to be used in combination with lowered levels of a halogen to provide the residual disinfection that UV does not. The lowered levels of halogen usage would in turn aid systems in meeting potential DBP Rule mandates.

Advances in UV technology, providing lower cost, more efficient lamps and more reliable equipment, are sustaining the current interest in UV disinfection. These advances have aided in the commercial application of UV for water treatment in the pharmaceutical, cosmetic, beverage and electronic industries. Its application is also increasing in both large-scale municipal water and wastewater disinfection (Oliver & Carey, 1976; White, Jernigan & Venosa, 1986; Whitby et al., 1984; Zukovs et al., 1986; Fahey, 1990; Maarschalkerweerd, Murphy & Sakamoto, 1990; Sommer & Cabaj, 1993) and small-scale household point-of-use and point-of-entry water disinfection (Kuennen et al., 1992). In small-scale situations, where municipally treated water may be unavailable, the number of reliable disinfection methods is limited. Here, economical and efficient UV systems are available that in combination with proper prefiltration act as an effective means of providing potable water with minimum operator involvement.

Additionally, concern exists as to the microbiological quality of drinking water treated by conventional methods (Craun, 1979, 1989, 1990; Hayes *et al.*, 1989; MacKenzie *et al.*, 1994), prompting the need to address alternative water and wastewater disinfection methods. Even ground water, often considered pristine and naturally protected, has been found to be microbiologically contaminated. A study currently in progress is finding that approximately 20% of environmental ground-water samples examined have tested positive for enteroviruses (M. Abbaszadegan, personal communication). Continual technological improvement in the methods used to recover and detect environmental pathogens will also increase the efficiency of finding these organisms in surface and ground water, and thus increase public awareness of the importance of the study of water and wastewater disinfection.

It is the purpose of this chapter to provide an insight into the use of UV for the disinfection of water and wastewater and to discuss the technical aspects involved with the measurement of UV dose, as this is the critical characteristic of UV disinfection systems. Initially, this chapter presents the history and biochemistry of UV disinfection, followed by a review of the literature on the UV susceptibility of waterborne pathogenic and indicator bacteria, viruses and protozoa. The fundamental inactivation kinetics for batch UV disinfection are then presented, including detailed descriptions of mixed second-order, multi-target and series-event inactivation models. Continuous-flow UV inactivation is then discussed at length. This discussion begins with a consideration of the flow dynamics and inactivation kinetics of idealized continuous-flow UV reactors, in which simplifying assumptions allow for reasonable estimations of UV dose. This is followed by an introduction to the use of biological and chemical actinometery to measure UV dose in complex, non-idealized continuous-flow UV reactors, in which the measurement of disinfection efficacy is much more difficult. The variables of UV intensity and contact time are discussed in detail, because of their importance in determining UV dose. The geometry of a typical continuous-flow UV reactor is presented as an introduction to the mathematical modeling of UV intensity distributions using the radial source and finite line source models. In addition, the importance of determining the residence time distribution of non-idealized continuous-flow reactors is discussed, including the presentation of indices used for hydraulic assessment of UV reactor design. Finally, an example is given to demonstrate the resolution of intensity and time distribution within a continuous-flow UV reactor. Interferences with UV disinfection are presented in terms of cellular repair, water quality and mechanical properties of UV systems. The end of the chapter contains a descriptive list of the mathematical nomenclature used.

History and biochemistry

The lethal effects of exposing microorganisms to UV have been known for decades. These effects were first reported by Downes & Blount (1877), who described the bactericidal effects of radiant energy from sunlight. They concluded that radiations of short wavelengths were responsible for the destruction of their mixed bacterial populations. The basic technology was established by the first decade of the 20th century with the development of the mercury vapor lamp by Hewitt in 1901 and enclosure of the lamp in a quartz sheath in 1906 to dampen the effects of temperature changes (Tonelli, Duff & Wilcox, 1978). For the next three decades, research was aimed at the identification of the optimum UV wavelength for disinfection. Initial studies (Barnard & Morgan, 1903; Bang, 1905; Newcomer, 1927; Wyckoff, 1932) narrowed the suspected optimum wavelength to a range of 250–266 nm. Clues to the mode of action of UV disinfection were not unveiled until the experiments of Gates (1928*a*), who published the first action spectrum for UV light, by demonstrating UV lethality against *Escherichia coli* and *Staphylococcus aureus* over a range of wavelengths. Gates (1928*b*) also demonstrated that the bactericidal action spectrum of UV closely resembled the absorption spectrum of nucleic acids or nucleic acid constituents, early evidence suggesting the key role that nucleic acids or their derivatives played in the disinfection of microorganisms by UV.

By the mid-1930s it was recognized that bacterial sensitivity to UV was species-dependent (Ehrismann & Noething, 1932; Duggar & Hollaender, 1934; Bucholz & Jeney, 1935). By 1941, it was discovered that UV sensitivity was strain-variable within a species (Rentschler *et al.*, 1941). In the 1940s, evidence relating UV inactivation to the structural alteration of genetic material and nucleic acids was reported (Hoellandar & Emmons, 1941; Loofhourow, 1948). Knowledge of the mechanism of UV inactivation had been expanded by the 1960s to identification of the specific reaction sites, i.e. intrastrand thymine dimer formation in DNA, and other reaction sites of lesser importance (Wacker, 1963; Setlow, 1964*a*,*b*; Smith, 1964). Enzymatic and photo-induced enzymatic repair mechanisms, or photoreactivation of UV-inflicted genetic damage, were discovered in the 1950s and 1960s (Kelner, 1949). Extensive reviews of these research areas are available (Smith & Hanawalt, 1969; Harm, 1980). The ability of organisms to recover from sublethal UV doses has been identified as a topic of practical significance to the wastewater disinfection field (Harris *et al.*, 1987*c*) and is discussed in more detail later in the chapter. In addition, a general relationship between the UV dose required to inactivate certain viruses and the mass of the viral genome has been proposed (Kallenbach *et al.*, 1989).

Ultraviolet susceptibility of waterborne pathogens and indicator organisms

The assessment of laboratory-generated inactivation data for waterborne microorganisms has been established for some time (Chick, 1908; Watson, 1908). The initial concepts put forth in these pioneering works remain in wide use today (Hoff, 1986*a,b*), although a number of modifications have been suggested to aid in the interpretation of results (Fair *et al.*, 1948; Hom, 1972). The interpretation of inactivation data is based on the use of the concentration–time (*Ct*) factor, i.e. the product of the disinfectant concentration and the contact time between the organism and the disinfectant. In UV technology, this has been termed the Bunsen–Roscoe Law (Jagger, 1967), in which UV dose (in mW s/cm^2), is the product of the UV intensity (*I*, in mW/cm^2), and the exposure time (*t*, in seconds), as given in Equation 11.1. (A list of mathematical nomenclature used in this chapter is given on page 359.)

$$\text{UV dose} = It \tag{11.1}$$

In most disinfection studies, it has been observed that the logarithm of the surviving fraction of organisms is nearly linear when it is plotted against the dose, where dose is the product of concentration and time (*Ct*) for chemical disinfectants, or intensity and time (*It*) for UV. A further observation is that constant dose yields constant inactivation. This is expressed mathematically in Equation 11.2:

$$\log \frac{N_S}{N_I} = function \ (It) \tag{11.2}$$

where N_S is the density of surviving organisms (number/cm^3) and N_I is the initial density of organisms pre-exposure (number/cm^3). The *function* of UV dose as implied in Equation 11.2 may be mathematically simple or dependent on a large number of factors that must each be considered in the equation. These factors will be discussed in more detail later in the chapter. The concept that constant dose yields constant inactivation is valid for conditions where the mechanism of inactivation is constant. In extreme cases, such as in the work reviewed by Kallenbach *et al.* (1989), high-intensity laser-generated UV applied in picosecond increments was found to be less effective than equivalent doses applied at low intensity over long exposure periods. This is presumably because the laser-applied UV caused excision of DNA fragments rather than the usually noted formation of DNA dimers. Further discussions in this text follow the concept that constant dose yields constant inactivation, as current technology is based on UV applied at lower intensities and longer exposure times relative to these laser systems.

Because of the logarithmic relationship of microbial inactivation versus UV dose, it is common to describe inactivation in terms of log survival, as expressed in Equation 11.3. For example, if one organism in 1000 survives exposure to UV, the result would be a -3 log survival, or a 3 log reduction.

$$\text{log survival} = \log \frac{N_S}{N_I} \qquad (11.3)$$

The determination of the UV susceptibility of various indicator and pathogenic waterborne microorganisms is fundamental in quantifying the UV dose required for adequate water disinfection. In making these determinations, the characteristics that affect UV dose output should be considered in laboratory experiments. These characteristics, discussed in detail in a further section of the chapter, include cell clumping and shadowing, suspended solids, turbidity and UV absorption. UV susceptibility experiments described in the literature are often based on the exposure of microorganisms under conditions optimized for UV disinfection, including filtration of the microorganisms to yield monodispersed, uniform cell suspensions and the use of buffered water with low turbidity and high transmission at 254 nm. UV susceptibility determinations made from these experiments may not reflect the disinfection that will be achieved under actual field conditions.

Waterborne bacteria

The UV susceptibility of waterborne *Escherichia coli* has been well documented (Zelle & Hollaender, 1955; Jepson, 1973; Rice & Hoff, 1981; Chang *et al.*, 1985; Harris *et al.*, 1987b; Wilson *et al.*, 1992b). The UV dose required for a 1 log reduction (10% survival) of *E. coli* is between 1.3 and 3.0 mW s/cm^2, whereas that required for a 3 log reduction (0.1% survival) is between 3.0 and 7.0 mW s/cm^2. In addition, the UV susceptibility of *E. coli* has been compared to that of other waterborne pathogenic bacteria, including *Legionella pneumophila* (Antopol & Ellner, 1979; Muraca, Stout & Yu, 1987), *Vibrio cholerae* (Roessler, Wilson & Kuennen, 1993), *Helicobacter pylori* (Roessler & Wilson, 1993), *Campylobacter jejuni*, *Yersinia enterocolitica* (Butler, Lund & Carlson, 1987) and others (Table 11.1). The susceptibility data in Table 11.1 were collected by adding uniform bacterial suspensions to high-quality water and then exposing the suspensions to UV. Although species-to-species variations do exist, the UV susceptibility of vegetative cells belonging to different waterborne bacterial species is of an equivalent magnitude (Chang *et al.*, 1985; Wolfe, 1990; Wilson *et al.*, 1992b). Under conditions of high water quality, waterborne vegetative bacteria appear to be highly susceptible to relatively low UV doses.

Table 11.1 *Ultraviolet dose required for 1 log and 3 log reductions of selected waterborne bacteria*

Bacteria	Ultraviolet dose (mW s/cm^2) required for reduction			
	1 log reduction	Reference	3 log reduction	Reference
Campylobacter jejuni	1.1	Wilson *et al.* (1992*b*)	3.8	Wilson *et al* (1992*b*)
			1.8	Butler *et al.* (1987)
Escherichia coli	1.3	Wilson *et al.* (1992*b*)	4.2	Wilson *et al.* (1992*b*)
	2.5	Harris *et al.* (1987*b*)	4.0	Harris *et al.* (1987*b*)
	3.0	Wolfe (1990)	3.0	Rice & Hoff (1981)
	2.1	Zelle & Hollaender (1955)	5.0	Butler *et al.* (1987)
			7.0	Chang *et al.* (1985)
Klebsiella terrigena	3.9	Wilson *et al.* (1992*b*)	9.1	Roessler *et al.* (1992)
Legionella pneumophila	2.5	Wilson *et al.* (1992*b*)	7.4	Wilson *et al.* (1992*b*)
	0.92	Antopol & Ellner (1979)	2.8	Antopol & Ellner (1979)
Salmonella typhi	2.3	Wilson *et al.* (1992*b*)	6.6	Wilson *et al.* (1992*b*)
	2.5	Wolfe (1990)	7.0	Chang *et al.* (1985)
	2.1	Zelle & Hollaender (1955)		
Shigella dysenteriae	0.89	Wilson *et al.* (1992*b*)	2.1	Wilson *et al* (1992*b*)
	2.2	Wolfe (1990)		
Vibrio cholerae	0.65	Wilson *et al.* (1992*b*)	2.2	Wilson *et al.* (1992*b*)
	3.4	Wolfe (1990)	2.9	Roessler *et al.* (1993)
Yersinia enterocolitica	1.1	Wilson *et al.* (1992*b*)	3.7	Wilson *et al.* (1992*b*)
			2.7	Butler *et al.* (1987)

Note: The chemical and physical conditions of water quality generally were optimized for ultraviolet disinfection (i.e. low absorptivity, low turbidity and filtration to minimize aggregation).

Waterborne enteric viruses

Enteric viruses can be anywhere from two to ten times more resistant to UV than *E. coli* (Chang *et al.*, 1985; Harris *et al.*, 1987*b*; Wolfe, 1990; Wilson *et al.*, 1992*b*). Hill *et al.* (1970) studied the UV inactivation of eight enteric viruses and reported no statistically significant difference ($p > 0.05$) in the susceptibility of poliovirus (types 1, 2 and 3), echovirus (type 1 and type 11) and coxsackievirus (type A-9). Hill reported that a UV dose of 35 mW s/cm^2 was capable of reducing these viruses by approximately 3 logs. Battigelli, Sobsey & Lobe (1993) could not statistically differentiate the responses of rotavirus SA11 and coxsackievirus B5 to UV and reported that UV doses of 42 and 29 mW s/cm^2, respectively, were required to reduce these viruses by approximately 4 logs. Chang *et al.* (1985) found similar UV inactivation curves for rotavirus SA11 and poliovirus type 1, and reported that a UV dose of approximately 30 mW s/cm^2 resulted in a 3–4 log reduction of the two viruses. In addition, Wilson *et al.* (1992*b*) reported that a UV dose of 39.4 mW s/cm^2 yielded a 4 log reduction in rotavirus SA11, poliovirus type 1 and hepatitis A virus. Harris *et al.* (1987*b*) found that a 4 log reduction of poliovirus type 1 required a UV dose of approximately 30 mW s/cm^2.

Reovirus type 1 has been found to be significantly ($p < 0.005$) more resistant to UV than to poliovirus type 1 (Harris *et al.*, 1987*b*), poliovirus types 2 and 3, echovirus type 1 and type 11, and coxsackievirus A9 (Hill *et al.*, 1970). Battigelli *et al.* (1993) demonstrated that hepatitis A virus was significantly less resistant to UV than were rotavirus SA11 and coxsackievirus B5, requiring a UV dose of 16 mW s/cm^2 for a 4 log reduction in the hepatitis A virus. Wiedenmann *et al.* (1993) reported a 4 log reduction in hepatitis A virus with a UV dose of approximately 20 mW s/cm^2 Q. S. Meng & C. P. Gerba (unpublished data) found that enteric adenovirus strains 40 and 41 required a UV dose of 90 and 80 mW s/cm^2, respectively, for a 3 log reduction. The investigators attributed this relatively low susceptibility to the double-stranded nature of the adenovirus genome.

An interesting observation concerning the relative resistance of viruses has been made by Kallenbach *et al.* (1989). They noted that viruses with high molecular weight, double-stranded DNA or RNA were easier to inactivate than those with low molecular weight, double-stranded genomes. Viruses with single-stranded nucleic acids of high molecular weight were easier to inactivate than those with single-stranded nucleic acids of low molecular weight. This is presumably true because the target density is higher in larger genomes. However, viruses with double-stranded genomes are less susceptible than those with single-stranded genomes, due to the ability of the naturally occurring enzymes within the host cell to repair damaged sections of the double-stranded genome, using the non-damaged strand as a template.

A compilation of the UV dose levels required for 1 and 3 log reductions in various waterborne enteric viruses and indicator organisms is listed in Table 11.2. The susceptibility data were generally collected by adding uniform viral suspensions to high-quality water and then exposing the suspensions to UV. The water conditions used for exposure to UV included the use of phosphate buffer at 0.05 mol/l with $A_{254} = <0.1$, pH 7.2 (Harris *et al.*, 1987*b*), filtered (0.08 μm) phosphate buffered saline (Chang *et al.*, 1985; Battigelli *et al.*, 1993) and filter-sterilized (0.45 μm) estuarine water, pH 8.0 (Hill *et al.*, 1970). The data given in Table 11.2 indicate that moderate UV doses effectively inactivate viruses of public health concern in high-quality water.

Table 11.2 *Ultraviolet dose required for 1 log and 3 log reductions of waterborne enteric viruses, bacterial spores and coliphage MS2*

Virus/indicator	1 log reduction	Reference	3 log reduction	Reference
Adenovirus strain 40	30.0	Q. S. Meng & C. P. Gerba	90.0	Gerba (unpublished data)
Adenovirus strain 41	23.6	(unpublished data)	80.0	
Coxsackievirus B1	15.6	Hill et al (1970)	46.8	Hill et al. (1970)
Coxsackievirus B5	11.9	Hill et al. (1970)	25	Battigelli et al. (1993)
Coxsackievirus A9			35.7	Hill et al. (1970)
Echovirus type 1	10.8	Hill et al. (1970)	32.5	Hill et al. (1970)
Echovirus type 11	12.1	Hill et al. (1970)	36.4	Hill et al. (1970)
Hepatitis A virus	3.7	Wolfe (1990)	15	Battigelli et al. (1993)
	5.5	Wiedenmann et al. (1993)	15.5	Wiedenmann et al. (1993)
	7.3	Wilson et al. (1992b)	21.9	Wilson et al. (1992b)
Polivirus type 1	5.0	Wolfe (1990)	23.1	Wilson et al. (1992b)
	7.7	Wilson et al (1992b)	24	Harris et al. (1987b)
	11.0	Hill et al. (1970)	30a	Chang et al. (1985)
			33.0	Hill et al. (1970)
Poliovirus type 2	12.0	Hill et al. (1970)	36.1	Hill et al. (1970)
Poliovirus type 3	10.3	Hill et al. (1970)	30.9	Hill et al. (1970)
Reovirus type 1	15.4	Hill et al. (1970)	46.3	Hill et al. (1970)
			45	Harris et al. (1987b)
Rotavirus SA11	8.0	Wolfe (1990)	25	Battigelli et al. (1993)
	9.9	Wilson et al. (1992b)	29.6	Wilson et al (1992b)
			30a	Chang et al. (1985)
Bacillus subtilis spores	14.2	Wilson et al (1992a)	39.9	Wilson et al. (1992a)
Coliphage MS2	18.6	Wilson et al. (1992b)	65.0	Havelaar et al. (1991)
			55.0	Wiedenmann et al. (1993)

Note: The chemical and physical conditions of water quality generally were optimized for ultraviolet disinfection (i.e. low absorptivity, low turbidity and filtration to minimize aggregation).
a A 3 to 4 log reduction was obtained at this ultraviolet dose (Chang et al., 1985).

Waterborne protozoa

In recent years, waterborne protozoan cysts and oocysts have received a great deal of attention, due to their role in waterborne disease outbreaks, especially species of *Giardia* and *Cryptosporidium* (Craun, 1990; Smith & Rose, 1990). Outbreaks of gastroenteritis attributed to these organisms most commonly occur in small systems using high-quality surface water as a source and chlorine as the only treatment method (Craun, 1979). Protozoan outbreaks, however, are not limited to these types of systems and have also occurred in filtered, chlorinated public water supplies (Craun, 1989; Hayes *et al.*, 1989; MacKenzie *et al.*, 1994). The growing use of UV as an alternative disinfectant has led researchers to investigate the potential of UV for treating water contaminated with these types of organisms.

Rice & Hoff (1981) found that a UV dose of 63 mW s/cm^2 achieved less than a 1 log reduction in *Giardia lamblia* cysts, and concluded that UV, at conventionally applied doses, is not a viable alternative for adequate treatment of water in the presence of *G. lamblia*. Wolfe (1990) reported that a 1 log reduction in cysts of *G. muris* could be achieved at a UV dose of 82 mW s/cm^2. Karanis *et al.* (1992) investigated the UV susceptibility of *Trichomonas vaginalis*, *G. lamblia* and dormant and vegetative stages of two different strains of *Acanthamoeba* and *Naegleria* spp. The results indicated that a dose of approximately 400 mW s/cm^2 was required to inactivate 3 logs of *T. vaginalis*, 180 mW s/cm^2 to inactivate 2 logs of *G. lamblia* cysts and approximately 72 mW s/cm^2 to inactivate 2 logs of *Acanthamoeba rhysodes* cysts and trophozoites. Chang *et al.* (1985) found that cysts of *Acanthamoeba castellani* required approximately 15 times the UV dose required to inactivate *E. coli* to the same extent (3 logs). Chang *et al.* (1985) indicated that a UV dose of approximately 80 mW s/cm^2 was required to inactivate 2 logs of *A. castellani* cysts. All data reported to date indicate that UV alone, at the doses applied in commercial systems, is insufficient for the control of waterborne protozoa. At present, the only effective mechanism for the control of these organisms is a multiple barrier design (Karanis *et al.*, 1992), which includes source protection, filtration, disinfection and protected distribution.

Indicators of ultraviolet disinfection

F-specific RNA coliphage such as MS2, f2 and Qβ have been suggested as bioassay indicator organisms for measuring the UV dose output of flow-through UV reactors (Havelaar *et al.*, 1990; Wilson *et al.*, 1992a). They have also been proposed as surrogate organisms to predict UV disinfection of viruses in water (Havelaar, 1987; Kamiko & Ohgaki, 1989; IAWPRC, 1991) and wastewater (Havelaar *et al.*, 1991). Coliphages have been recommended because of an ecological and biological similarity to the enteric

322   P. F. ROESSLER AND B. F. SEVERIN

viruses, a linear logarithmic inactivation response, ease of recovery and enumeration, high numbers naturally occurring in wastewater prior to disinfection, nonpathogenicity and relatively low UV susceptibility. Havelaar *et al.* (1991) found similar UV inactivation rate constants for F-specific coliphages as for reoviruses, and reported that a 3 log reduction of coliphage MS2 required a UV dose of approximately 65 mW s/cm^2 (Table 11.2). Similar reductions in coliphage MS2 have been reported to require a UV dose of 55 mW s/cm^2 (Wilson *et al.*, 1992*b*; Wiedenmann *et al.*, 1993). Wilson *et al.* (1992*b*) reported that coliphage MS2 was approximately twice as resistant to UV as were rotavirus, poliovirus and hepatitis A virus.

The National Sanitation Foundation (1991) Standard 55 recommends the use of *Bacillus subtilis* spores and *Saccharomyces cerevisiae* for certifying the disinfection performance of Class 'A' and Class 'B' point-of-use and point-of-entry UV water treatment devices. Class A systems are designed to disinfect bacteria and viruses from contaminated water to a safe level, and require a UV dose output of 38 mW s/cm^2. Class B systems are designed for reducing nonsporeforming heterotrophic bacteria commonly found in treated and disinfected public drinking water and require a UV dose output of 16 mW s/cm^2. For comparative purposes, 3 log reductions in *B. subtilis* spores and *S. cerevisiae* require UV doses of 39.9 and 24.6 mW s/cm^2, respectively (Wilson *et al.*, 1992*a*).

Klebsiella terrigena has also been proposed as a UV water disinfection surrogate, to predict the level of inactivation of vegetative waterborne bacterial pathogens, because of its relatively low UV susceptibility (Roessler *et al.*, 1992). *K. terrigena* was found to be from 1 to 4 times as resistant to UV as the other bacterial strains tested, including *Vibrio cholerae*, *Shigella dysenteriae*, *Salmonella typhi*, *Escherichia coli* and *Aeromonas hydrophila*.

Batch ultraviolet inactivation

Batch UV inactivation data are usually collected in stirred, flat, thin-layer, closed reactors with the use of water with low UV absorbance. In batch UV reactors, uniform UV intensities exist and contact time can be strictly controlled. With these two variables controlled, the determination of UV dose in batch reactors is simplified. To deliver UV to these types of reactors, a collimating beam apparatus has been suggested (Qualls & Johnson, 1983). The light emitted at the end of the collimating beam is perpendicular to the batch reactor surface, thus creating a uniform, constant irradiation field that can be accurately quantified by means of a radiometer and photodetector calibrated for detecting 254 nm light (Johnson & Qualls, 1984; Havelaar *et al.*, 1990). To conduct UV exposures, a pure culture cell suspension of the microorganism of interest is prepared and dispensed into a stirred quartz petri dish reactor designed to fit at the end of the collimating beam. The UV susceptibility of specific microorganisms may then be determined by exposing the organism to a range of UV doses and using a fixed intensity and varying contact time. A timer regulating a shutter positioned at the end of the collimating beam may be used for precise control of exposure time.

Modeling ultraviolet intensity

Of significant importance in the batch inactivation experiments described above is the calculation of the average UV intensity applied to the stirred cell suspension ($\bar{I}$). The modeling technique for determining the average UV intensity in batch reactors is termed the linear light attenuation model. The model is based on the Beer–Lambert law, which states that the reduction of light intensity is caused only by absorption as given in Equation 11.4,

$$\frac{dI}{dH} = -EI \tag{11.4}$$

where H is the depth of the water film (in cm), E is the attenuation in $base_e$ (/cm), and I is the intensity (mW/cm^2). This equation may be integrated across the depth of the water film using the boundary conditions $I = I_0$ when $H = 0$, where I equals the UV intensity at any point within the reactor and I_0 is the actual UV intensity measured at the end of the collimating beam. The resulting Equation 11.5 yields an expression for the intensity at any point throughout the depth of the water film.

$$\frac{I}{I_0} = e^{-EH} \tag{11.5}$$

Equation 11.5 is often written in $base_{10}$ with reference to output from spectrophotometers, in which case the absorbance is given as A (/cm). The absorbance in this numerical system is defined by Equation 11.6. Most spectrophotometers utilize a 1.0-cm path length, such that $H = 1.0$. The relationship between the different absorbance factors is $A = E/2.303$.

$$A = -\frac{\log_{10}(I/I_0)}{H} \tag{11.6}$$

Another commonly used term in the study of UV disinfection is transmittance, T. Transmittance is the fraction of light transmitted across a 1.0-cm path length ($H = 1.0$), as expressed in Equation 11.7.

$$T = \frac{I}{I_0} = 10^{-AH} \qquad (11.7)$$

The average UV intensity at any point within the stirred cell suspension in the batch reactor, $\bar{I}$, may be calculated by integrating Equation 11.5 and averaging over the pathlength H, and is presented in Equation 11.8 (Morowitz, 1950).

$$\bar{I} = I_0 \left(\frac{1 - e^{-EH}}{EH} \right) \qquad (11.8)$$

If the surface intensity I_0 is measured with a radiometer, then I_0 is multiplied by a correction factor of 0.96 to account for the reflection of light at the surface of the water film (Jagger, 1967; National Sanitation Foundation, 1991). If a biological or chemical actinometer is used to measure I_0, the correction factor is eliminated.

Kinetics of batch ultraviolet inactivation

Batch inactivation curves are generated by exposing cell suspensions to a range of UV doses. This methodology allows for the determination of the UV susceptibility of various microorganisms, including waterborne pathogens, under a variety of controlled conditions. Data are usually presented as the logarithm of the fraction of survival versus the UV dose applied, as described in Equation 11.3. Presentation in this fashion allows for comparison of the UV susceptibility of different strains, species and genera of microorganisms over a range of UV doses (Figure 11.1).

Figure 11.1 and Equations 11.1–11.3 imply that exposure to a constant dose yields a constant inactivation. Many authors have approached this phenomenon using mechanistic (Haas, 1980) or statistical models (Kimball, 1953; Gurian, 1956; Hiatt, 1964; Wei & Chang, 1975; Severin, Suidan & Engelbrecht, 1983a). The events leading to inactivation are organism-specific, disinfectant-specific, water quality-specific and reactor design-dependent. It is these non-idealities that dictate the mathematical conditions of the *function* term in Equation 11.2 and rule out the use of a specific modeling technique as a 'best model'. The best model for a given set of conditions has the following criteria: (1) specifically defined assumptions, (2) ease of data interpretation, and (3) ease of application to specific reactor designs. The following section reviews three commonly used models in the study of UV disinfection: the mixed second-order model, the multi-target model and the series-event model.

Batch disinfection curves typically fall into three classes, often referred to as straight (Type A), shouldered (Type B) and tailed (Type C) curves (Figure 11.2). Type A inactivation curves are usually attributed to first-order kinetics with respect to both the number of viable organisms (Chick,

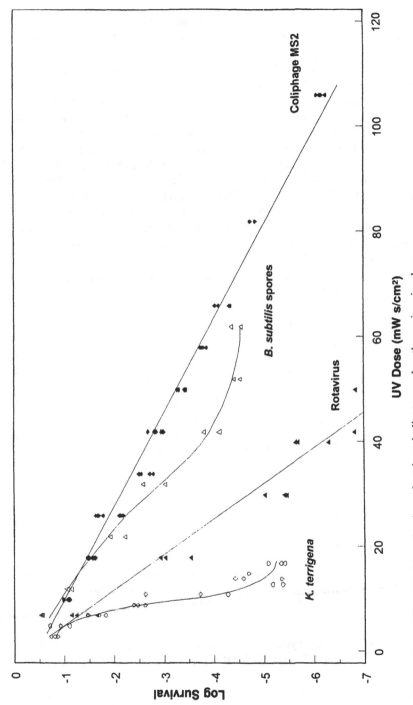

Figure 11.1   Ultraviolet inactivation of various indicator and pathogenic microbes.

1908) and the concentration of disinfectant (Watson, 1908) or intensity. Thus, the overall reaction is termed 'mixed second order'. Type B curves are called initial resistance curves and are commonly attributed to a requirement for multiple detrimental reactions within a target prior to inactivation. Type B curves can be modeled by the use of two common kinetic expressions: (1) series-event and (2) multi-target kinetics. Type C curves are called 'tailed curves' and are justified by differential resistance within a mixed-strain population or by the existence of several populations of different clump sizes within a cell suspension. The clumped cell populations are differentiated by the number of cells per clump. Modeling of Type C curves is much more complex than Types A or B curves, and requires identification of the initial numbers of cells in each population. Each population must then be tracked individually by means of series-event, multi-target, or mixed second-order kinetics. A model for tracking clumps with mixed second-order kinetics has been described by Poduska & Hershey (1972).

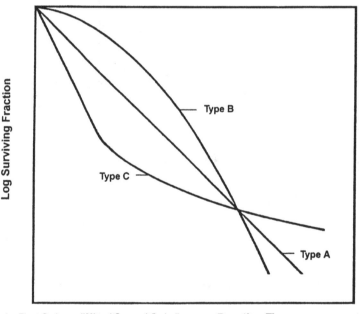

A: First Order or "Mixed Second Order"   **Reaction Time**
B: Initial Resistance
C: Tailing

Figure 11.2 Typical inactivation curves. type A, first order or 'mixed second order'; Type B, initial resistance; Type C, tailing.

Mixed second-order kinetics (Type A inactivation curves)

Some organisms, especially the bacteriophages, exhibit logarithmic inactivation that is linear with time, as exemplified by the linear response of coliphage MS2 in Figure 11.1 and by the Type A curve in Figure 11.2. It is assumed that this response reflects the rate of inactivation, which is first-order-dependent on the number of surviving organisms and on UV intensity, and is expressed in Equation 11.9.

$$\frac{dN_S}{dt} = -KIN_S \tag{11.9}$$

In this equation, N_S is the surviving fraction (number/cm^3), t is the exposure time, K is the inactivation rate constant (cm^2/mWs) and I is the UV intensity (mW/cm^2). Integration of Equation 11.9 across the limits $N_S = N_I$ at $t = 0$, where N_I is the initial density of organisms, yields Equation 11.10. This is the typical expression for mixed second-order inactivation kinetics, and the simplest inactivation model.

$$\frac{N_S}{N_I} = e^{-KIt} \tag{11.10}$$

Multi-target model (Type B inactivation curves)

The resistance of microorganisms that exhibit a Type B response to UV inactivation has, in the past, been modeled with multi-target kinetics. This model was first developed for types of radiation other than UV, and current observations on the mechanism of UV inactivation do not generally support the model. However, it has still found general use (Smith & Hanawalt, 1969; Harm, 1980), because of its simplicity of logic, simplicity of mathematics and ability to fit batch data. In this model, it is assumed that a particle contains a finite number, q, of discrete critical targets. Each target must be hit prior to full inactivation of the particle. Since the number of targets is finite, the probability of attaining a hit on the next target is decreased as the reaction proceeds. A particle may represent an organism with q targets or a clump of organisms possessing a total of q targets. Due to the enumeration methods commonly used in disinfection studies, however, it is impossible to differentiate between the survival of a clump of cells or that of a single organism. Therefore, the terms 'organism' or 'particle' are used interchangeably for this model.

The multi-target model is most easily derived from a statistical approach, presented as follows. A kinetic rate derivation is also available (Severin *et al.*, 1983*a*). Assuming that mixed second-order kinetics describe the rate of inactivation of a critical target, the probability, $P(0)$, of a specific target surviving attack is given by Equation 11.11:

$$P(0) = e^{-KIt} \tag{11.11}$$

where K is a rate constant (cm²/mW s) and t is the exposure time (s) to UV of intensity I (mW/cm²). The probability of inactivating the first target is $(1 - e^{-KIt})$. Assuming that all targets are equivalent and that damage is randomly distributed among the targets, then the probability of survival of a particle with q critical targets is given by Equation 11.12:

$$\frac{N_S}{N_I} = 1 - (1 - e^{-KIt})^q \tag{11.12}$$

where N_S is the density of surviving organisms and N_I is the density of organisms initially present. It should be noted that when the number of hits is equal to one ($q = 1$), Equation 11.12 reduces to the simple mixed second-order model as defined in Equation 11.10.

Series-event model (Type B inactivation curves)

Another approach to modeling UV inactivation is with series-event kinetics. An event is assumed to be the accumulation of a unit of damage. Events occur in a stepwise fashion and each step is assumed to be an integer function. The rate at which an organism passes from one event level to the next is first-order with respect to UV intensity and independent of the event level occupied by the organism. As long as an organism is exposed to UV it continues to collect damage. However, a threshold number (n) exists, whereby organisms that have reached an event level greater than or equal to the threshold number are inactivated. All organisms that have accumulated less than n events survive. This model assumes that there is a mathematically infinite number of reaction sites such that the reaction of one site does not influence the rate of the reaction of the next site. In addition, the threshold number may vary, depending on the culturing conditions used, both prior to and after irradiation. For a given set of culturing conditions, however, the threshold number remains constant. The above is expressed in Equation 11.13 as a series of chemical reactions based upon a single organism:

$$M_0 \xrightarrow{kI} M_1 \xrightarrow{kI} \ldots M_i \xrightarrow{kI} \ldots M_{n-1} \xrightarrow{kI} M_n \xrightarrow{kI} \ldots \tag{11.13}$$

where k is the mixed, second-order reaction rate constant (cm²/mW s), I is the UV intensity (mW/cm²), M_i is an organism that has reached event level i, and n is the threshold number of the organism.

For a batch reactor operated under the conditions described in Batch ultraviolet inactivation, the rate at which organisms pass through event level i, dN_i/dt (number/cm³ s) is given by Equation 11.14,

$$\frac{dN_i}{dt} = -kIN_{i-1} - kIN_i \tag{11.14}$$

where N_i is the population of organisms with i units of damage. Equation 11.14 may be incorporated into a material balance for a batch reactor, and the resulting equation is solved sequentially from event level $i = 0$ to $i = n-1$. The general expression for the time-varying density of organisms occupying event level i at any time is given by Equation 11.15:

$$N_i = - \frac{N_I (\mathrm{k}It)^i}{i!} \, e^{-\mathrm{k}It} \tag{11.15}$$

where N_I is the initial density of organisms prior to exposure to UV. The density of surviving organisms N_S is the sum of all organisms that have not reached event level n and is shown in Equation 11.16 (Severin *et al.*, 1983*a*).

$$N_S = \sum_{i=0}^{n-1} N_i = N_I e^{-\mathrm{k}It} \sum_{i=0}^{n-1} \frac{(\mathrm{k}It)^i}{i!} \tag{11.16}$$

The series-event model has also been useful in the study of chemical disinfection, where I is replaced by the concentration of the disinfectant (Severin, Suidan & Engelbrecht, 1984*b*). When the threshold number is equal to one ($n = 1$), Equation 11.16 reduces to the mixed second-order model, as defined in Equation 11.10.

Comparison of series-event and multi-target models

The comparison of these two reaction models has been made elsewhere (Severin *et al.*, 1983*a*). Both models were found to adequately fit the batch inactivation curves for three organisms, *E. coli*, *Candida parapsilosis*, and f2 bacteriophage. Furthermore, both models were useful in the extrapolation of batch data to results from a completely mixed flow-through reactor. Figure 11.3 is an example of the inactivation data fitted to the series-event model. A plot of these data fitted to the multi-target model is not presented here, as both models yield similar curves. Table 11.3 is a summary of the kinetic constants found for the inactivation of these organisms with the two models. Since the assumptions behind each model are fundamentally different, the amount of damage required to inactivate the organisms as quantified by the threshold number (n) in comparison to the number of critical targets (q) differs greatly. The kinetic constants also differ in magnitude. For example, *C. parapsilosis* requires an accumulation of 15 events (n) or 1187 targets (q) to be destroyed before inactivation occurs. Likewise, the kinetic constants are 0.891 and 0.433 cm²/mW s for series-event and multi-target models, respectively. It is noteworthy that both models reduce to mixed second-order kinetics when the threshold number (n) or target number (q) equals one, such as with the Type A curve exhibited by the f2 bacteriophage.

The major difference in the two models is the mathematical perception of the targets or events. The logic of the multi-target model necessitates that

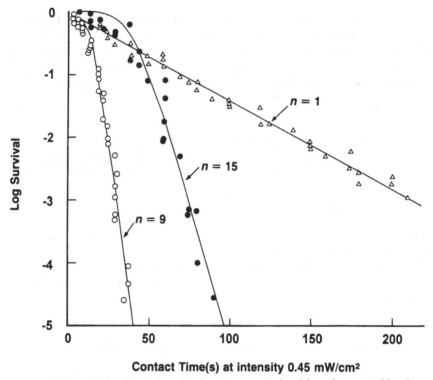

Figure 11.3   Batch inactivation data: analysis with series-event kinetics.
Open circles, *E. coli*; filled circles, *C. parapsilosis*; triangles, f2 bacterial
virus.

the number of targets (q) be finite and discrete, such that the probability of
attaining successive hits within an organism is reduced as the number of
available targets is depleted. The number of targets (q) is the total extent
to which an organism may be damaged, because it represents the total
number of sites that exist. The assumption behind the multi-target model
would break down if more than q targets existed. By comparison, the
threshold number (n) in the series-event model represents a finite number
of hits or events required from a mathematically large (infinite) number
of sites available. The rate of obtaining the last event prior to reaching
the threshold (n) is therefore equivalent to the rate of obtaining the first
event. Since there are many more sites available for reaction than required
for inactivation, the reactivations may proceed well beyond the threshold
number (n).

   The mathematical difference between the two models becomes apparent
in the interpretation of photoreactivation data. In an analysis of photoreactivation, Stevens (1980) showed that photoreactivated samples of *E. coli*
exhibited similar, yet broader initial shouldering in the inactivation curve

relative to the curves exhibited by non-photoreactivated samples. When it was analyzed by both multi-target and series-event models (Severin *et al.*, 1983*a*), this increased resistance in the photoreactivated samples was evident as a larger number of targets (q) and a larger threshold number (n), respectively. The mathematical interpretation of the series-event model allows for the collection of damage beyond the threshold number (n). However, the increased number of targets (q) caused by photoreactivation is a mathematical paradox for the multi-target model, as this model dictates that only a finite, lesser number of targets exist, as depicted in the non-photoreactivated samples. The authors have a preference for the series-event model, since it seems better to represent the known chemistry of UV inactivation and photoreactivation than does the multi-target model.

Continuous-flow ultraviolet inactivation

In contrast to batch inactivation conditions, most UV inactivation systems are designed to expose a continuous flow of water to UV. Both contact and non-contact continuous flow reactor types exist. Contact reactors are designed to use quartz-enclosed UV lamps immersed in the water to be disinfected. The size of these reactors vary, utilizing as many as 500 lamps in open channel reactors with loosely defined inlet and outlet ports, to as few as one lamp in a well-defined annular reactor system. In non-contact reactors, the UV lamp does not directly come into contact with the water to be disinfected. Older non-contact reactor designs consisted of a shallow channel of water situated below a parabolic reflector, housing the UV lamps. Newer non-contact reactors, termed coaxial reactors, consist of tubes constructed of UV transmittable material which carry the water past any number of UV lamps, using several different designs. In addition, newer reactors often utilize reflectors of various designs and reflective materials to increase disinfection efficacy.

Table 11.3. *Multi-target and series–event model kinetic constants calculated for* Escherichia coli, Candida parapsilosis *and f2 bacteriophage*

	E. coli	*C. parapsilosis*	f2 bacteriophage
Multi-target model			
Number of targets, q	201	1187	1
K (cm^2/mW s)	0.893	0.433	0.0724
Regression R^2	0.958	0.953	0.986
Series–event model			
Threshold events, n	9	15	1
k (cm^2/mW s)	1.538	0.891	0.0724
Regression R^2	0.967	0.965	0.986

As demonstrated by Equation 11.1, the UV dose applied to a unit of water is a function of UV intensity and contact time. In continuous-flow reactor systems, the path taken by an organism when passing through the reactor determines its contact time with UV. In addition, the path flows through sections within the reactor irradiated with different UV intensities. If multiple flow paths are hypothesized to make up the entire flow field of a reactor, it follows that a population of microorganisms entering a continuous-flow reactor will be exposed to a range of UV intensities, contact times and thus UV doses. Unfortunately, the means of measuring contact times and UV intensity within continuous-flow reactors are at present insufficient for proper prediction of the range of UV doses applied to a population of organisms. This is in sharp contrast to batch UV reactors, in which a uniform intensity field is delivered to the water to be disinfected for a precise, controlled exposure time.

The topic of continuous flow UV inactivation is presented by a discussion of the flow dynamics of idealized and non-idealized reactors and the resulting effect on UV intensity and contact time, followed by a discussion of the inactivation kinetics of idealized flow reactors. This sets a foundation for discussing the complexities of non-idealized continuous-flow reactors. The best available methods for determination of average UV dose output of non-idealized reactors will be discussed both in terms of the use of biological and chemical actinometry, and by calculation from measurements of UV intensity and contact time distributions. Finally, an example demonstrating the effect of intensity and time variation on the determination of UV dose in non-idealized, continuous-flow UV reactors is presented.

Flow dynamics and idealized inactivation kinetics

Mixing in continuous-flow UV reactors is considered to occur in the longitudinal and radial directions. Longitudinal mixing, as defined here, is the dispersion of a group of microorganisms in the direction of water flow, whereas radial mixing is the dispersion of microorganisms perpendicular to the direction of water flow. In a UV reactor demonstrating longitudinal mixing, some fractions of the microbial population will pass rapidly through the reactor relative to other fractions, which remain in the reactor for longer periods of time. In severe cases, short-circuiting of flow will occur. Here, some fractions of the population will exit the reactor prior to receiving the minimum UV dose required to achieve the level of inactivation desired. Under these conditions, UV reactor design becomes costly and inefficient (Thampi & Sorber, 1987). For optimal design of UV reactors, longitudinal mixing should be minimized, ensuring that microorganisms passing through the reactor are exposed to UV for a uniform time period.

In simple reactor systems, radial mixing dictates the distance at which the microbes are located from the UV source as they pass through the reactor, thus defining the UV intensity to which a group of microorganisms is exposed. Haas & Sakellaropoulos (1979) suggested that, at high flow velocities, turbulent conditions exist and all microorganisms have an equal probability of being located at any given radial distance from the UV source. Thus, the organisms are exposed to all intensities present through the cross section of the reactor. At low flow velocities, laminar flow conditions may develop in which the water is considered to segregate into layers. Each layer remains at a constant distance from the UV source throughout the disinfection process and thus is exposed to a different UV intensity. This is termed radial stratification, in which the UV intensity received is based on the location of the microorganism within the reactor or distance from the UV source. Under optimal UV reactor design conditions, radial mixing is maximized, ensuring that all microorganisms travelling through the reactor will be exposed to a uniform UV intensity.

Flow dynamics

As previously mentioned, two types of flow can be considered to occur in continuous-flow UV reactors: idealized and non-idealized flow. Under conditions of ideal flow, simplified assumptions allow for reasonable estimates of UV dose. However, in practice, the flow is non-ideal and simplified assumptions concerning UV dose determination cannot be readily made. The optimal hydraulic pattern in UV reactor design is a condition described as 'plug-flow'. Idealized or perfect plug-flow has been defined as water flow with no longitudinal mixing and perfect radial mixing (Severin, Suidan & Engelbrecht, 1984*a*). Although idealized or perfect plug-flow cannot be created in practice, it may be encouraged by proper hydraulic design. Some attempts to encourage proper hydraulic design include elimination of short-circuiting by ensuring complete use of the entire volume of the reactor. This can be accomplished by designing the reactor so that the inlet and outlet ports yield an even distribution of water flow across the entire reactor. Deflecting water flow as it enters the reactor and inserting baffles are other means to accomplish an even water distribution. In addition, the reactor should be designed for high-velocity water flow to create radial turbulence (Scheible, 1987) even at the cost of increasing longitudinal mixing (Severin *et al.*, 1984*a*). Simple closed reactor systems should also be designed with a high aspect ratio, defined as the ratio of reactor length (l) to reactor hydraulic radius (R). For l/R values greater than 50, plug-flow characteristics dominate, due to the development of a more uniform cross-sectional velocity, creating less longitudinal dispersion (Thampi, 1990).

The hydraulic condition opposite from plug-flow is described as complete mixing. Here, complete mixing occurs in both the direction of flow and perpendicular to flow. In idealized complete mixing, a population of micro-organisms entering a UV reactor immediately becomes completely mixed into the total volume of the reactor. Reactor designs demonstrating charac-teristics of complete mix hydraulics are undesirable from a disinfection standpoint, as, in theory, idealized complete mixing indicates that a portion of the microbial population would exit the reactor with a contact time equal to zero.

Kinetics of idealized continuous-flow ultraviolet inactivation

The UV inactivation kinetics of batch reactors having a uniform UV inten-sity can be modeled with the use of Equations 11.10, 11.12 and 11.16. These same Equations are also applicable to continuous-flow reactors with perfect plug-flow. Likewise, the mixed second-order, multi-target and series-event inactivation models are applicable to completely mixed UV reactors. When complete mixing conditions are present, two assumptions guide the deri-vation of the kinetics: (1) the density of surviving organisms in the effluent is exactly equal to the density at any point within the reactor, and (2) there is no change in the surviving cell density with respect to time within the reactor. The simplest case is that of a single-hit or single-event inactivation, equivalent to the mixed second-order model. The change in density of surviving organisms with time within the reactor is zero, and the overall inactivation rate is equivalent to the rate of input of organisms (QN_I) minus the rate at which surviving organisms leave the reactor (QN_S), as shown in Equation 11.17:

$$V \frac{dN_S}{dt} = QN_I - QN_S - KIVN_S = 0 \tag{11.17}$$

where V is the liquid volume of the reactor (cm^3), Q is the volumetric flow rate (cm^3/s), N_s is the density of surviving organisms within the reactor (number/cm^3), N_I is the density of viable organisms in the influent water flow (number/cm^3), K is the mixed second-order reaction rate constant (cm^2/mW), and I is the intensity within the reactor (mW/cm^2). Replacing the quotient Q/V with t and rearranging yields an expression for the surviv-ing fraction based on mixed second-order kinetics for a completely mixed reactor, as given in Equation 11.18 (Severin et al., 1983a):

$$\frac{N_S}{N_I} = \frac{1}{(1+KIt)} \tag{11.18}$$

Similar derivations may be made for other kinetic expressions as well. For example, the multi-target model for completely mixed reactors is given in Equation 11.19 (Severin *et al.*, 1983*a*).

$$\frac{N_S}{N_I} = 1 - \prod_{i=1}^{q} \frac{i\mathrm{K}It}{(1+\mathrm{K}It)} \tag{11.19}$$

The series-event model for completely mixed reactors is given in Equation 11.20.

$$\frac{N_S}{N_I} = 1 - \left[1 + \left(\frac{1}{\mathrm{k}It} \right) \right]^{-n} \tag{11.20}$$

In Equations 11.10, 11.12, 11.16–11.20, the UV intensity (I) can be replaced with $\bar{I}$, the average UV intensity within the reactor. The use of the average intensity within the equations comes directly from the derivations, not an a priori assumption that the average is the proper unit of measure. Calculation of the average intensity, $\bar{I}$, in the special case of single lamp annular reactor is derived later in the chapter. The use of the average intensity for any UV reactor configuration, however, must be carefully evaluated on the basis of known mixing or intensity distributions. Its use is valid only under the constraints imposed by the mathematical definitions of perfect plug-flow and complete mixing.

Actinometry for measuring average ultraviolet dose in non-idealized reactors

One of the major challenges facing researchers in the field of UV disinfection of water and wastewater is the measurement of the UV dose output of non-idealized, continuous-flow UV reactors (Jagger, 1967). This measurement is complicated by the complex design geometries and flow patterns of practical UV reactors (Severin *et al.*, 1984*a*), which, as previously mentioned, create non-uniform UV intensity gradients (Suidan & Severin, 1986) and a distribution of contact times within the reactor (Scheible, 1983). In addition, variations in chemical and physical water quality parameters also interfere with UV dose output of continuous-flow reactors (Qualls, Flynn & Johnson, 1983; Johnson & Qualls, 1984; Qualls *et al.*, 1985; Whitby & Palmateer, 1993). To effectively measure the UV dose output of these systems, a test method must prove capable of considering all elements from each of the above-mentioned factors. Optimally, the measurement system should be relatively simple, accurate, reliable and reproducible. The method should also be sensitive enough to detect changes in reactor UV dose output that occur with design alterations during

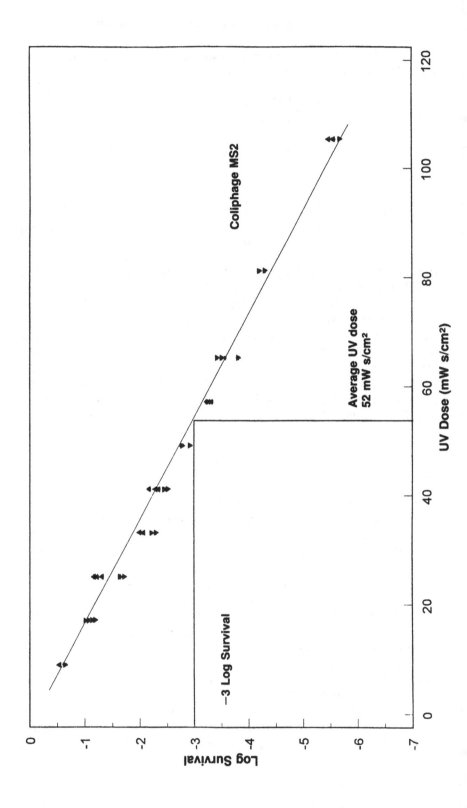

the reactor development and performance optimization process, in response to changes in water quality and with alterations in UV intensity with regard to the mechanical efficiency of the lamp/ballast combination. For the measurement of the average UV dose in non-ideal flow reactors, two approaches have been used. The first is to use actinometric methods, either biological (Qualls & Johnson, 1983; Qualls, Dorfman & Johnson, 1989; Havelaar *et al.* 1990; Wilson *et al.*, 1992*a*) or chemical (Harris *et al.*, 1987*a*; Mark *et al.*, 1990*a,b*; Hoyer *et al.*, 1992). Secondly, UV dose may be estimated from measurements of UV intensity and retention time distributions (Qualls & Johnson, 1985; Scheible, Casey & Forndran, 1986; Suidan & Severin, 1986), with results substantiated by biological or chemical actinometry. However, it has been recognized that, by themselves, neither of the two approaches is completely satisfactory. The best method of estimating the UV dose output of continuous-flow reactors will be a combination of biological or chemical actinometry along with mathematical modeling (Darby, Snider & Tchobanoglous, 1993). It may be best concluded that there is no protocol currently available that exactly delineates the applied UV dose in any reactor system with both non-ideal retention times and non-uniform UV intensities.

Biological actinometry

Biological actinometric measurement of UV dose (Qualls & Johnson, 1983) involves initially determining the UV susceptibility of a selected bioassay 'indicator' microorganism under controlled batch conditions. This is conducted by determining the level of bioassay microorganisms surviving (N_S, in number/cm^3) from a known initial concentration (N_I, in number/cm^3) after exposure to a range of UV doses. A standard UV susceptibility curve is then generated by plotting the log survivors over the range of UV doses applied (See above, under Batch ultraviolet inactivation). To measure the UV dose of a continuous-flow UV reactor, the bioassay organism is continuously injected at a known concentration (N_I) into water entering the reactor and its concentration determined in water exiting the reactor (N_S). The log survival is then calculated from previously constructed UV susceptibility curves with dose plotted against log survival. An extrapolation using the bioassay indicator organism coliphage MS2 is depicted in Figure 11.4. In the figure, the log survival was found to be −3.0, which extrapolated to a UV dose output of 52 mW s/cm^2. The average UV intensity (mW/cm^2) from the reactor may be determined by dividing the assayed UV dose (mW s/cm^2) by the average contact time (seconds) as determined from residence time distribution analysis. The application of biological actinometry to flow-through UV reactors should as much as possible be a controlled assay, as deviations in voltage supply to UV reactor, water flow rate, water chemical and physical conditions and lamp age greatly affect the UV dose value measured.

Alternatively, biological actinometry can be used to measure the UV dose received by separate fractions of the indicator organisms as a function of time after injection. Qualls & Johnson (1983) describe the use of *Bacillus subtilis* spores in a method analogous to a tracer study by injecting the spores as a single pulse into the water stream entering a continuous flow reactor. With this method, fractions of water exiting the reactor are collected in a rotating sampling tray as a function of time after injection. The procedure is conducted with the UV lamps 'on' and again with the lamps 'off', to generate data from pulse injections of irradiated and unirradiated *B. subtilis* spores. The distribution of the unirradiated spores represents the residence time distribution (RTD) curve of the reactor (Figure 11.5), i.e. the amount of time during which fractions of the *B. subtilis* spores are in contact with UV. The survival rate (N_S/N_I) of each flow fraction is determined from corresponding sampling points (based on time from injection) from irradiated and unirradiated spore pulse injections. The average UV intensity for each fraction is then determined by dividing the assayed dose by the time after injection at which the particular fraction was collected.

F-specific RNA coliphages (MS2 and Qβ) have been suggested as bioassay indicator organisms for measuring the UV dose output of continuous-flow UV reactors (Kamiko & Ohgaki, 1989; Havelaar *et al.*, 1990). Coliphage MS2 has been suggested because of (1) first-order inactivation kinetics, (2) relatively low susceptibility towards UV, (3) non-pathogenicity towards humans, (4) the ability to generate titers of up to 10^{12} PFU/ml, and (5) no innate ability to photoreactivate. In addition, the assay is highly consistent and reproducible (Kamiko & Ohagaki, 1989). Havelaar *et al.* (1990) used coliphage MS2 to verify predicted UV dose as determined from mathematical models. Wilson *et al.* (1992*a*) proposed the use of coliphage MS2 as a replacement of *B. subtilis* spores and *S. cerevisiae* as recommended in the National Sanitation Foundation Standard 55, due to the linearity of the MS2 dose–response curve over a wide range of UV doses (0–140 mW s/cm^2), its accurate enumeration at both low and high UV doses, enumeration in as little as 6 h and ease of attaining high titers in stock cultures (10^{10} to 10^{11} PFU/ml). Due to its low susceptibility towards UV, coliphage MS2 has also been proposed as a UV water disinfection surrogate for prediction of the level of inactivation of several waterborne bacterial and viral pathogens (Roessler *et al.*, 1992; Wilson *et al.*, 1992*b*).

Chemical actinometry

In addition to biological actinometers, several chemical actinometers are available for quantifying the levels of 254 nm light (UV intensity) emitted

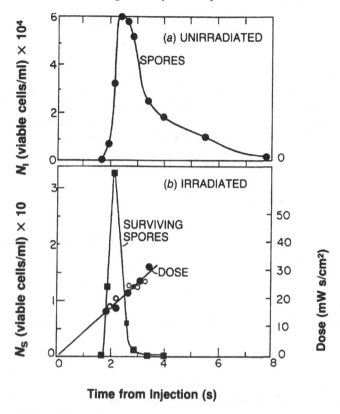

Figure 11.5   Biological actinometry assay to determine ultraviolet intensity output.

by the low-pressure mercury arc (Kuhn, Braslavsky & Schmidt, 1989). Chemical actinometers undergo photochemical decomposition of a known quantum yield when exposed to UV, with the quantification of reaction product used to estimate UV dose. Therefore, the photoproduct must be stable and readily measured (Harris *et al.*, 1987*a*). One such example of a chemical actinometer is potassium ferrioxalate (Hatchard & Parker, 1956), which undergoes photochemical decomposition upon exposure to light over a wide range of wavelengths. The change in ferrous iron concentration before and after irradiation, often measured spectrophotometrically with phenanthroline solution, is used to calculate the dose within the reactor (Harris *et al.*, 1987*a*). Another chemical actinometer is the peroxydisulfate-*t*-butanol system (Mark *et al.*, 1990*a*). This actinometer uses an oxygen-saturated solution of potassium peroxydisulfate with *t*-butanol. Photolysis of the peroxydisulfate anion ($S_2O_8^{2-}$) by 254-nm light creates the reactive SO_4^- radical anion, which attacks *t*-butanol, yielding a proton and the free radical $\cdot CH_2C(CH_3)_2OH$. The reaction of this free radical with dissolved

O_2 produces further protons, forming an acid yield of 1.8 mol/einstein (± 0.2) at 20 °C. In the photolysis of peroxydisulfate (10^{-2} mol/dm^3) and t-butanol (10^{-1} or 1 mol/dm^3) solutions, Mark *et al.* (1990*a*) found that the formation of H$^+$ and SO_4^{2-} increased linearly with respect to the UV dose applied. An advantage of the system is that proton formation can be followed with a pH meter or by titration, methods both routinely performed by even small waterworks. This same actinometer system was used by Hoyer *et al.* (1992) to measure the UV dose in optical cells and continuous-flow UV reactors, and for calibrating UV sensors.

Modeling ultraviolet intensity distribution in non-idealized reactors

A second method proposed for determining the UV dose of non-ideal, continuous-flow UV reactors is separate assessment of the UV intensity and contact time distributions to which microorganisms are exposed as they pass through the reactor. Knowledge of the independent variations within the time and intensity distributions provides some insight into the range of variations expected for UV dose. The following discussion focuses on modeling the distribution of UV intensity within a UV reactor.

The geometric relationship of the UV lamp to the water determines which mathematical model is to be used to estimate UV intensity distribution. For collimated beams emitting UV light into a thin water film (batch inactivation), the attenuation of light is assumed to follow the linear model, as previously discussed under Batch ultraviolet inactivation. The linear model is also applicable to parabolic reflector systems situated above a shallow channel of water (Roeber & Hoot, 1975). For modeling the UV intensity emitted from cylindrical lamps immersed in water, two models are available. The simpler but less accurate model is the radial source or infinite line source model. In this model, all light is assumed to be emitted radially, i.e. perpendicularly, from the UV lamp surface. The second more complex model for contact reactor systems is the finite line source model, based on the premise that the UV line source is actually a series of point sources. Here, UV is assumed to be emitted from each point of the line source.

Geometry of line sources

In contact reactor systems, the lamp is protected from the water to be disinfected by being housed within a quartz tube. To facilitate calculation of UV intensity distributions, the effects of the air gap between the lamp and quartz tube on light transmission are neglected. Conditions are further simplified by the assumption that the reflection or refraction of light at all

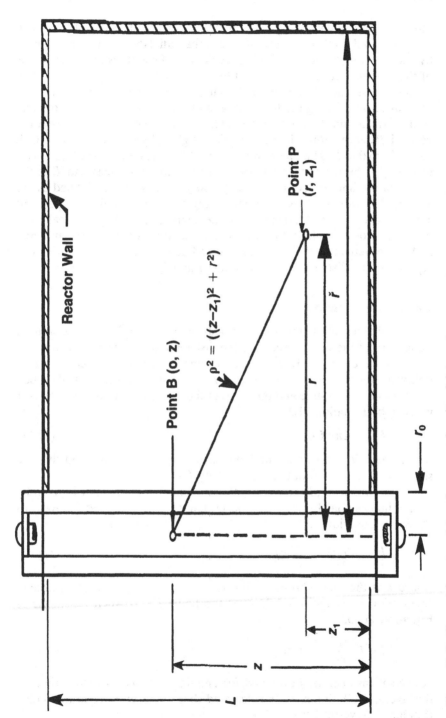

Figure 11.6 Geometric nomenclature for an annular ultraviolet reactor.

interfaces is negligible. Figure 11.6 is a schematic drawing of a typical, contact, annular UV reactor. Distance in the radial direction is represented by r where r_0 is the outer radius of the quartz tube and r is the inner radius of the reactor wall. UV is assumed to be emitted from a hypothetical line source located at the center of the lamp. The variable z represents the vertical distance along the line source, and L is the total length of the line source. In the finite line source model, it is also necessary to define the vertical distance within the reactor. This is given by the variable z_1, which varies from 0 to L. The line source has a total energy emission of $I_1 L$ (mW) where I_1 is the linear intensity of the line source (mW/cm) and L is the length of the line source (cm). The quantity I_0 (mW/cm^2) is defined as the average UV intensity, i.e. the total energy emitted per unit time per unit area averaged for the total surface of the quartz tube. A hypothetical point P (r, z_1) is located within the reactor volume at location r between r_0 and $\check{r}$. A hypothetical point B $(0, z)$ is located at $r = 0$ and z. The variable ϱ represents the distance between points B and P.

Radial source model

In the radial or infinite line source model, all light is assumed to be emitted radially from the line source; i.e., the distribution is one-dimensional and intensity varies as $1/r$. Due to axial symmetry, it is recognized that I_0 is uniform everywhere at the quartz surface. Thus, if no light is absorbed between the line source and the quartz tube, energy emitted to the reactor is given by Equation 11.21:

$$I_1 L = 2\pi r_0 L I_0 \tag{11.21}$$

where the UV intensity at the lamp surface is I_0 (mW/cm^2) and the line source is theorized to output an intensity I_1 (mW/cm).

The UV intensity, I (mW/cm^2), at any point P in the reactor is given by a one-dimensional form of Lambert's law (Jacob & Dranoff, 1966), as shown in Equation 11.22:

$$\frac{1}{r} \frac{d(rI)}{dr} = -EI \tag{11.22}$$

where E is the monochromatic absorbance at 254 nm reported in base e. Integration of Equation 11.22 over the limits $I = I_0$ when $r = r_0$ yields Equation 11.23.

$$I = I_0 \frac{r_0}{r} e^{-E(r-r_0)} \tag{11.23}$$

A special case can be developed for annular reactors. In such a reactor, the average UV intensity is calculated, beginning with Equation 11.24, (Suidan & Severin, 1986):

$$\bar{I} = \frac{2I_0 r_0}{(\bar{r}^2 - r_0^2)} \int_{r_0}^{\bar{r}} e^{-E(r-r_o)} \, dr \qquad (11.24)$$

The integrated result is given as Equation 11.25 (Suidan & Severin, 1986):

$$m = \frac{\bar{I}}{I_0} = \frac{2r_0}{E(\bar{r}^2 - r_0^2)} \left[1 - g^{-E(\bar{r}-r^0)} \right] \qquad (11.25)$$

where m is defined as the intensity factor and represents the ratio of the average UV point intensity in the reactor, I, to the UV intensity at the lamp surface, I_0. A very rudimentary method for estimating I in multi-lamp reactors could be developed by use of Equation 11.23. In this situation, each point in the reactor would be irradiated by many lamps, thus the total irradiation at point P would be estimated as the sum of the intensities from each lamp with the use of Equation 11.23.

Finite line source model

A second model for estimating UV intensity is the finite line source model. This model is based upon the premise that a lamp source may be approximated by a series of point sources located along a line segment, and that UV is emitted spherically from all points on the lamp axis. The geometry of the finite line source is shown in Figure 11.6. A differential element, dz, is located on the line source. The line source has a total length, L, and an intensity I_1. A point located at dz emits a point intensity of I, dz in units of mW. The energy of the total line source is $I_1 L$.

Therefore, the practical approximation of energy of $I_1 \, dz$ is created by dividing L into S segments, such that $I_1 L = I_1 S \Delta z$, and Δz approaches dz. Within the reactor, each point P is subjected to UV from all points on the line, and the UV intensity at P is given as the sum of the contributions from all of these points. The distance between a point in the reactor and a point on the line source is given as ϱ in Equation 11.26.

$$\varrho = \sqrt{(z-z_1)^2 + r^2} \qquad (11.26)$$

The dissipation of light from a point on the line source is given by Lambert's law of absorption in Equation 11.27.

$$\frac{d(\varrho^2 I)}{\varrho^2 d\varrho} = -EI \qquad (11.27)$$

Equation 11.27 is further modified to account for no absorbance of light between a point on the line source and the quartz tube and integrated to yield Equation 11.28 (Jacob & Dranoff, 1970).

$$I = \frac{I_l \, dz}{4\pi\varrho^2} \, e^{-E\varrho[(r-r_0)/r]} \tag{11.28}$$

Equation 11.28 is the basic equation for the point source summation method for estimating intensities in complex reactor configurations (Jacob & Dranoff, 1970). By this method each lamp is divided into a large number of point sources and the reactor volume is divided into a larger number of light receptor points. Each point in the reactor receives energy from each point source along each lamp. The total UV intensity at each point in the reactor is the summation of the individual contributing sources, irrespective of the direction of the originating light. Use of the point source summation method to calculate the average UV intensity has been described in detail in the United States Environmental Protection Agency (1986) *Design Manual: Municipal Wastewater Disinfection.*

A cautionary note on the point source summation method must be expressed concerning the origin of the value of the point source intensity, $I_l \, dz$. Consider the problem where the intensity of the line source, I_l is not known, and is to be measured with a chemical actinometer such as potassium ferrioxalate. For these chemical methods, the energy is absorbed very near to the lamp surface. If UV were emitted radially, as described by the radial source model, no light would be lost out of the top or the bottom of the quartz tube. However, if light is emitted in all directions, as in the case of a finite line source model, then energy is lost from the top and bottom of the quartz tube. The chemical actinometer does not account for this lost energy. Any approximation of I_l for modeling purposes must include this energy, or the line source intensity as calculated will be artificially low. Therefore, the value of I_l used in the finite line model must be greater than I_l for the radial model to obtain the same energy input to the reactor. The value of I_l for the infinite line or radial source model is related to the average surface intensity, I_0, by Equation 11.21. The value of I_l for the finite line model, however, is related to the average surface intensity, I_0, by Equation 11.29:

$$I_l = 2\pi r_0 \, I_0 / \beta \tag{11.29}$$

where β is a transmission factor relating the ratio of energy transmitted to the reactor to the total energy emitted by the finite line source. The value of β can be derived from energy balances to give Equation 11.30 (Suidan & Severin, 1986):

$$\beta = \sqrt{1 + \left\langle \frac{r_0^2}{L^2} \right\rangle} - \frac{r_0}{L} \qquad (11.30)$$

The transmission factor, β, has practical ramifications for the geometry of reactor design, as is seen in Figure 11.7, where the effective transfer of energy to the reactor is plotted against the lamp aspect ratio, i.e. the ratio of lamp length (L) to the quartz tube radius (r_0). The figure demonstrates that long, narrow lamps are theoretically more efficient than short, wide lamps, as more than 95% of the energy from a finite line source enters the reactor if L/r_0 is greater than 20. For L/r_0 greater than 100, more than 98% of the energy enters the reactor. The finite line source model also predicts that the UV intensity at the lamp surface will vary along the lamp surface, with the highest intensity at the midpoint of the lamp. Thus, it may be seen that in any reactor system that involves intimate contact between water and the quartz tube, there will be both radial and longitudinal intensity gradients.

For the special case of the annular reactor, the average UV intensity as estimated with the use of the finite line source model is given by Equation 11.31:

$$x = \frac{\bar{I}}{I_0} \qquad (11.31)$$

where the value of x is the average of the intensity within the annular reactor normalized to the surface intensity of the radial model, and is functionally equivalent to the expression for m given in Equation 11.25. However, the calculation of I for the finite line source model is very detailed (Suidan & Severin, 1986), involving the integration of the point intensity from Equation 11.28 over the reactor radius, the reactor length and the lamp length. The calculation also includes correction for β, depending on the origin of the value of I_1.

The relationship between x and m relating to average UV intensities has previously been discussed for annular reactor systems with lamp aspect ratios of 24, 72 and 120 (Suidan & Severin, 1986). These aspect ratios were chosen because of the availability of UV lamps in lengths of 30.5, 91.4 and 152.4 cm. A typical radius for quartz tubes is 1.27 cm (Suidan & Severin, 1986). Figure 11.8 represents the case where the lamp aspect ratio is 24. The figure also depicts cases in which the ratio of the reactor radius to lamp radius, $\bar{r}/r_0$, is changed from 2 to 20. Note that for the aspect ratio of 24, x is greater than m for values up to 10. When $\bar{r}/r_0$ exceeds 10, m is greater than x. When it is valid to use the average intensity to calculate UV dose, such as in idealized flow conditions, values of x and m are indistinguishable for all cases where the absorbance is high, i.e. where the

Lamp Aspect Ratio (L/r_0)

Transmission Factor β

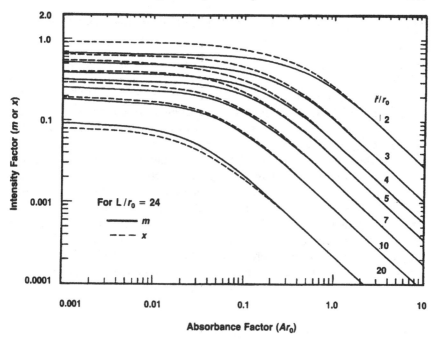

Figure 11.8   Intensity factors for an annular reactor with $L/r_0 = 24$.

product of absorbance (A) and radius (r_0) is greater than 0.5. This is consistent with a conservation of energy input in the two models.

Some practical observations can be made from these figures. Lukiesh & Holladay (1944) reported a range of absorbencies of $A = 0.016$ to 0.024/cm for waters from three public drinking water supplies. Severin (1980) indicated a range of $A = 0.07$ to 0.10/cm for secondary treated wastewater effluents. Assuming $r_0 = 1.27$ cm, expected values of Ar_0 for typical UV disinfection reactor applications range from approximately 0.02 to 0.03 for drinking water. As seen in Figure 11.8, there is a sufficient difference between x and m to warrant the use of the more complex model. Therefore, for application to drinking waters, the finite line source model is preferred over the radial source model for the assessment of the average UV intensity within an annular reactor. For treated wastewaters, the approximate range of Ar_0 of 0.09 to 0.127 represents a transition range where m may be used as a replacement for x. In highly absorbing liquids, the radial and finite line source models are considered equivalent in the calculation of average UV intensity.

Determining residence time distribution in non-idealized reactors

In addition to mathematical modeling of the UV intensity distribution within a reactor, the amount of time in which microorganisms passing through the reactor are in contact with UV, or the reactor RTD, must also be determined. The most simplistic measurement of contact time within a continuous flow UV reactor is division of the reactor volume (V) by the flow rate (Q) of the water to be treated. This determines the average time in which the microorganisms are in contact with UV. However, the geometry of non-idealized reactors generate conditions in which actual contact time is inconsistent with the predicted or theoretical contact time (Qualls & Johnson, 1983; Qualls *et al.*, 1985). Thus, a group of microorganisms travelling through a UV reactor are often exposed to UV over a distribution of contact times. In addition, the importance of precise measurements of the relatively short contact times related to UV disinfection (seconds) are magnified when compared to the longer exposure times often required for chemical disinfection (minutes).

Residence time distribution curves

Longitudinal mixing and actual contact time in UV reactors can be partially described by generating RTD curves. This type of analysis is based on a stimulus–response experimental design, in which a non-reactive tracer of known concentration is injected into the water stream immediately before it enters the reactor. The tracer concentration is then quantitatively monitored over time in water immediately it exits the reactor. The RTD curve may then be generated by plotting the concentration of tracer against time.

The applicability of tracers to measurement of the RTD of a continuous-flow UV reactor requires that the tracer pulse be completely mixed in the full water stream immediately before it enters the UV lamp portion of the reactor, and that representative water samples be taken immediately before and after the water passes through the lamp portions of the reactor (Scheible, 1983). Thampi & Sorber (1987) developed a method for generating RTD curves to determine the mixing characteristics of reactors with relatively short retention times, utilizing a conductivity cell positioned at the effluent end of the UV reactor to monitor test water conductivity after the injection of a salt tracer. The signal response was captured on a chart recorder, which instantaneously gave RTD data for any desired flow setting. Other tracer techniques have also been discussed and proposed for use in evaluating reactor performance (Severin, 1980; Nieuwstad, Havelaar & van Olphen, 1991). In addition, Scheible (1987) developed a step input method in order to conduct RTD analysis in open-channel UV reactors with less defined inlet and outlet ports.

Residence time distribution curve analysis

Reactor hydraulics can be visually and mathematically described by both the shape of the RTD curve and the distribution of the area under the curve (Figure 11.9). Several hydraulic indices are shown on the RTD curve, including the time at which the tracer initially appears (t_i), the time at which the concentration of the tracer peaks (t_p) and the time at which 10 (t_{10}), 50 (t_{50}) and 90% (t_{90}) of the tracer has passed through the reactor. In addition, the mean retention time of the reactor (θ) can also be calculated from the RTD (US Environmental Protection Agency, 1986). Several researchers have used this type of analysis to define the hydraulic performance of UV reactors (Severin, 1980; Harris *et al.*, 1987a; Scheible 1987; Thampi & Sorber, 1987; Nieuwstad *et al.*, 1991). From these indices, a number of calculations are possible to assist in characterization, improvement and quantitative evaluation of UV reactor hydraulic design (Scheible, 1983; US Environmental Protection Agency, 1986). In addition, guidelines have been suggested for specification of RTD indices to allow for the quantitative assessment of plug-flow conditions (Scheible, 1983; US Environmental Protection Agency, 1986) and are listed in Table 11.4.

Dispersion

The RTD curve can also be used to calculate the dispersion (D_x), or longitudinal mixing that occurs in flow-through UV reactors (Levenspiel & Smith, 1957; US Environmental Protection Agency, 1986). The value of (D_x) can be defined as a measure of the spread of the RTD about the average retention time, as in Figure 11.9 (Scheible, 1983). Scheible (1987) indicates that if the RTD curve closely resembles a normal Gaussian distribution, then an estimation of dispersion can be made from Equation 11.32.

$$\sigma_\theta^2 = \frac{\sigma^2}{\theta^2} = \frac{2(D_x)}{vx} \tag{11.32}$$

where $\sigma_{\theta 2}$ is the dimensionless variance, σ^2 is the variance of the RTD curve, θ^2 is the mean residence time (s) as calculated from the RTD, D_x is the dispersion coefficient (cm^2/s), v is the velocity of the water in the longitudinal direction (cm/s) and x is the average distance (cm) travelled by the water while under the influence of UV. It has been suggested that a dispersion value of less than 50 cm^2/s indicates reactor designs approaching plug-flow conditions (Thampi, 1990). The value of (D_x)/vx has been termed as the dispersion number (d), and has also been used as an index for proper UV reactor design. A value of 0 would indicate no dispersion, less than 0.01 would indicate low dispersion, between 0.01 and 0.1 would denote moderate dispersion, and high dispersion would be indicated by d values above 0.1 (US Environmental Protection Agency, 1986).

Table 11.4. *Description of hydraulic indices from ultraviolet reactor residence time distribution analysis*

Hydraulic index	Index description
t_i/T	This index is defined as the severe short-circuiting index and is the ratio of the time at which the tracer initially appears (t_i) to the theoretical retention time of the reaction (T). T is determined by dividing the void volume of the reactor by the flow rate of the water to be treated. For perfect plug-flow reactors, the ratio is 1.0, and approaches 0 with increased mixing. The t_i/T value should be greater than 0.5 for effective design (United States Environmental Protection Agency, 1986).
t_p/T	This index is defined as the average short-circuiting index and will indicate the presence of dead spots within the reactor and estimate the effective volume of the UV reactor. The index is the ratio of the time required for the tracer concentration to peak (t_p) to the theoretical retention time of the reactor. For perfect plug-flow reactors, the ratio is 1.0, with the index decreasing with increased mixing and short-circuiting. The t_p/T value should be greater than 0.9 for effective design (United States Environmental Protection Agency, 1986).
$1-(t_p/t_{50})$	This is a third index of short-circuiting and is based on the ratio of the time required for the tracer concentration to peak (t_p) to the time required for 50% of the tracer to pass. The index value is 0 for ideal plug-flow and 1.0 for ideal complete mixing (Thampi & Sorber, 1987)
(t_{90}/t_{10})	This is defined as the Morrill Dispersion Index and is a measure of the spread of the residence time distribution curve (Morrill, 1932). The index is the ratio of the time required for 90% of the tracer to pass to the time required for 10% of the tracer to pass. A value of 1.0 would indicate ideal plug-flow. This value should be less than 2.0 for effective design (Scheible, 1987).
t_i/T	This index is also a measure of effective use of the entire volume of the reactor and is determined by dividing the mean residence time by the theoretical residence time of the reactor. A value of 1.0 would indicate that full use is being made of the reactor volume. Values less than 1.0 indicate the effective volume is less than the actual volume of the UV reactor. The t_i/T value should be close to 1.0 for effective design (Scheible, 1987).
t_{50}/t_i	This index measures the skewness of the residence time distribution curve and is a ratio of the center of gravity of the curve (time in which 50% of the tracer had passed) to the mean residence time of the reactor. A perfectly symmetrical residence time distribution curve will have an index of 1.0. If the curve is skewed to the left, the index will be less than 1.0. For effective reactor design, this value should be between 0.9 and 1.0 (United States Environmental Protection Agency, 1986).

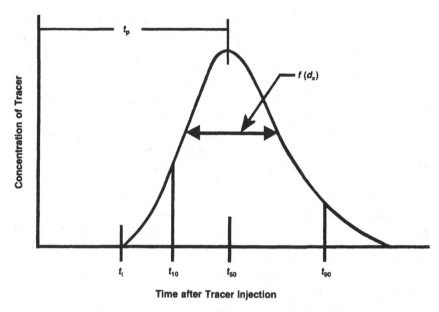

Figure 11.9   Nomenclature for hydraulic indices of residence time distribution curves.

Resolution of ultraviolet intensity and residence time variations

As previously mentioned, the estimation of the UV dose in a continuous-flow reactor is extremely difficult, due to the range of UV intensities and retention times likely to be encountered within the reactor. It is therefore unlikely that, even given full understanding of intensity and time variation within a particular UV reactor, the true extent of UV dose variation can be predicted. However, knowledge of these two independent parameters is useful in describing the boundary conditions for UV dose within a reactor. In discussing the variations in intensity and time within a reactor, the flow path concept may be used. By this concept, microorganisms travel along flow paths as they pass through a UV reactor. The flow path taken determines the UV intensity and contact time of exposure to UV. Tracing of flow paths and the organisms within them yields the boundary expectations for overall effectiveness of the reactor.

Effects of intensity and time variation on ultraviolet dose

The following is a model proposed by the authors for predicting the range of applied UV dose from intensity and time distribution data. To develop the following example, a reactor described by Darby *et al.* (1993) was used. Darby *et al.* described a four-lamp reactor in which the lamps were on

7.5-cm centers within a 15 × 15-cm face. Flow was directed along the length of the lamps. To estimate the UV intensity distribution for the example, a single intensity gradient analysis of the plane perpendicular to flow was conducted with the use of the radial source model, with water absorbance assumed to be equal to zero. Use of the model allowed calculation of the UV intensity at 3200 points within the reactor. The results were normalized to give an average intensity of 1.0 intensity units, and the UV intensity distribution was found to be 1.0 ± 0.18 intensity units (Figure 11.10a).

The US Environmental Protection Agency (1986) recommends that the Morrill dispersion index (t_{90}/t_{10}) be less than 2.0 to ensure that no short-circuiting occurs within a reactor. This boundary condition was used to approximate RTD for the hypothetical reactor used in this exercise. With the assumption that the distribution of retention times within the reactor was normal about an average retention time of 1.0, the distribution of contact times within the reactor was set at 1.0 ± 0.29 time units, reflecting the condition t_{90}/t_{10} ± 2.0 (Figure 11.10b).

From these two distributions, special cases may be established to determine the effects of intensity and time variation on a hypothetical organism passing through the reactor. For this exercise, the bioassay organism was assumed to be susceptible to UV in a fashion consistent with mixed second-order kinetics, as is often observed with coliphages. A postulated UV susceptibility curve was established for the organism so that a UV dose of 1.0 intensity time units resulted in a 0.0001 surviving fraction, equivalent to a −4.0 log survival. The kinetic constant K calculated for this hypothetical organism by Equation 11.10 was 9.21. This effectively set a hypothetical biological actinometery standard constant with batch inactivation in a uniform intensity field, such as is generated with a collimated beam apparatus.

Consider the idealized case where the continuous flow reactor has no longitudinal mixing and a uniform intensity field. The expected average UV dose would be 1.0 + 0.0 intensity time units, and the hypothetical organism would yield a 0.0001 surviving fraction, as summarized in Table 11.5, Case 1. If the reactor has no longitudinal mixing and no radial mixing, then the water entering the reactor would be exposed to a UV dose of 1.0 + 0.18 intensity time units, based soley on the variation in intensity. The surviving fraction would be 0.00039, and the calculated UV dose as extrapolated from the bioactinometric assays would be 0.85 intensity time units (Table 11.5, Case 2). If the intensity is uniform, or complete radial mixing is assumed, but the flow is dispersed to the extent of 1.0 ± 0.29 time units (t_{90}/t_{10} <2.0, as in Figure 11.10b), then the surviving fraction would be 0.0027 and the back-calculated UV dose would be 0.66 intensity time units (Table 11.5, Case 3).

One possible case is that a portion of the water enters a flow path that is exposed to a random intensity and random time within the set boundary conditions of I = 1.0 ± 0.18 intensity units and t = 1.0 ± 0.29 time units.

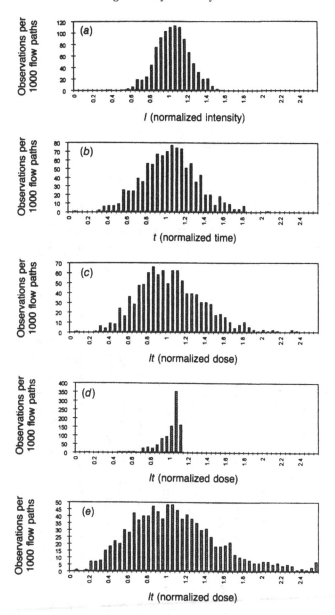

Figure 11.10   Effect of intensity and time distribution on simulated dose responses. (*a*) Intensity distribution; (*b*) time distribution; (*c*) Intensity and time are random; (*d*) best case; short exposure at high intensity; (*e*) worst case: short exposure at low intensity.

Table 11.5. *An exercise in simulation of perceived ultraviolet dose based on a hypothetical reactor and bioactinometer*

	Intensity distribution $(I/\bar{I})$ (average ± SD)	Time distribution $(t/\bar{t})$ (average ± SD)	Dose distribution $(It/\bar{I}\bar{t})$ (average ± SD)	Surviving fraction[a]	Calculated dose[b] $(It/\bar{I}\bar{t})$
Case 1: No longitudinal dispersion, uniform intensity	1.00 ± 0.00	1.00 ± 0.00	1.00 ± 0.00	0.0001	1.0
Case 2: No longitudinal dispersion, intensity gradients[c]	1.00 ± 0.18	1.00 ± 0.00	1.00 ± 0.18	0.00039	0.85
Case 3: Longitudinal dispersion[d], uniform intensity	1.00 ± 0.00	1.00 ± 0.29	1.00 ± 0.29	0.0027	0.66
Case 4: Longitudinal dispersion, random intensity gradient	1.00 ± 0.18	1.00 ± 0.29	1.00 ± 0.34	0.00292	0.63
Case 5: Longitudinal dispersion, intensity gradients: BEST CASE	1.00 ± 0.18	1.00 ± 0.29	1.00 ± 0.12	0.00104	0.75
Case 6: Longitudinal dispersion, intensity gradients: WORST CASE	1.00 ± 0.18	1.00 ± 0.29	1.00 ± 0.47	0.00681	0.54

[a] The average from 1000 flow paths passing through the reactor, based on mixed second-order inactivation kinetics. The inactivation constant K for the hypothetical organism is 9.21 (intensity/time units), where a dose of 1.0 yields a 0.0001 surviving fraction (−4 log survival).

[b] Extrapolated with the surviving fraction and the kinetic relationship described in footnote a above.

[c] The standard deviation of intensity is based on a normal distribution applied to an intensity field with the use of a four-lamp reactor (Darby *et al.*, 1993) and with the radial source model and the assumption of no absorbance.

[d] The standard deviation of retention time is based on a normal distribution with $t_{90}/t_{10} = 2.0$ (United States Environmental Protection Agency, 1986).

The distribution of dose applied to this double random event is $It = 1.0 \pm 0.34$ (Figure 11.10c). The expected surviving fraction measured using biological actinometery would be 0.00292 and the back-calculated UV dose from the actinometric measurement would be 0.63 intensity time units, as in Table 11.5, Case 4.

The best case scenario for radially stratified flow would be that in which the water that spent the longest time in the reactor passed through areas within the reactor with the lowest intensity, and water exiting the reactor in the shortest time passed through the areas of highest intensity. This scenario is represented in Figure 11.10d, where the UV dose distribution is 1.0 ± 0.12. The expected surviving fraction, if biological actinometery were used, would be 0.00104, and the back-calculated UV dose would be 0.75 intensity time units, as in Table 11.5, Case 5.

The worst possible case for radially stratified flow would be one in which the water with the longest retention time passed through reactor areas with the highest intensity, and water exiting the reactor in the shortest time was exposed to the lowest intensity. On the basis of the initial time and intensity distributions, this case would result in a UV dose distribution of 1.0 ± 0.47 intensity time units (Figure 11.10e). The surviving fraction of the bioassay organism would be 0.00681 and the extrapolated dose would be 0.54 intensity time units, as in Table 11.5, Case 6.

This example is presented to alert the reader to the complex codependence of UV intensity with retention time in a simple hypothetical UV reactor. Even in the simple system presented, the perceived dose as back-extrapolated with bioactinometery varied from 1.0 to 0.54 intensity time units, even though in all cases the average UV dose was 1.0 intensity time units. While the information presented in Table 11.5 provides insight into the relationships of time and intensity to the UV dose as measured by biological actinometery, the model is not universally applicable to reactor systems, in that the true nature of the distributions are probably not normally distributed, the organisms tested may not show mixed second-order inactivation kinetics and the level of inactivation of the test organism would affect the UV dose calculated from survival data. However, this exercise demonstrates the effects of intensity and time variation on microbial inactivation and calculated UV dose levels.

Interferences with ultraviolet disinfection efficiency

The effectiveness of UV disinfection under field conditions is dictated by several factors, including repair or reactivation of microorganisms exposed to UV, chemical and physical water quality parameters and mechanical properties of UV disinfection systems. The ability of a microorganism to repair itself after exposure to UV is dependent not only on the physiology of the particular organism, but also on the history of the organism, both before and after exposure to UV. Water quality parameters affecting UV disinfection include absorbance at 254 nm, suspended solids and turbidity. Finally, the major mechanical properties determining UV effectiveness include electrical ballast integrity, the cleanliness of the lamp quartz surfaces and the temperature of the UV lamps. This section of the chapter is a brief overview of these important parameters.

Cellular repair

The phenomena of cellular repair following sublethal exposure to UV have been studied by many investigators (Kelner, 1949; Carson & Petersen, 1975; Salaj-Smic et al., 1980; Knudson, 1985; Harris et al., 1987c, Simonson, Kokjohn & Miller, 1990; Mechsner et al., 1991; Bridges, 1992; Lindenauer & Darby, 1994). Two major repair mechanisms have been identified (Smith & Hanawalt, 1969), with either of the two enzyme systems capable of restoring cell viability to UV-irradiated organisms. The first repair system, termed 'dark repair', is a mechanism that excises dimerized pyrimidine base pairs and allows the reinsertion of undimerized bases by other enzymes. A second system, termed 'photoreactivation', is stimulated by exposure of injured cells to near-UV light (340–380 nm). Upon activation, the photoreactivating enzyme recognizes, binds and splits thymine dimers. The regenerative capacity of any organism type is related to its inventory of repair enzymes, with some organisms known to be more efficient than others. In addition, reactivation is speculated to be a universal phenomenon with evolutionary importance (Raunbler & Margulis, 1980).

Sunlight is known to contain sufficient near-UV light to induce the reactivation of coliforms in UV-irradiated secondary effluents (Johnson et al., 1979; Scheible, Binkowski & Milligan, 1979; Wolf, Petrasek & Esmond, 1979). One hour of exposure to sunlight typically increases the surviving fraction by a factor of 10. In an analysis of reactivation data from Stevens (1980), Severin et al. (1983a) showed that the effect of near-UV exposure on a pure culture of E. coli was to increase the threshold number (n) in the series-event model. The inactivation constant (k) was found to remain the same. Mechsner et al. (1991) stated that the UV dose applied for disinfection must be sufficient to counteract the effects of the various repair mechanisms that have evolved in microorganisms, as regrowth of injured cells in these experiments reached approximately 30% of the pre-UV exposure levels. In addition, Lindenauer & Darby (1994) found that an increase in UV dose is extremely valuable in minimizing photoreactivation events, and that increased levels of suspended solids increased photoreactivation by decreasing the UV dose delivered to the cells to be disinfected. The practical effects of photoreactivation deserve more study, especially in view of the fact that certain pathogens are known to photoreactivate (Chanda & Chatterjee 1976).

Water quality parameters

One of the most important water quality parameters affecting UV disinfection is the absorbance of the water at 254 nm (see Equations 11.5 and 11.6), which has been used to predict disinfection efficiency (Petrasek *et al.*, 1980; Severin, 1980; Scheible, 1987). Most waters to be treated have absorbances in the range of $E = 0.02$ to 0.1/cm. Many naturally occurring chemicals, such as phenolic compounds, humic acids from decaying vegetation, lignin sulfates from pulp and paper production, and ferric iron, have been determined to interfere with UV transmission (Yip & Konasewich, 1972).

Suspended solids within the water to be treated are also known to effect UV disinfection efficiency. Beyond a potential increase in absorbance, the effect of these particles on UV disinfection is three-fold (US Environmental Protection Agency, 1986; Darby *et al.*, 1993): (1) clumps of organisms skew the kinetic response, due to the method by which survival is measured, i.e. the plate count method (Galasso & Sharp, 1965); (2) organisms in a clump tend to shadow each other, a typical situation for UV (Qualls *et al.*, 1983) as well as other disinfectants (Hoff, 1978; Brigano *et al.*, 1980); and (3) particles scatter UV (Qualls & Johnson, 1983). These mechanisms may partially explain why studies comparing sand-filtered and unfiltered wastewater effluents have shown superior UV performance with the filtered water (Severin, 1980; Johnson & Qualls, 1984; Qualls *et al.*, 1985; Darby *et al.*, 1993).

The effects of particles on cell clumping and shadowing have an observable impact on the efficacy of UV disinfection of wastewaters. Oliver & Cosgrove (1975) found that sonication of wastewaters to break up clumps resulted in large increases in disinfection efficiency. Qualls *et al.* (1983) found that the major obstacle to inactivating more than 3 or 4 log units of fecal coliforms in wastewater effluents was clumps larger than 70 μm in diameter. In addition, organisms deep within suspended solids may be partially protected from UV exposure (Qualls *et al.*, 1983). This means of protection was first identified by Roeber & Hoot (1975), who found that coliform levels in UV-irradiated, high-turbidity samples increased after blending. Low-turbidity samples did not show increased counts after blending. To consider the effects of particle shielding in predicting UV disinfection, Scheible *et al.* (1986) and Scheible (1987) proposed a disinfection model that incorporates the protected fraction or particle-associated microorganisms into the prediction of survival.

Suspended solids are also known to scatter UV. In one early study, fuller's earth and koalinite clay did not adsorb UV equally when equal turbidities were suspended in water (Huff, 1965). Apparently not all particles adsorb UV and it is surmised that clays act to scatter light rather than to adsorb it (Qualls *et al.*, 1983). Qualls *et al.* (1983) determined that scattering accounted for 12% of the absorbance in wastewater effluent samples. They concluded that the average intensity in a reactor is usually higher than predicted from photometric methods.

Mechanical properties of the ultraviolet system

Electrical ballasts are available in two styles, mechanical and electronic. Currently, there is a lack of information available to substantiate the superiority of either design, as controversy exists as to the effects of higher outputs, cooler temperatures and control of output pulsing by electronic ballasts on the rate of lamp solarization (i.e. chemical changes within the quartz caused by reaction of UV with impurities within the quartz). UV lamps are available in two types, with low and medium intensity. US manufacturers of UV large-scale equipment tend to prefer the low-intensity lamps, whereas European manufacturers prefer the medium-intensity lamps. Questions currently exist over the advantages of higher point intensity versus the number of lamps. Manufacturers utilizing low-intensity lamps claim that dispersement of energy by many low-intensity lamps is more uniform, and therefore more efficient, than the use of fewer, more intense lamps. Secondary claims surround the observation that the higher-intensity lamps operate at higher temperatures, both at the lamp and at the ballast, therefore increasing the potential for increased rates of solarization and chemical precipitation on the quartz surface. In addition, UV lamps need replacement periodically due to solarization of the quartz. The typical life expectancy of lamps is around 2000 h of burn time.

The quartz sheaths of contact UV reactors, which come in direct contact with water, must periodically be cleaned, to maximize disinfection efficiency. Cleaning frequency is of the order of once or twice per year, depending on the quality of the water to be treated. Collection of suspended solids and chemical precipitation of iron and calcium salts on the exposed quartz surfaces are the major reasons for the need for cleaning. Reactors with few lamps may include mechanical wiper systems, while reactors with many lamps do not include this feature. In this case, cleaning may include taking the reactor out of service and soaking the quartz sleeves in citric acid or a commercial lime remover.

The temperature of the UV lamp has a large bearing on the effectiveness of UV lamp output. The highest efficiency occurs at 105 °F (41 °C) while lamps operating at 80 or 150 °F (27 or 66 °C) are only about 50% as efficient (Luckiesch, Taylor & Kerr, 1944). Control of the surface temperatures of the lamp is often accomplished by cooling the lamps with pure nitrogen (Scheible, 1983). Otherwise, the lamp temperature is maintained by heat loss through the quartz sheath.

On the same note, one of the engineering advantages speculated for UV disinfection is the insensitivity of photochemical reactions to changes in temperature (Reinsisch, Gloria & Androes, 1970; Yip & Konasewich, 1972; Tonelli *et al.*, 1978). However, few data are available to substantiate this claim. Severin, Suidan & Engelbrecht (1983*b*) measured the activation energy of the inactivation of f2 virus, *C. parapsilosis* and *E. coli*. Activation energies were very low (>1100 cal/g mol); nearly an order of magnitude less than observed with other disinfectants. Thus, from a biological standpoint water temperature is not a major factor in UV disinfection.

Summary

UV is well documented as an effective disinfectant for the control of waterborne bacteria and viruses in water and wastewater, and is gaining popularity in larger-scale municipal applications, small water-treatment systems serving a limited number of customers and household point-of-use/ point-of-entry applications. This has in part been due to concerns with the adverse health effects related to halogenated disinfection practices and the development of efficient, economical designs for UV lamps, ballast and reactors. Of the disinfectants available for water treatment, UV is probably the most well understood with regard to the biochemistry of cellular inactivation. However, UV disinfection remains a challenging field for the academic as well as the practitioner. The challenge stems from the need to understand UV dose measurement more comprehensively and the effect of residence time variation and non-uniform UV intensity fields within the complex, non-ideal UV reactors in use today.

Mathematical nomenclature

A = monochromatic absorbance at 254 nm using base$_{10}$ (/cm); $A = E/2.303$
D_x = dispersion, or longitudinal mixing that occurs in flow-through UV reactors (cm^2/s)
d = the dispersion number, an index for proper UV reactor design (dimensionless); $d = (D_x)/vx$
dH = differential depth in a thin film, flat reactor (cm)
dr = differential radius along a radius (cm)
dz = differential length along a line source (cm)
E = monochromatic absorbance at 254 nm using base$_e$ (/cm); $E = 2.303A$
H = depth of liquid in a thin film, flat reactor (cm)
$\bar{I}$ = average UV intensity (mW/cm^2)
I_0 = intensity at the surface of a quartz tube, or at the surface of a thin-film batch reactor (mW/cm^2)
I = intensity at any point ϱ in the reactor (mW/cm^2)
I_1 = intensity of a line source (mW/cm)

$I_1 \, dz$ = intensity of a single point source along a line (mW)
K = kinetic constant for the multi-target model ($cm^2/mW \, s$)
k = kinetic constant for the series-event model ($cm^2/mW \, s$)
L = total length of line source (cm)
L = length of the UV lamp for calculation of lamp aspect ratio (cm)
l = length of UV reactor for calculation of reactor aspect ratio (cm)
M_i = microorganisms having reached event level i
m = intensity factor in an annular reactor using the radial source model (dimensionless); $m = \bar{I}/I_0$
N_i = density of organisms with i hits or density at event level i (number/cm^3)
N_I = initial density of organisms (number/cm^3)
N_s = density of surviving organisms (number/cm^3)
n = threshold for inactivation in the series-event model
ϱ = distance between a point in the reactor and a point on a finite line source (cm)
Q = liquid flow rate (cm^3/s)
q = number of critical sites in the multi-target model
R = reactor hydraulic radius (cm), for calculation of reactor aspect ratio
r = radial distance in an annular UV reactor (cm)
r_0 = outer radius of quartz tube in an annular UV reactor (cm)
$\check{r}$ = the inner radius of an annular UV reactor wall (cm)
t = contact time in batch reactor (s)
$\bar{t}$ = average contact time in a flow-through reactor (s)
t = theoretical contact time in a flow-through reactor (s); $t = Q/V$
t_i = the time at which the tracer initially appears in residence time distribution studies (s)
t_{10} = the time at which 10% of the tracer has passed through the reactor (s)
t_{50} = the time at which 50% of the tracer has passed through the reactor (s)
t_{90} = the time at which 90% of the tracer has passed through the reactor (s)
t_p = the time at which the concentration of the tracer peaks in residence time distribution studies (s)
θ = the mean retention time of a flow-through reactor (s)
V = the volume of a flow-through UV reactor (cm^3)
v = the velocity of water in the longitudinal direction, used in residence time distribution calculations (cm/s)
x = the average distance traveled by the water while under the influence of UV (cm)
x = the intensity factor for the finite line source model (dimensionless); $x = \bar{I}/I_0$
z = the vertical distance along a line source (cm)
z_1 = the vertical distance within an annular reactor (cm)

References

Antopol, S. C. & Ellner, P. D. (1979). Susceptibility of *Legionella pneumophila* to ultraviolet radiation. *Applied and Environmental Microbiology*, **38**, 347–8.

Bang, S. (1905). Die Wirkungen des Lichtes auf Mikoorganismen. *Mitt. Finsens. Med. Lysinst*, **9**, 164.

Barnard, J. E. & Morgan, H. (1903). On the bactericidal action of some ultraviolet radiations as produced by the continuous current arc. *Proceedings of the Royal Society (London)*, **72**, 126.

Battigelli, D. A., Sobsey, M. D. & Lobe, D. C. (1993). The inactivation of hepatitis A virus and other model viruses by UV irradiation. *Water Science and Technology*, **27**, 339–42.

Bridges, B. A. (1992). Mutagenesis after exposure of bacteria to ultraviolet light and delayed photoreversal. *Molecular and General Genetics*, **233**, 331–6.

Brigano, F. A. O., Scarpino, P. V., Cronier, S., Zink, M. L. & Hoff, J. C. (1980). Effect of particles on inactivation of enteroviruses in water by chlorine dioxide. In *Progress in Wastewater Disinfection Technology*, EPA-600/9–79-018, ed. A. D. Venosa. Cincinnati: Environmental Protection Agency.

Brungs, W. A. (1973). Effects of residual chlorine on aquatic life. *Journal, Water Pollution Control Federation*, **45**, 2180–93.

Bucholz, J. & Jeney, A. U. (1935). Bacterial effect of ultra-violet light. *Zentr. Bakteriol.*, **5**, 229.

Butler, R. C., Lund, V. & Carlson, D. A. (1987). Susceptibility of *Campylobacter jejuni* and *Yersinia enterocolitica* to UV radiation. *Applied and Environmental Microbiology*, **53**, 375–8.

Carson, L. A. & Petersen, N. J. (1975). Photoreactivation of *Pseudomonas cepacia* after ultraviolet exposure: a potential source of contamination in ultraviolet-treated waters. *Journal of Clinical Microbiology*, **1**, 462–4.

Chanda, P. K. & Chatterjee, S. N. (1976). Photoreactivating property of *Vibrio cholerae* cell systems. *Canadian Journal of Microbiology*, **22**, 1186.

Chang, J. C. H., Ossoff, S. F., Lobe, D. C., Dorfman, M. H., Dumais, C. M., Qualls, R. G. & Johnson, J. D. (1985). UV inactivation of pathogenic and indicator organisms. *Applied Environmental Microbiology*, **49**, 1361–5.

Chick, H. (1908). An investigation of the laws of disinfection. *Journal of Hygiene*, **8**, 92–158.

Craun, G. F. (1979). Waterborne giardiasis in the United States; a review. *American Journal of Public Health*, **69**, 817–19.

Craun, G. F. (1989). Recent statistics of waterborne disease outbreaks. In *Waterborne Disease in the United States*, ed. G. F. Craun. Boca Raton, Florida: CRC Press.

Craun, G. F. (1990). Waterborne giardiasis. In *Giardiasis*, ed. E. A. Meyer. Amsterdam: Elsevier.

Darby, J. L., Snider, K. E. & Tchobanoglous, G. (1993). Ultraviolet disinfection for wastewater reclamation and reuse subject to restrictive standards. *Water Environment Research*, **65**, 169–80.

Downes, A. & Blount, T. (1877). Research on the effect of light upon bacteria and other organisms. *Proceedings of the Royal Society (London)*, **26**, 488.

Duggar, B. M. & Hollaender, A. (1934). Irradiation of plant viruses and of microorganisms with monochromatic light, I and II. *Journal of Bacteriology*, **37**, 219–41.

Ehrismann, O. & Noething, W. (1932). Uber die bactericide wirkung monochromatischen lichtes. *Z. Hyg. Infektionskrankh*, **113**, 597.

Fahey, R. J. (1990). The UV effect on wastewater. *Water Engineering Management*, **137**, 15–20.

Fair, G. M., Morris, J. C., Chang, S. L, Weil, I. & Burden, R. P. (1948). The behavior of chlorine as a water disinfectant. *Journal, American Water Works Association*, **40**, 1051–61.

Galasso, G. J. & Sharp, D. G. (1965). Effect of particle aggregation on the survival of irradiated vaccina virus. *Journal of Bacteriology*, **4**, 1138.

Gates, F. L. (1928a). *Journal of General Physiology*, **13**, 231.

Gates, F. L. (1928b). *Science*, **68**, 479.

Gurian, J. M. (1956). Note on fitting the multi-hit survival curve. *Biometrics*, **12**, 123–6.

Haas, C. N. (1980). A mechanistic model for chlorine disinfection. *Environmental Science and Technology*, **14**, 339–40.

Haas, C. N. & Sakellaropoulos, G. P. (1979). Rational analysis of u.v. disinfection reactors. *American Society of Chemical Engineers National Conference of Environmental Engineering*, San Francisco, July.

Harm, W. (1980). *Biological Effects of Ultraviolet Radiation*. Cambridge: Cambridge University Press.

Harris, G. D., Adams, V. D., Moore, W. M. & Sorenson, D. L. (1987a). Potassium ferrioxalate as chemical actinometer in ultraviolet reactors. *Journal of Environmental Engineering*, **113**, 612–27.

Harris, G. D., Adams, V. D., Sorenson, D. L. & Curtis, M. S. (1987b). Ultraviolet inactivation of selected bacteria and viruses with photoreactivation of the bacteria. *Water Research*, **21**, 687–92.

Harris, G. D., Adams, V. D., Sorenson, D. L. & Dupont, R. R. (1987c). The influence of photoreactivation and water quality on ultraviolet disinfection of secondary municipal wastewater. *Journal, Water Pollution Control Federation* **59**, 781–7.

Hatchard, C. G. & Parker, C. A. (1956). A new sensitive chemical actinometer. II. Potassium ferroxalate as a standard chemical actinometer. *Proceedings of the Royal Society (London)*, **A235**, 518–36.

Havelaar, A. H. (1987). Bacteriophages as model organisms in water treatment. *Microbiology Science*, **4**, 362–4.

Havelaar, A. H., Meulemans, C. C. E, Pot-Hogeboom, W. M. & Koster, J. (1990). Inactivation of bacteriophage MS2 in wastewater effluent with monochromatic and polychromatic ultraviolet light. *Water Research*, **24**, 1387–93.

Havelaar, A. H., Nieuwstad, T. J., Meulemans, C. C. E. & van Olphen, M. (1991). F-specific RNA bacteriophages as model viruses in UV disinfection of wastewater. *Water Science Technology* **24**, 347–52.

Hayes, E. D., Matte, T. D., O'Brien, T. R., McKinley, T. W., Logsdon, G. S., Rose, J. B., Ungar, B. L. P., Word, D. M. & Juranek, D. D. (1989). Large community outbreak of cryptosporidiosis due to contamination of a filtered public water supply. *New England Journal of Medicine*, **320**, 1372–6.

Hiatt, C. W. (1964). Kinetics of the inactivation of viruses. *Bacteriology Review*, **28**, 150–63.

Hill, W. F., Hamblet, F. E., Benton, W. H. & Akin, E. W. (1970). Ultraviolet devitalization of eight selected enteric viruses in estuarine water. *Applied Microbiology*, **19**, 805–12.

Hoff, J. C. (1978). The relationship of turbidity to disinfection of potable water. In *Evaluation of the Microbiology Standards for Drinking Water*, EPA 570/9–78-00C, ed. C. Hendricks. Cincinnati: Environmental Protection Agency.

Hoff, J. C. (1986a). Inactivation of microbial agents by chemical disinfectants.

Project Summary, EPA/600/52-86/067. Cincinnati: Environmental Protection Agency.

Hoff, J. C. (1986*b*). Strengths and weaknesses of using Ct values to evaluate disinfection practice. *American Water Works Association Water Quality Conference XIV*, Portland, Oregon, November.

Hollaendar, A. & Emmons, C. W. (1941). Wavelength dependence of mutation production with special emphasis on fungi. *Cold Spring Harbor Symposium*, **9**, 179.

Hom, L. W. (1972). Kinetics of chlorine disinfection in an eco-system. *Journal of the Sanitary Engineering Division of the American Society of Civil Engineers*, **98**, 183-94.

Hoyer, O, Kryschi, R., Piecha, I., Mark, G., Schuchmann, M. N., Schuchmann, H. P. & von Sonntag, C. (1992). UV fluence rate determination of the low-pressure mercury arc in the disinfection of drinking water. *Journal of Water Supply Research and Technology*, **41**, 75-81.

Huff, C. B. (1965). Study of ultraviolet disinfection of water and factors in treatment efficiency. *Public Health Report*, **80**, 695.

IAWPRC Study Group on Health Related Water Microbiology (1991). Bacteriophages as model viruses in water quality control. *Water Research*, **25**, 529-45.

Jacob, S. M. & Dranoff, J. S. (1966). Radial scale-up of perfectly mixed photochemical reactors. *Chemical Engineering Progress Symposium Series*, **62**, 47-55.

Jacob, S. M. & Dranoff, J. S. (1970). Light intensity profiles in a perfectly mixed photoreactor. *Journal of the American Institute of Chemical Engineers*, **16**, 359-63.

Jagger, J. H. (1964). Photoprotection from far ultraviolet effects in cells. *Advances in Chemical Physics*, **7**, 584.

Jagger, J. H. (1967). *Introduction to Research in Photobiology*. Englewood Cliffs, New Jersey: Prentice Hall.

Jepson, J. D. (1973). Disinfection of water supplies by ultraviolet radiation. In *Proceedings of Water Treatment Examination*, **22**, 175-91.

Johnson, J. D., Aldrich, K., Francisco, D. E., Wolff, T. & Elliot, M. (1979). UV disinfection of secondary effluent. In *Proceedings of the National Symposium on Progress in Wastewater Disinfection*, EPA-600-7/79-018. Cincinnati: Environmental Protection Agency.

Johnson, J. D. & Qualls, R. G. (1984). Ultraviolet disinfection of a secondary effluent: measurement of dose and effects of filtration. EPA-600/2-84-160, NTIS No. PB85-114023, Cincinnati: Environmental Protection Agency.

Jolley, R. L. (1975). Chlorine-containing organic constituents in chlorinated effluents from sewage treatment plants. *Journal, Water Pollution Control Federation*, **47**, 601.

Kallenbach, N. R., Cornelius, P. A., Negus, D., Montgomerie, D. & Englander, S. (1989). Inactivation of viruses by ultraviolet light. In *Virus Inactivation in Plasma Products*, ed. J. J. Morgenthaler, *Current Studies in Hematology and Blood Transfusions*, vol. 56, pp. 70-82. Basel: Karger.

Kamiko, N. & Ohgaki, S. (1989). RNA coliphage Qβ as a bioindicator of the ultraviolet disinfection efficacy. *Water Science Technology*, **21**, 227-31.

Karanis, P., Maier, W. A., Seitz, H. M. & Schoenen, D (1992). UV sensitivity of protozoan parasites. *Journal of Water Supply Research and Technology*, **41**, 95–100.

Kelner, A. (1949). Photoreactivation of ultraviolet-irradiated *Escherichia coli* with special reference to the dose-reduction principle and to ultraviolet-induced mutation. *Journal of Bacteriology*, **58**, 611.

Kimball, A. W. (1953). The fitting of multi-hit survival curves. *Biometrics*, **9**, 201–11.

Knudson, G. (1985). Photoreactivation of uv-irradiated *Legionella pneumophila* and other *Legionella* species. *Applied Environmental Microbiology*, **49**, 975–80.

Kuennen, R., Taylor, R., Wilson, B., Roessler, P. & VanDellen, E. (1992). Performance of a point-of-use water treatment system for treating chemically and microbially contaminated drinking water. *First International Conference on the Safety of Water Disinfection: Balancing Chemical and Microbial Risk*. Washington, DC, September.

Kuhn, H. J., Braslavsky, S. E. & Schmidt, R. (1989). Chemical actinometery. *Pure and Applied Chemistry*, **61**, 187–210.

Levenspiel, O. & Smith, W. K. (1957). Notes on the diffusion-type model for the longitudinal mixing of fluids in flow. *Chemical Engineering Society*, **6**, 227–33.

Lindenauer K. G. & Darby, J. L. (1994). Ultraviolet disinfection of wastewater: effect of dose on subsequent photoreactivation. *Water Research*, **28**, 805–17.

Loofhourow, J. R. (1948). The effects of ultraviolet radiation on cells. *Growth*, **12**, 75.

Luckiesh, M. & Holladay, L. L. (1944). Disinfecting water by means of chemical lamps. *General Electric Review*, **4**, 14.

Luckiesch, M., Taylor, A. & Kerr, G. (1944). Germicidal energy. *General Electric Review*, **9**, 7.

Maarschalkerweerd, J., Murphy, R. & Sakamoto, G. (1990). Ultraviolet disinfection in municipal wastewater treatment plants. *Water Science Technology*, **22**, 145–52.

MacKenzie, W. R., Hoxie, N. H., Proctor, M. E., Gradus, M. S., Blair, K. A., Peterson, D. E., Kazmierczak, J. J., Addiss, D. G., Fox, K. R., Rose, J. B. & Davis, J. P. (1994). A massive outbreak in Milwaukee of *Cryptosporidium* infection transmitted through the public water supply. *New England Journal of Medicine*, **331**, 161–7.

Mark, G., Schuchmann, M. N., Schuchmann, H. P. & von Sonntag, C. (1990a). The photolysis of potassium peroxodisulphate in aqueous solution in the presence of *t*-butanol: a simple actinometer of 254 nm radiation. *Journal of Photochemistry and Photobiology A: Chemistry*, **55**, 157–68.

Mark, G., Schuchmann, M. N., Schuchmann, H. P. & von Sonntag, C. (1990b). A chemical actinometer for use in connection with UV treatment in drinking water processing. *Journal of Water Supply Research and Technology*, **39**, 309–13.

Martens, D. W. & Servizi, J. A. (1974). *Acute Toxicity of Municipal Sewage to Fingerling Sockeye Salmon*. *Progress Report 30*. New Westminister, British Columbia, Canada: International Pacific Salmon Fisheries Commission.

Mechsner, K., Fleischmann, T., Mason, C. A. & Hamer, G. (1991). UV disinfection: short term inactivation and revival. *Water Science Technology*, **24**, 339–42.

Morowitz, H. J. (1950). Absorption effects in volume irradiation of microorganisms. *Science*, **111**, 229–30.

Morrill, A. B. (1932). Sedimentation basin research and design. *Journal, American Water Works Association*, **24**, 1442–63.

Muraca, P., Stout, J. E. & Yu, V. L. (1987). Comparative assessment of chlorine, heat, ozone and uv light for killing *Legionella pneumophila* within a model plumbing system. *Applied and Environmental Microbiology*, **53**, 447–53.

Murphy, K. L., Zaloum, R. & Fulford, D. (1975). The effect of chlorination practice on soluble organics. *Water Research*, **9**, 389.

National Sanitation Foundation (1991). *Standard 55: Ultraviolet Microbiological Water Treatment Systems*. Ann Arbor, Michigan: NSF Joint Committee on Drinking Water Treatment Units.

Newcomer, H. S. (1927). The abiotic action of ultraviolet light. *Journal of Experimental Medicine*, **26**, 841.

Nieuwstad, T. J., Havelaar, A. H. & van Olphen, M. (1991). Hydraulic and microbiological characterization of reactors for ultraviolet disinfection of secondary wastewater effluent. *Water Research*, **25**, 775–83.

Oliver, B. G. & Carey, J. H. (1976). Ultraviolet disinfection as an alternative to chlorination. *Journal, Water Pollution Control Federation*, **48**, 2619.

Oliver, B. G. & Cosgrove, E. G. (1975). The disinfection of sewage treatment plant effluents using ultraviolet light. *Canadian Journal of Chemical Engineering*, **53**, 170.

Petrasek, A. C., Wolf, H. W., Esmond, S. E. & Andrews, D. C. (1980). *Ultraviolet Disinfection of Municipal Wastewater Effluents*, EPA-600/2–80-102. Cincinnati: Environmental Protection Agency.

Poduska, R. A. & Hershey, D. (1972). Model for virus inactivation by chlorine. *Journal, Water Pollution Control Federation*, **44**, 738–45.

Qualls, R. G., Dorfman, M. H. & Johnson, J. D. (1989). Evaluation of the efficiency of ultraviolet disinfection systems. *Water Research*, **23**, 317–25.

Qualls, R. G., Flynn, M. P. & Johnson, J. D. (1983). The role of suspended particles in ultraviolet disinfection. *Journal, Water Pollution Control Federation*, **55**, 1280–5.

Qualls, R. G. & Johnson, J. D. (1983). Bioassay and dose measurement in UV disinfection. *Applied and Environmental Microbiology*, **45**, 872–7.

Qualls, R. G. & Johnson, J. D. (1985). Modeling and efficiency of ultraviolet disinfection systems. *Water Research*, **19**, 1039–46.

Qualls, R. G., Ossoff, S. F., Chang, J. C. H., Dorfman, M. H., Dumais, C. M., Lobe, D. C. & Johnson, J. D. (1985). Factors controlling sensitivity in ultraviolet disinfection of secondary effluents. *Journal, Water Pollution Control Federation*, **57**, 1006–11.

Raunbler, M. B. & Margulis, L. (1980). Bacterial resistance to ultraviolet irradiation under anaerobiosis: implications for pre-phanerozoic evolution. *Science*, **7**, 638.

Reinisch, R. F., Gloria, H. R. & Androes, G. M. (1970). *Photochemistry of Macromolecules*. New York: Plenum Press.

Rentschler, H. C. et al. (1941). *Journal of Bacteriology*, **41**, 745.

Rice, E. W. & Hoff, J. C. (1981). Inactivation of *Giardia lamblia* cysts by ultraviolet irradiation. *Applied and Environmental Microbiology*, **42**, 546–7.

Roeber, J. A. & Hoot, F. M. (1975). *Ultraviolet Disinfection of Activated Sludge Effluent Discharging to Shellfish Waters*, Report No. EPA-600/2–75-060, NTIS No. PB-249460. Cincinnati: United States Environmental Protection Agency.

Roessler, P. F. & Wilson, B. (1993). Ultraviolet light susceptibility of *Helicobacter pylori* and comparison to waterborne pathogenic bacteria and indicator organisms. *93rd General Meeting of the American Society for Microbiology* (abstract N-62), Atlanta, May.

Roessler, P. F., Wilson, B. & Kuennen, R. (1993). Ultraviolet light susceptibility of *Vibrio cholerae* O1 classical and el tor biotypes. *American Society for Microbiology Conference: Water Quality in the Western Hemisphere*, San Juan, Puerto Rico, April.

Roessler, P. F., Wilson, B., VanDellen, E., Abbaszadegan, M. & Gerba, C. P. (1992). Ultraviolet light susceptibility comparison of waterborne pathogens and indicator microorganisms. *92nd General Meeting of the American Society for Microbiology* (abstract Q-9), New Orleans, May.

Rook, J. J. (1974). Formation of haloforms during chlorination of natural waters. *Water Treatment Examination*, **23**, 234.

Salaj-Smic, E., Petranovic, D., Petranovic, M. & Trgovcevic, Z. (1980). Relative roles of *uvrA* and *recA* genes in the recovery of *Escherichia coli* and phage Λ after ultraviolet irradiation. *Radiation Research*, **83**, 323–9.

Scheible, O. K. (1983). Design and operation of uv disinfection systems. In *Proceedings of the Preconference Workshop on Wastewater Disinfection, 56th Annual Water Pollution Control Federation Conference*, Atlanta, October.

Scheible, O. K. (1987). Development of a rationally based design protocol for the ultraviolet light disinfection process. *Journal, Water Pollution Control Federation*, **59**, 25–31.

Scheible, O. K., Binkowski, G. & Milligan, T. J. (1979). Field scale evaluation of UV disinfection of a secondary effluent. In *Proceedings of the National Symposium on Progress in Wastewater Disinfection*, EPA-600–7/79–018. Cincinnati: Environmental Protection Agency.

Scheible, O. K., Casey, M. C. & Forndran, A. (1986). *Ultraviolet Disinfection of Wastewaters from Secondary Effluent and Combined Sewer Overflows*, EPA-600/2–86/005, NTIS No. PB86–145182. Cincinnati: Environmental Protection Agency.

Setlow, R. B. (1964a). Physical changes and mutagenesis. *Journal of Cellular and Comparative Physiology*, **64**, 51.

Setlow, R. B. (1964b). *Mammalian Cytogenics and Related Problems in Radiobiology*, ed. C. Paven, C. Chagas, D. Frosta-Pessoa & L. R. Caldas. Oxford: Pergamon Press.

Severin, B. F. (1980). Disinfection of municipal wastewater effluents with UV light. *Journal, Water Pollution Control Federation*, **52**, 2007.

Severin, B. F., Suidan, M. T. & Engelbrecht, R. S. (1983a). Kinetic modeling of uv disinfection of water. *Water Research*, **17**, 1669–78.

Severin, B. F., Suidan, M. T. & Engelbrecht, R. S. (1983b). Effects of temperature on ultraviolet light disinfection. *Environmental Science and Technology*, **17**, 717–21.

Severin, B. F., Suidan, M. T. & Engelbrecht, R. S. (1984a). Mixing effects in uv disinfection. *Journal, Water Pollution Control Federation*, **56**, 881–8.

Severin, B. F., Suidan, M. T. & Engelbrecht, R. S. (1984b). Series event kinetic model for chemical disinfection. *Journal of Environmental Engineering*, **110**, 430–9.

Simonson, C. S., Kokjohn, T. A. & Miller, R. V. (1990). Inducible UV repair potential of *Pseudomonas aeruginosa* PAO, *Journal of General Microbiology*, **136**, 1241–9.

Smith, K. C. (1964). *Photophysiology*, 2nd edn, ed. A. C. Giese. New York: Academic Press.

Smith, K. C & Hanawalt, P. C. (1969). *Molecular Photobiology*. New York: Academic Press.

Smith, V. H. & Rose, J. B. (1990) Waterborne cryptosporidiosis. *Parasitology Today*, **6**, 11–12.

Sommer, R. & Cabaj, A. (1993). Evaluation of the efficiency of a UV plant for drinking water disinfection. *Water Science Technology*, **27**, 357–62.

Stevens, W. H. (1980). Quantitative aspects of photoreactivation in *Escherichia coli* following exposure to ultraviolet light. Master of Science Special Problem Report. Urbana: University of Illinois.

Suidan, M. T. & Severin, B. F. (1986). Light intensity models for annular u.v. disinfection reactors. *Journal of the American Institute of Chemical Engineering*, **32**, 1902–9.

Thampi, M. V. (1990). Basic guidelines for specifying the design of ultraviolet disinfection systems. *Pollution Engineering*, **22**, 65–9.

Thampi, M. V. & Sorber, C. A. (1987). A method for evaluating the mixing characteristics of u.v. reactors with short detention times. *Water Research*, **21**, 765–71.

Tonelli, F. A., Duff, R. & Wilcox, B. (1978). *Ultraviolet Disinfection of Domestic Sewage and Stormwater – a Literature Evaluation*, Research Paper 2046. Canada: Ministry of the Environment.

United States Environmental Protection Agency (1986). *Design Manual: Municipal Wastewater Disinfection*, EPA/625/1–86/021. Cincinnati: Environmental Protection Agency.

Wacker, A. (1963). Molecular mechanisms of radiation effects. *Progress in Nucleic Acid Research*, **1**, 369.

Ward, R. W. & DeGrave, G. M. (1978). Residual toxicity of several disinfectants in domestic wastewater. *Journal, Water Pollution Control Federation*, **50**, 46.

Watson, H. E. (1908). A note on the variation of the rate of disinfection with change in the concentration of the disinfectant. *Journal of Hygiene*, **8**, 536.

Wei, J. H. & Chang, S. L. (1975). A multi-poisson distribution model for treating disinfection data. In *Disinfection, Water and Wastewater*, ed. J. D. Johnson. Ann Arbor, Michigan: Ann Arbor Science.

Whitby, G. E. & Palmateer, G. (1993). The effect of uv transmission, suspended solids and photoreactivation on microorganisms in wastewater treated with uv light. *Water Science Technology*, **27**, 379–86.

Whitby, G. E., Palmateer, G., Cook, W. G., Maarschalkerweerd, J., Huber, D. & Flood, K. (1984). Ultraviolet disinfection of secondary effluent. *Journal, Water Pollution Control Federation*, **56**, 844–50.

White, S. C., Jernigan, E. B. & Venosa, A. D. (1986). A study of operational ultraviolet disinfection equipment at secondary treatment plants. *Journal, Water Pollution Control Federation*, **58**, 181–92.

Wiedenmann, A., Fischer, B., Straub, B., Wang, C. H., Flehmig, B. & Schoenen, D. (1993). Disinfection of hepatitis A virus and ms-2 coliphage in water by ultraviolet irradiation: comparison of uv susceptibility. *Water Science Technology*, **27**, 335–8.

Wilson, B. R., Roessler, P. F., Abbaszadegan, M., Gerba, C. P. & Van Dellen, E. (1992*a*). UV dose bioassay using coliphage MS2. *92nd General Meeting of the American Society for Microbiology* (abstract Q-10), New Orleans.

Wilson, B. R., Roessler, P. F., VanDellen, E., Abbaszadegan, M. & Gerba, C. P. (1992*b*). Coliphage MS2 as a uv water disinfection efficacy test surrogate for bacterial and viral pathogens. *Proceedings of the Water Quality Technology Conference, American Water Works Association*, Toronto, Ontario, Canada, May.

Wolf, H. W., Petrasek, A. C. Jr & Esmond, S. E. (1979). Utility of uv disinfection of secondary effluent. In *Proceedings of the National Symposium on Progress in Wastewater Disinfection*, EPA 600-7/79-018. Cincinnati: Environmental Protection Agency.

Wolfe, R. L. (1990). Ultraviolet disinfection of potable water. *Environmental Science Technology*, **24**, 768–73.

Wyckoff, R. W. C. (1932). The killing of colon bacilli by ultraviolet light. *Journal of General Physiology*, **15**, 351.

Yip, R. W. & Konasewich, D. E. (1972). Ultraviolet sterilization of water; its potential and limitations. *Water Pollution Control (Canada)*, **14**, 14.

Zelle, M. R. & Hollaender, A. (1955). Effects of radiation on bacteria. In *Radiation Biology, Ultraviolet and Related Radiations*, vol. 2, ed. A. Hollaender. New York: McGraw-Hill.

Zukovs, G., Kollar, J., Monteith, H. D., Ho, K. W. A. & Ross, S. A. (1986). Disinfection of low quality wastewaters by ultraviolet light irradiation. *Journal, Water Pollution Control Federation*, **58**, 199–206.

Chapter 12

Thermal inactivation of microorganisms

GUY LE JEAN and GÉRARD ABRAHAM

Introduction

As emphasized by Lee & Guilbert (1918), thermal disinfection is not a sudden event that would obey an 'on–off' law. In fact, at moderated lethal temperatures, population dynamics may effectively be measured. At very high lethal temperature, as the microbial death velocities increase very greatly, the period of observability becomes so short that the phenomenon can be considered to be instantaneous. Conversely, for temperatures at which the microbial population is able to grow, thermal death does not occur. It is the breadth of the range of temperatures between no death and quasi-instantaneous death that makes it possible to say that the phenomenon is not sudden. Of course, this temperature range depends mainly on the nature of the target microorganism. Also, for many reasons, most thermal disinfection treatments are carried out within this moderate temperature range.

When the product that has to be disinfected is not temperature-sensitive, it is possible to apply thermal treatments at very high temperatures for a long time, but this is most often not the case. When the product is temperature-sensitive, the treatment must be carried out at moderate temperatures, and models of thermal death kinetics are needed to identify temperatures that satisfy the disinfection goal but minimize the alteration of the product. Also, the more realistic and reliable a model, the fewer safety coefficients have to be applied and the more negative effects of thermal treatment minimized.

The most currently used mathematical model was developed at the beginning of this century. Although it has been shown to be quite efficient, since its design largely evolved from thermal disinfection processes (especially in food processing), important simplifications have been conceded that may be avoided by the use of more elaborate models. Moreover, the widespread usage of computers, coupled with increasing automation, allows consideration of the use of more complex models.

The aim of this chapter is to provide basic information on models of microbial thermal death. However, the diversity of applications of thermal disinfection processes and the variety of target microorganisms render it impossible to address all the aspects of the subject, and only a few items relative to specific applications appear in this chapter. The discussion is directed toward the more common problems and the mathematical treatment is developed through examples we have selected as particularly representative. The first section presents the experimental methods applied to the generation of data on thermal death kinetics and the results they generally yield. Models developed to fit experimental observations are presented in the second section. The third section deals with the application of models in the determination of kinetics parameters and the prediction of the dynamics of microbial populations in the general case of non-isothermal treatment.

Thermal death kinetics: experiments

Kinetics data generation

Thermal treatments for kinetics data generation may be carried out with two different goals:

(1) Determination of the numerical values of the parameters of a mathematical model that is known to be satisfactory in its description of microbial death dynamics;
(2) Formulation of a model for a microbial population whose death kinetics in some given conditions is not known.

For the first goal, any thermal treatment (whether isothermal or non-isothermal) may apply, provided that a suitable numerical method is available to estimate the necessary parameter values. It is noteworthy that, for this aim, the collection of non-isothermal population dynamics data would be an interesting approach, since it would minimize errors due to thermal transfer lag times. Moreover, as the whole temperature range may be studied in one treatment, it is also an economical method. However, this method has rarely been used (Ingraham & Marier, 1964, 1965; Hayakawa, Schnell & Dick, 1969; Matsuda, Kamaki & Matsunawa, 1981; Reichart, 1981; Swartzel, 1984), and there are no consistent reports on the reliability of kinetics parameter values determined from non-isothermal treatments.

   For the second goal, it is necessary to differentiate the respective effects of treatment times and temperature variations by performing isothermal treatments at different temperature levels for various times. The success of isothermal treatments requires particular attention to thermal transfers occurring at the heating and cooling steps. Indeed, to get correct data, the experimental method should be designed so that delays due to thermal transfers between the microbial population sample and the hot source and/or cold source are negligible. Also, if heat sources are made of temperature-varying apparatus, the transient temperatures during heating and cooling should be checked. Unfortunately, the occurrence of heating and/or cooling delays are unavoidable and impose a limit on the reliability of data for the length of treatment. As short treatment times have very often been shown by experience to provide essential information, many experimental methods have been proposed that minimize thermal delays to allow high temperature–short time treatments.

Experimental methods

Experimental methods depend largely on the temperature range of treatment. Treatments at temperatures under 100 °C are carried out to test the thermoresistancy of bacterial vegetative forms, yeasts and molds, and can consist of simply immersing tubes containing microorganism suspensions into a water bath set at the required temperature. However, special apparatus is sometimes designed to shorten the lag time in heating the sample to the experimental temperature, for example, the reactor developed by Doutsias (1974) for testing the thermoresistancy of yeasts. The sterile medium (50–100 ml) is heated at the required temperature by circulation of the heating fluid in the double wall of the reactor. It is then inoculated with a small volume (10–100 μl) of a concentrated suspension of yeasts. A magnetic rod placed in the reactor permits the mixing of the yeasts in the medium. A hole in the lid allows sampling at the desired time.

Temperatures higher than 100 °C are used to study only the highly thermoresistant microorganisms, as bacterial spores. In this case, an oil bath is needed or special apparatus in which heating is insured by steam injection. Two different kinds of methods are available: batch methods and continuous-flow methods. Bigelow & Esty (1920) used a batch method that consisted of immersing sealed test tubes ('thermal death time' tubes) containing 1–4 ml of an inoculated sample in a suitable heating medium. After the desired time of treatment, the sample was removed and rapidly chilled in cold water. A similar method is currently applied using flame sealed capillary tubes containing about 50 μl of microbial suspension and a temperature-regulated oil bath (Stern & Proctor, 1954; Davies *et al.*, 1977; Michels, Spiliotis & Etoa, 1985; Abraham *et al.*, 1990; Le Jean *et al.*, 1994). This method does not allow isothermal treatments shorter than 10 s. A second type of batch method is derived from the thermoresistometer designed by Stumbo (1948), in which samples of a spore suspension or homogenate are placed in a small reactor chamber, in which temperature is controlled by quick admission and release of steam. The recent development of a reactor in which hydrophilic disc containing 10 μl of a microbial suspension is admitted and removed automatically from a steam chamber allowed David & Merson (1990) to perform isothermal treatments as short as 0.1 s.

Continuous-flow methods appeared in about 1950 (Tobias, Heirried & Ordal, 1953; Tobias, Kaufman & Tracy 1955) with the use of small-bore plate-type heat exchangers to study thermal death at high temperatures. As this equipment still induced large heating and cooling delays, Wang Scharer & Humphrey (1964) designed an original apparatus in which controlled flows of microbial suspension and hot water were mixed at the entry of an adiabatic horizontal reactor whose length determined the isothermal exposure time according value of to flow rate. They determined the equilibrium temperature to be reached in less than 0.001 s and performed treatments as short as 0.2 s. Swartzel (1984) developed a continuous-flow non-isothermal reactor for the study of the kinetics of liquid food constituents (microbial contamination included), for which kinetics models were previously known.

This is not an exhaustive review, but rather a short historical panorama of the experimental methods developed for studies of microbial death kinetics. The advantages and disadvantages of these methods are various. Capillary tube experiments are still much in use, certainly because of low costs and ease of implementation, but the manipulations are cumbersome, and are convenient only for a temperature range in which the microbial death velocities are moderate. High investment costs represent the major inconvenience of batch reactors. At least, continuous-flow reactors require large volumes of microbial suspension and knowledge of residence time distributions (Heppel, 1985). It must be noted that extrapolation of results from capillary experiments to temperatures exceeding those studied experimentally presents important risks of error on estimating the lethal effect of a thermal treatment (Wang *et al.*, 1964; Davies *et al.*, 1977; Jonsson *et al.*, 1977).

Typical survival curves

Since the earliest efforts (Chick, 1910; Bigelow, 1921; Esty & Meyer, 1922), in which authors showed experimentally that microbial isothermal death was an exponential phenomenon, survival curves have commonly been represented on semilogarithmic graphs, giving the logarithm of the surviving fraction versus the time of treatment. A strictly exponential death rate is then represented by a straight line. In fact, important deviations from the straight line are very often observed. Deviations appear for short as well as long heating times, and some efforts have generated experimental survival curves that contained nothing comparable to a straight line. Figures

12.1, 12.2 and 12.3 illustrate the typical shapes of microbial survival curves shown in published works.

Figure 12.1 illustrates exponential death (curve A) and typical short-time deviations. Deviations of type B (bilinear) and C (shoulder without increase) have been observed for vegetative cells as well as for bacterial spores. Type D deviations are specific to bacterial spores and are associated with the presence of dormant spores that are able to grow on a suitable medium only after activation by heat at high temperatures. Type D curves may be obtained for vegetative cells at temperatures for which growing and death occur simultaneously, but disinfection processes normally do not deal with such temperatures. Figure 12.2 shows a sigmoidal survival curve that may appear for any type of microorganism. Tailing (Figure 12.3) seems to appear independently of the initial shape of survival curve and only for extended treatments leading to extremely low survival fractions (less than 10^{-6}). Table 12.1 provides references contributing such observations for various microorganisms.

It is now well documented that the shape of a survival curve is not only an intrinsic characteristic of the microorganism tested but also depends on various factors including the strain, the age of the culture, the compositions

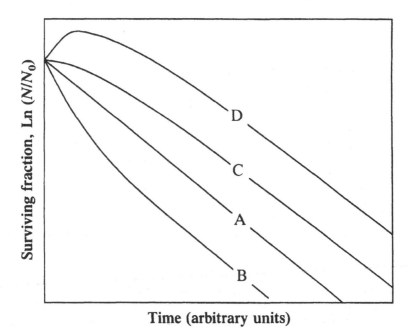

Time (arbitrary units)

Figure 12.1 Typical shapes of isothermal survival curves. Exponential death (A) and deviations appearing for short treatment times. B, bilinear curves; C, shoulder without increase; D, shoulder with increase (specific to thermal death of bacterial spores).

Table 12.1. *Examples of published works reporting nonlinear isothermal survival curves for various microorganisms*

Reference	Microorganism	Shape of isothermal survival curve
Briggs (1960)	B. licheniformis spores	B
	B. pantothenticus spores	S
	B. subtilis spores	B
Shull & Ernst (1962)	B. stearothermophilus spores	D
Moats et al. (1971)	Salmonella anatum	C
	Streptococcus faecalis	T
	Escherichia coli	T
	Salmonella senthenberg	B
Van Uden & Vidal-Leira (1976)	S. cerevisiae	C
Davies et al. (1977)	B. stearothermophilus spores	A, B, C, S
Berry & Bradshaw (1980)	B. stearothermophilus spores	B
Spiliotis (1983)	B. stearothermophilus spores	B
Barillère et al.(1985a,b)	S. bayanus	S
	S. cerevisiae	B
	S. rouxii	S
Ababouch et al. (1987)	C. sporogenes spores	B
	C. botulinum spores	T
	B. cereus spores	A
Feehery et al. (1987)	B. stearothermophilus spores	C
Mackey & Derrick (1987)	Salmonella thompson	B
Abraham et al. (1990)	B. stearothermophilus spores	D
David & Merson (1990)	B. stearothermophilus spores	D
Fernandez et al. (1994)	B. stearothermophilus spores	C
Le Jean et al. (1994)	B. stearothermophilus spores	D

Note: A, linear; B, bilinear; C, shoulder without increase; D, shoulder with increase; S, sigmoid; T, tailing.

of media used for initial culture, heating conditions and microbial enumeration method (water activity, pH, nutrients, pH indicators, temperatures, etc.) Factors affecting thermal death characteristics are a wide subject and each particular case needs to be considered separately. These problems will not be considered in this chapter, but workers who perform thermal death experiments for specific goals should pay attention to these potential problems when defining their experimental methods. Isothermal survival curves are generally reproducible for fixed experimental conditions and may be extrapolated to elaborate predictive models.

Thermal death kinetics: modeling

Basic concepts in thermal death modeling

The subject of microbial thermal death kinetics has been approached through both vitalistic and mechanistic concepts. The vitalistic theory considers that individuals of a population are not identical, but instead possess different degrees of resistance such that they need different treatment times to be killed. In this case, the shape of survival curves would depend on the distribution of survival times within the microbial population. The mechanistic theory is based on the assumption that all individuals possess similar thermal resistance properties and that thermal death is due to subjacent biochemical reactions. Thermal death models may then be derived from various hypotheses according to the general laws of chemical kinetics theory. The vitalistic theory has been poorly supported experimentally, and a large majority of authors prefer to interpret their data according to the mechanistic theory. Barillère, Dubois & Bidan (1985*a*) noted that

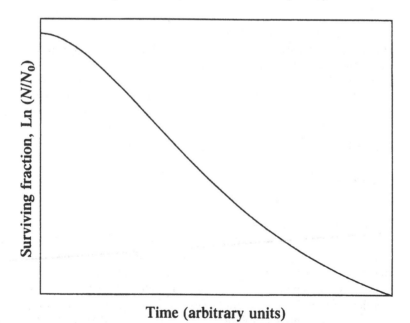

Time (arbitrary units)

Figure 12.2   Typical shape of isothermal survival curve, showing deviations from linearity for short and long heating time: sigmoid.

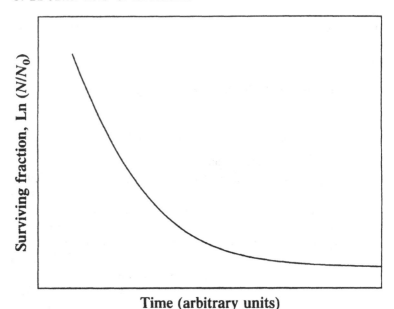

Time (arbitrary units)

Figure 12.3   Typical shape of isothermal survival curve. Deviation observed for extended treatments, bringing strong destruction levels: tailing.

it was very difficult to verify the vitalistic theory mathematically. Also, the recent use of advanced experimental techniques such as differential scanning calorimetry (DSC) (Maeda, Teramoto & Koga, 1975; Maeda, Nogushi & Koga, 1978; Miles *et al.*, 1986; Anderson *et al.*, 1991; Mackey *et al.*, 1991, 1993; Belliveau *et al.*, 1992), which enable researchers to correlate rate constant values and other properties associated with microorganisms undergoing thermal death with the occurrence of thermographic peaks, may give more experimental support to the mechanistic theory.

Exponential death models

Conventional model

To quantify the isothermal exponential death, the earliest authors such as Bigelow (1921) and Katzin, Sandholdzer & Strong (1943) introduced the currently used decimal reduction time (D_θ) *and* z parameter. D is the time needed at a particular temperature to reduce the microbial population to 10% survival. The parameter z is the increase of temperature required to divide D_θ by 10 (the variation of D with temperature is usually represented on a semilogarithmic graph of log (D) versus $\theta\,°C$ which is called a thermal death time (TDT) curve). According to these definitions, isothermal survival curves are calculated from the following equations:

$$\frac{N}{N_0} = 10^{\frac{-t}{D\theta}}$$ (12.1)

$$D_\theta = D_{\theta_r} 10^{\frac{\theta_r - \theta}{z}}$$ (12.2)

where: t = duration of treatment (min); N = number of surviving spores after a treatment of t min; N_0 = initial spore count; θ = temperature of treatment (°C); θ_r = reference temperature (°C); D_θ = decimal reduction time at θ°C (min); z = inverse of the slope index of the TDT curve (°C).

It is still usual to determine D_θ and z parameters (by linear regressions of log (N/N_0) versus time at various temperatures, and log (D_θ) versus θ) to interpret thermal death kinetics data. Thus, workers dealing with food preservation processes generally employ these parameters to evaluate the lethal effects of non-isothermal treatments through the numerical calculation of a sterilizing (or pasteurizing) value:

$$F_{\theta_r}^z \left(\text{or } P_{\theta_r}^z \right) = \int_0^\infty 10^{\frac{\theta(t) - \theta_r}{z}} \, dt$$ (12.3)

This value represents the time of exposure required at the reference temperature θ_r that would produce a lethal effect equivalent to that of a treatment whose temperature history is $\theta(t)$. The survival ratio is then calculated by ln (N/N_0) = $-F_{\theta_r}^z/D_\theta$.

These equations have been established to allow a quantification of the phenomenon, but did not account for any theoretical assumption. Now, according to the mechanistic concepts, it may be assumed that exponential death is ruled by a subjacent molecular reaction of the first order.

First-order reaction kinetics

Equations of the conventional model are in fact very similar to those derived from a first-order unimolecular reaction mechanism using the Arrhenius relationship (Equation 12.5), and the exponential survival curves established at various temperatures can also be well described as follows:

$$\frac{N}{N_0} = e^{-kt}$$ (12.4)

$$k = A e^{\frac{-E_a}{RT}}$$ (12.5)

where k = rate constant (/s); t = time of treatment (s); T = absolute temperature (K); R = constant of ideal gas (J/mol per K); E_a = energy of activation of the reaction (J/mol); A = constant.

Once the parameter values are determined (by linear regressions on ln (N/N_0) versus time for various temperatures, and ln (k) versus $1/T$) one can calculate the lethal effect of any non-isothermal treatment $T(t)$ by integrating numerically the differential form of Equation 12.4:

$$\text{Ln } (N/N_0) = \int_0^\infty - Ae^{\frac{-E_a}{RT(t)}} d_t \tag{12.6}$$

This integral (where $T(t)$ is the temperature history of the treatment expressed in kelvin) provides the same information as the conventional sterilizing (or pasteurizing) value.

Advantage of the Arrhenius relationship

Equation 12.4 is equivalent to Equation 12.1 when the experimental D and k values are considered, and these parameters are linked by: D = ln (10)/k. This equality is no longer valid when D and k are recalculated from Equations 12.2 and 12.5, and several authors have examined the adequacy of these equations (Wang et al., 1964; Scharer, 1965; Jones, 1968; Jonsson et al., 1977; Burton et al., 1977; Davies et al., 1977; Perkin et al., 1977; Spiliotis, 1983; David & Merson, 1990; Ocio et al., 1994). As the superiority of any one equation has not been clearly established, safety can be taken as the criterion for proposal of a theoretical discrimination. In fact, Equation 12.2 can be obtained from a first-order limited development of Equation 12.4. Accordingly, many authors (Cerf, 1977; Spiliotis, 1983) have shown that the parameter z is linked to the activation energy of the reaction by:

$$z = 2.303 \frac{RT_m^2}{E_a} \tag{12.7}$$

where T_m = mean temperature in the range wherein z and E_a were estimated.

Therefore, under the assumption of a subjacent molecular reaction, the parameter z is a function of the temperature and, theoretically, the approximation of the Arrhenius relationship by Equation 12.2 would be applicable only for a limited range of temperature. When both equations are applied to an extended range of temperature, values from Equation 12.2 appreciably exceed values calculated from Equation 12.4. This is illustrated in Figure 12.4. From a safety standpoint, the use of Equation 12.4 is preferable, since it leads to conservative estimations.

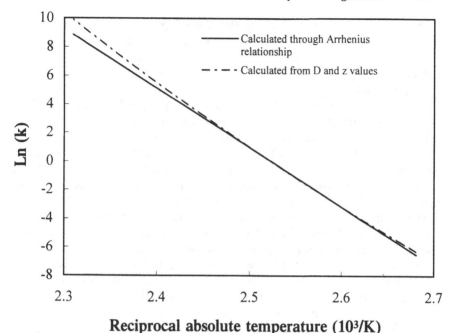

Figure 12.4 Comparison of rate constant values calculated from Arrhenius relationship and from D_θ and z values. Data from Le Jean *et al.* (1994) established for *B. stearothermophilus*.

Non-exponential death models

General issues

Most often, non-exponential death is due to short time deviations, and isothermal survival curves present a linear asymptote. If the deviation is weak enough, conventional and first-order models may be useful from a practical standpoint, since they give predictions with negligible errors by simple calculations. Unfortunately, strong deviations are currently observed. However, the method is still used and, most often, the initial portions of curves are totally neglected. A few authors took initial phenomena into account by a 'constant intercept ratio' (intersection of the linear portion of the survival curve with the ordinate axis), which was equivalent to the logarithm of a so-called 'phantomatic' initial concentration of spores at a reference temperature (Hayakawa *et al.*, 1969; Matsuda *et al.*, 1981). Equivalently, Berry & Bradshaw (1986) took the initial shoulder of a type D curve into account with a lag time prior to first-order inactivation. Such procedures may allow acceptable predictions only if the intercept ratio or the lag time does not vary with the temperature. That has never been determined experimentally. On the contrary, most data available in the literature show that it is largely temperature-dependent. For example, Le

Jean et al. (1994) measured lag times varying from 50 min (at 111 °C) to 1 min (at 127 °C) on type D survival curves established for Bacillus stearothermophilus spores. Also, since the real shapes of the isothermal curves are neglected, the actual effect of heat treatments of short duration remains totally unpredictable. Now, models assuming either successive or competing first-order reactions, as well as multiple heat-sensitive sites, account for the initial shapes of survival curves.

The adequacy of conventional and first-order models was also seriously criticized by Moats et al. (1971), who emphasized that they were inconsistent with the occurrence of tailing effects (Figure 12.3). These authors noticed that because most investigators did not extend their study to strong destruction levels, they might not be able to observe the tailing that could be a normal characteristic of bacterial populations. According to observations of thermal injury (revealed by the selectivity of microbial recovery media or different optimal growth temperatures), the authors insisted that the death process was due to the activation of multiple critical sites in the cell; and that use of a unimolecular reaction, which assumes the existence of an underlying single sensitive site, is not realistic. In a previous paper, Moats (1971) developed a model based on multiple site theory but did not consider its convenience for tailing effects. In fact, among the hypotheses advanced to explain tailing (see Moats et al., 1971; Cerf, 1977) no single hypothesis has been shown (to our knowledge) to be efficient at elaborating a satisfying predictive model. Thus, workers dealing with disinfection goals that need strong destruction should account for tailing effects only through direct experimental verification of thermal treatment, as applied through numeration procedures allowing accurate low-level counting of surviving microbial populations.

Models accounting for multiple reactions and a single critical site

The assumption that thermal death is not a direct single first-order process but a more complex mechanism composed of successive and competing reactions enables the development of mathematical models that do well in terms of accounting for short temporal deviations. The occurrence of multiple reactions depends upon the concept that microorganisms can be found in different hypothetical states, and that they occur in these states initially, or that these states are induced by the thermal treatment. These hypothetical states depend on the microbial population considered, but let us illustrate this through the particular case of bacterial spores. These spore forms are known to be particularly resistant to lethal agents and their thermal destruction has been extensively studied to establish the type of reliable predictive models required for food sterilization process control and optimization.

Example of models developed for bacterial spores

All of the aforementioned types of isothermal survival curves (Figures 12.1 to 12.3) have been reported by authors studying the thermal death of bacterial spores. Bilinear survival curves (Figure 12.1, type B) are most often attributed a mixture of two species, or strains that possess different thermal resistance. This is also often considered as tailing. Cerf (1977) discussed a number of publications that report such a phenomenon, which can be described as two parallel first-order destructions, each of which is characterized by its own reaction rate constant (k_1 or k_2):

$$N_1 = N_{01}e^{-k_1 t} \tag{12.8}$$

$$N_2 = N_{02}e^{-k_2 t} \tag{12.9}$$

Feehery *et al.* (1987) proposed the occurrence of heat injury to explain type C survival curves for *B. stearothermophilus* spores. This requires that viable spores submitted to heat would have to be injured prior to destruction. The use of two media (one selective stopping injured spores growing) permitted the authors to establish survival curves, showing that the proportion of heat-injured spores increased with time of treatment at a given temperature. The authors did not elaborate any model, but their assumption can be represented schematically by the following mechanism:

non-injured spores $\longrightarrow$ injured spores $\longrightarrow$ inactivated spores

Injury may contribute to type C deviations (and perhaps for type B if the initial population is a mix of injured and non-injured spores that have different thermal resistance and if injury is slower than inactivation), but this cannot explain the apparent increase in the initial population that is observed for type D deviations.

Type D survival curves are particular to thermal death experiments performed on bacterial spores, and bring to the fore the occurrence of a mechanism incorporating at least two successive reactions. Shull, Cargo & Ernst (1963) associated type D survival curves with the presence of a mixed population of dormant and activated spores. According to this approach, there are dormant spores that should be heat-activated in order that they may develop the ability to grow in a suitable medium, and those activated spores then become thermally vulnerable. The activation phenomenon is supposed to obey first-order kinetics, so that the destruction process is assimilated into the succession of two first-order reactions, which is represented schematically by:

dormant spores $\xrightarrow{k_a}$ activated spores $\xrightarrow{k_d}$ inactivated spores

at $t = 0$	M_0	N_0
at $t > 0$	M	N	$(M_0 + N_0) - (M + N)$

k_a and k_d being respectively, the first-order activation and inactivation rate constants. The authors translated this mechanism into an isothermal death equation similar to the following:

$$N = N_0 \left[\alpha e^{-k_a t} + (1 - \alpha) e^{-k_d t} \right] \tag{12.10}$$

Where $\alpha = \dfrac{M_0}{N_0} \dfrac{k_a}{k_d - k_a}$

To verify the validity of the hypotheses, Equation 12.6 was fitted to the experimental plots. For that, the initial shoulder was associated with simultaneous initial activation of dormant spores and destruction of activated spores. The linear asymptote was associated with the exponential death of activated spores by imposing $k_d < k_a$. The destruction rate constant (k_d) was then determined through a linear regression on the linear asymptote and the other parameters (M_0 and k_a) were estimated by successive adjusments to fit the initial portion of the curve. As they failed to get a satisfying fit, they assumed that the activation was either not a first-order phenomenon or not the only reason for the occurrence of the initial shoulder.

Rodriguez *et al.* (1988, 1992) developed a model incorporating competing activation and inactivation of dormant spores, and parallel inactivation of a mix of two populations of spores differing by their heat resistance ('activated spores' and 'less heat-resistant fraction'). This model was elaborated to account for complex survival curves showing type D deviation superimposed on a fast linear decay, due to inactivation of a feebly resistant population. The destruction process was then represented by the following mechanism:

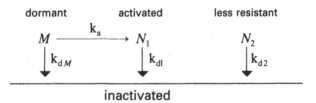

As the authors imposed the condition $k_{dM} = k_{d1} = k_{d2}$ the isothermal dynamics of a numerable population was given by:

$$N(t) = M_o \left(1 - e^{-k_a t} \right) e^{-k_d t} + N_{10} e^{-k_d t} + N_{20} e^{-k_{d2} t} \tag{12.11}$$

The model was used (Rodriguez *et al.* 1992) to interpret type D survival curves obtained for *B. subtilis* spores (there was no fraction N_2). Each reaction was supposed to be ruled by first-order kinetics and the inactivation of both dormant and activated spores was characterized by the same velocity constant (k_d). As done by Shull *et al.* (1963), the authors assumed $k_a > k_d$ and determined k_d by linear regression on the linear asymptote of survival curves. They noticed that small discrepancies appeared early in isothermal treatments between curves generated by the model as compared to the experimental data, and pointed out the importance of frequent accurate sampling for short duration time treatments.

Sapru *et al.* (1992) considered the same mechanism without a less resistant fraction (N_2), but judged it more prudent to hypothesize that inactivation of viable dormant spores differs from that of activated spores. The isothermal response equation was then:

$$N(t) = N_{10}{}^{-k_{dl}t} + \frac{k_a}{k_a + k_{dM} - k_{dl}} M_0 \left(1 - e^{-(k_a + k_{dM} - k_{dl})t} \right) e^{-k_{dl}t} \qquad (12.12)$$

They determined the parameters of the model for *B. stearothermophilus* spores from isothermal dynamics of activated numerable spores, but estimated the initial dormant population by direct microscopic count, assuming that the whole population was 100% viable (the dormant fraction was then the difference between microscopic count and initial plate count). Using this model, the authors accurately predicted the non-isothermal activated population dynamics for two sub-ultra-high temperature (UHT) heat treatments. In a subsequent paper (Sapru *et al.*, 1993), the same authors compared these predictions to those calculated from three other models – the conventional, the Rodriguez and the Shull models – which were identified by the consideration that these three models were derived as special cases from the Sapru model. In this way, the inactivation of activated spores was imposed a priori to be the slowest reaction in the case of Shull's model ($k_d < k_a$). However, this was not verified by the final parameter values, so that the comparison was not valid. This point was not discussed by the authors and the alternative case ($k_a < k_d$) was not checked.

Abraham *et al.* (1990) developed mathematical aspects of Shull's model and checked the theoretical shapes of the survival curves it would generate under both alternative cases defined by $k_a < k_d$ and $k_d < k_a$. According to the relative values of k_a, k_d and τ (M_0/N_0 dormancy = ratio), the authors showed that when $k_d < k_a$, type A, C and D curves could be generated, whereas when $k_a < k_d$, all curves illustrated in Figure 12.1 (A, B, C and D) could be generated. Now there is no special reason to reject a priori (as is usually done) the case where $k_a < k_d$, and the observation of type B curves by several authors (see Table 12.1) is consistent with this assumption. Moreover, when testing the model for a suspension of *B. stearothermophilus* spores, the authors analyzed experimental data through a procedure that enabled them to reject the case $k_d < k_a$ and determined parameters that generated isothermal survival curves which fitted well with the experimental data plots. Le Jean *et al.* (1994) corroborated these results using additional data, and the identified model, allowed an accurate prediction of non-isothermal dynamics for the activated spores. The authors also pointed out that, if activation is really the slowest reaction then a dormant fraction must exist for as long as activated spores are present, and that depending on the ratio k_d/k_a, the dynamic dormancy ratio (M/N) may remain important even at the end of the thermal treatment

Discussion

Important questions arise with the use of these kinetics models developed for thermal inactivation of bacterial spores. For example, it is a fact that obtaining a good fit to experimental data is an important and attractive aspect of a mathematical model. This goodness of fit is in fact the major interest behind model development. It is evident that increasing the number of model parameters may be an efficient response to a lack of fit, but then it also must be noticed and understood that having a good mathematical fit does not guarantee the validity of the hypotheses upon which the model is based. The first-order model accounts for only one transition (activated to inactivated) and, because it neglects the dormant fraction, may not be conservative. The Shull model accounts for two transitions that are well supported experimentally (activation of dormant spores and inactivation of activated spores) and, although it does not deal with heat injury, it seems to be the more prudent model when the interpretability of experimental facts is considered. Unsatisfied with the fit obtained with Shull's model, Rodriguez, and later Sapru, introduced an additional hypothetical transition, allowing for a direct inactivation of dormant spores. However, the resulting goodness of fit of their model to their experimental data does not prove that this transition actually exists. Indeed, it is probable that some other possible assumptions regarding an additive hypothetical transition leading to the same increase in the number of parameters (adding one rate constant) could allow the same accuracy of fit. If so, were another transition to be assumed, then it would be interesting to check the alternative possibilities and choose (if it cannot be verified experimentally) the more conservative one. Moreover, although the linear asymptote of survival curves indicates that the limiting phenomenon is ruled by quasi-first-order kinetics, nothing proves that all transitions are of the same order, and until it is verified experimentally, assumption of first-order kinetics must be considered as an implicit simplification adopted as a concession, because of insufficient experimental data.

Models accounting for multiple critical sites

The representation of microbial death as a mechanism which relies upon one single reaction is certainly an idealization of thermal inactivation processes. Such a representation supposes that a transition from one state to another would be induced by thermal inactivation of a single sensitive site per cell. This idea is perhaps too simplistic, because in reality the loss of viability of a cell is accompanied by various phenomena (effects upon proteins, RNA, DNA, ribosome denaturation, cell membrane damage, etc.), so that thermal death could be due to competing and possibly complementary actions at multiple critical sites. Moreover, the theories that suggest the occurrence of varying degrees of injury do not agree with the idea of direct death by action at a single sensitive site. It is in part for this reason that models accounting for thermal behavior have been derived from models developed for the effects of ionizing radiation (Alper, Gillies & Elking, 1960; Powers, 1962; Tyler & Dipert, 1962).

For example, Moats (1971) developed a model suggesting that a cell is inactivated when a given proportion X of N present critical sites, all having the same heat resistance, are inactivated (multiple targets, single-hit model). The inactivation of individual sites was characterized by a first-order inactivation rate constant, k. From these parameters, the author calculated the probability of cell survival as a function of time. They consequently determined values of X, N and k for *Salmonella anatum* and *Pseudomonas viscosa* that fit the experimental data. Such models could effectively account for varying degrees of injury by the assumption that inactivation at each critical site contributed to heat injury, and that the number of sites inactivated corresponded with a particular degree of injury. This approach seems to be attractive, but no supportive information exists which would prove that all sites either are of the same nature or have the same heat resistance. Without such supportive knowledge, their definition of heat injury would remain a theoretical concept, and would not represent actual states of the cell. In fact, the development of models that accurately account for inactivation as a multi-step process with intermediate states requires more specific information than is now available on the nature and thermal properties of the critical heat-sensitive sites.

Application of models

Once a model is known to well describe the thermal inactivation dynamics of a microorganism under some given set of conditions, the model can be used to characterize the microbial population by determination of appropriate parameter values from isothermal treatment data. The model can then be used to predict the population dynamics for any thermal treatments performed upon that organism. As experimental data provided by classical enumeration procedures generally show a large amount of scattering, particular attention should be paid to the experimental design whenever data are collected for parameter estimation. Also, the estimation of prediction errors when population dynamics are simulated provides important information regarding the capacity of a model to generate an accurate prediction when previously determined parameter values are used. It is essential to check the validity of any identified model. Unfortunately, authors generally are not very thorough when reporting the results from validity testing.

Determination of rate constants

Rate constants are generally determined through the estimation of Arrhenius relationship coefficients. The first need is to determine isothermal rate constant values from experimental isothermal survival curves. Proceeding from an approach by Guttman, Wilks & Hunter (1971), Lenz & Lund (1980) developed an experimental plan that minimizes the amount of experimental work needed to reach acceptable accuracy. The authors stated that 5–6 isothermal survival curves at regular temperature intervals ($\Delta T = z/2 = R T^2_m/E_a$; see Equation 12.5) was a good compromise. With the use of this approach, the temperature range should be as large as possible within the limitations imposed by the thermal treatment experimental procedure. They also indicated that data should be collected for three time points during the period of heat exposure, and that these three time points should be regularly spaced within the time period that the lowest microbial counts are observed, and that this process should be repeated for each temperature. Now, three time points may supply adequate information for simple first-order kinetics, but this would surely not be the case when the inactivation is described by a mechanism incorporating several transitions with different rate constants. Indeed, such composite mechanisms lead to equations expressed as sums of exponential terms (Equations 12.10, 12.11 and 12.12). With the use of models that have large rate constants (or those that employ the sum of several rate constants) it is important to perform sampling at short time intervals within that portion of the survival curve where the deviation from linearity is observed. This knowledge was emphasized by Rodriguez *et al.* (1992).

When a first-order kinetics model is used, isothermal rate constant values are given by the slopes of experimentally determined survival curves (graphed as $\ln(N/N_0)$ versus time) and are easily determined by a simple linear regression. When a model with several rate constants is used, it is usual to proceed by successive regressions on residuals obtained by successively substracting the previously determined terms, and then adjusting the rate values on the basis of non-linear regression. To this end, Rodriguez *et al.* (1988) used the Levensberg–Marquardt non-linear regression method (referred to Press *et al.*, 1986), and Sapru *et al.* (1992) used a modified Gauss–Newton iterative method (Sapru *et al.* referred to information from the SAS Institute, 1985).

Once the necessary isothermal values are available, Arrhenius relationship coefficients are determined by linear regressions on plots that give the logarithm of the rate constants versus the reciprocal of the absolute temperature. Then, the Arrhenius relationship (Equation 12.5) is most often written as:

$$\text{Ln}(k) = a + b/T \tag{12.13}$$

Where $a = \ln(A)$, and $b = -E_a/RT$. Although it is rarely done, it is easy graphically to show the accuracy of the developed regression line. Guttman *et al.* (1971) showed that for a given temperature, the statistical behavior of ln (k) determined by linear regression can be estimated by a Student law $T_{n-2}(m(T), \sigma^2(T))$, where n is the number of experimental values of ln (k) used for the regression, $m(T)$ the value of ln (k) calculated from the regression line equation at temperature T and $\sigma(T)$ the standard error of $m(T)$. Thus, the 95% confidence limits for Arrhenius plots are given by the particular value t_{n-2}; 0.025 of the Student variable:

$$95\% \text{ confidence limits} = m(T) \pm t_{n-2;\,0.025}\,\sigma(T) \tag{12.14}$$

$$\text{with } \sigma(T) = \sigma_\varepsilon \sqrt{\frac{1}{n} + \frac{\left(\frac{1}{T} - \left\langle\frac{1}{T}\right\rangle\right)^2}{\sum_{i=1}^{n}\left(\frac{1}{T_i} - \left\langle\frac{1}{T}\right\rangle\right)}} \tag{12.15}$$

Where σ_ε = standard error of the regression line

$$\left\langle\frac{1}{T}\right\rangle = \frac{1}{n}\sum_{i=1}^{n}\frac{1}{T_i}$$

This is illustrated in Figure 12.5 for activation and inactivation rate constants of Shull's model for *B. stearothermophilus* spores (data from Le Jean *et al.*, 1994). In this study, the model parameters were identified by assuming that k$a \neq$ kd (according to the procedure proposed by Abraham *et al.*, 1990) which is a condition for obtaining a linear asymptote in iso-thermal survival curves. The calculated confidence intervals superimposed in Figure 12.5 show that this condition was clearly verified for nearly all except the lowest temperatures. However, the application of Abraham's procedure was allowed by the high correlation coefficients for linear regressions on the asymptote for 111 and 115 °C. This case illustrates well

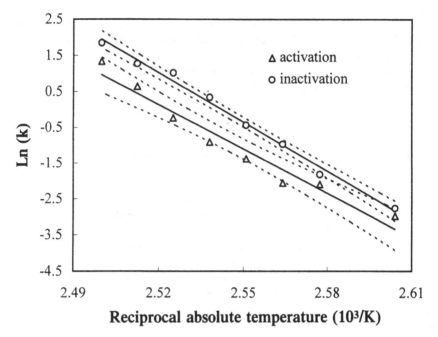

Figure 12.5   Arrhenius relationships for Shull's model as applied to
B. steatothermophilus spores. Data from Le Jean *et al.* (1994). Continuous
lines, linearization by the least-square method; dashed lines, 95%
confidence intervals determined from least-square linearization residuals.

the necessity of performing isothermal treatments over as large a temperature range as possible; and, for experiments performed at extreme temperatures, the widening of the confidence intervals at those extreme
temperatures expresses the increasing importance of the uncertainties
associated with the model parameters. Particularly at high temperatures,
when the reaction velocities are high, any error in rate constant value
would lead to important errors in predicting the surviving population. At
the very least, nonlinearity of experimental Arrhenius plots may be
observed. This would indicate that the employed kinetic model was poorly
chosen. It is also possible that a collision effect (Labuza, 1980) may cause
deviation from linearity at high temperatures.

The knowledge of Arrhenius coefficients for any rate constant of a given
model allows the prediction of population dynamics for any isothermal
treatment. Deriving this information generally requires numerically integrating a system of differential equations that describe instantaneous population dynamic velocity. Let us take the example of Shull's models, for
which dormant and activated populations vary instantaneously according to
the following system of differential equations derived from the inactivation
mechanism:

$$\frac{dM}{dt} = -k_a \, M \tag{12.16}$$

$$\frac{dN}{dt} = -k_d \, N + k_a \, M \tag{12.17}$$

$$k_a = \exp\left(\alpha_a + \frac{\beta_a}{T} \right) \tag{12.18}$$

$$k_d = \exp\left(\alpha_d + \frac{\beta_d}{T} \right) \tag{12.19}$$

The solution is calculated for a given temperature history (recorded in a numerical format (t_i; T_i)) and fixed initial conditions:

$$N(t_0) = N_0; \; M(t_0) = M_0 = t \, N_0$$

For that, the previous differential equations (Equations 12.16 and 12.17) are numerized and can be integrated numerically through a standard procedure, with the rate constant values recalculated at each time step from the Arrhenius relationships (Equations 12.18 and 12.19).

Now, let us assume that the applied model is 100% valid. Then, the reliability of the prediction depends mainly on the accuracy of the Arrhenius relationships that were determined by linear regressions. The uncertainties of the results from numerical integration may be calculated on the basis of the statistical data from linearizations. This can be performed by the use of a Monte-Carlo simulation method, which simulates a large number of predictions (for the heat treatment considered) from the generation of numerous pseudo-random coefficients of Arrhenius relationships. This technique employs the following procedure:

(1) Repeat the following three steps a great number of times:
   (a) For each rate constant generate one pseudo-random isothermal value for each of the n temperatures that were studied experimentally; this requires the realization of n independent Student variables with $n - 2$ degrees of freedom (T_{n-2}). For the constant k_i, the pseudo-random value at temperature T_j is given by:

$$\mathrm{Ln}\,[k_i(T_j)] = m_i(T_j) + t_{ij}\,\sigma_i(T_j) \tag{12.20}$$

Where $T_j = j$st temperature (K) at which experimental value of k_i was determined for the initial linear regression (Arrhenius plot linearization), $j = 1 \ldots n$; $m_i(T_j)$ = value of k_i calculated from the initial regression line at temperature T_j; t_{ij} = realization of Student variable for k_i at temperature T_j; $\sigma_i(T_j)$ = standard error of $m_i(T)$

at temperature T_j determined from initial least-squares linearization (see Equation 12.15).

(b) For each rate constant, perform linear regression on pseudo-
random isothermal values: this is done to derive the corresponding
pseudo-random Arrhenius relationship coefficients.

(c) Proceed to integral calculation of the population dynamics with
the use of these new coefficients.

(2) Determine the confidence interval of the prediction by checking the
statistical representing distribution values for the surviving (results of
step 1c, at time t_i). For example, 95% confidence limits $CL_{min}(t_i)$
and $CL_{max}(t_i)$ are found by determination of values for the surviving
fractions to verify the respective probabilistic conditions:

$$P((N/N_0) < CL_{min}(t_i)) = 0.025 \qquad (12.21)$$

$$P((N/N_0) < CL_{max}(t_i)) = 0.975 \qquad (12.22)$$

The prediction of non-isothermal dynamics, with confidence limits, is illustrated in Figure 12.6 (data from Le Jean *et al.*, 1994), which compares
predictions from the conventional and Shull models developed for an aqueous suspension of *B. stearothermophilus* spores. The widening of confidence
intervals along the time course of thermal treatment probably expresses
an accumulation of error at each step of the calculation, so that the final
uncertainty of predictions at the end of the treatment becomes very large.
Therefore, it is more likely that a lack of fit will occur at the end of the
thermal treatment than at the beginning. The failure of Shull's model to
fit the last two points carries less weight in terms of validation, than does
the good fit occurring at the beginning of the treatment. The calculation
of confidence limits enables us to quantify the predictive accuracy of an
identified model, and can be used to determine safety coefficients when
the model is applied to the design of thermal disinfection processes.

Conclusion

This chapter provided basic information on the experimental procedures
used for studying thermal inactivation, data interpretation and the application of models of kinetics. The importance given to bacterial spores comes
from our own limited laboratory experience, and also from the large
number of published works that have relied upon the use of bacterial
spores when the dynamics of microbial thermal inactivation were studied.
We did not discuss at length the important topics of heat injury and the
biochemical processes involved in thermal inactivation of microorganisms.
The actual causes of thermal inactivation depend upon several factors,

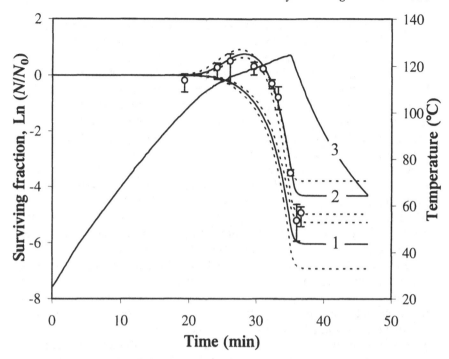

Figure 12.6 Comparison of prediction by the conventional model (1) and the Shull model (2) applied to *B. steatothermophilus* spores submitted to a non-isothermal treatment (3). Dashed lines, 95% confidence intervals determined through a Monte-Carlo simulation method, accounting for rate constant residuals from least-square linearization. Open circles, average of the 3 most probable number (MPN) values. Error bars, maximum and minimum MPN values.

among which are the nature of the target microorganisms and the heat vectors (moist or dry heat, heating medium, etc.). This subject is simply too wide to have been adequately addressed in this chapter. The development of kinetic parameters based upon numerical data also was not mentioned, and for the same reason. Instead, we preferred to develop the overall concept of kinetic modeling, assuming that readers interested in other aspects of thermal inactivation could refer to the cited literature.

Thermal inactivation kinetics is most often approached through the conventional model that uses z and D_θ values, thus yielding important simplifications. One could ask: is it possible, for a given target species of microorganism, to propose a kinetics model that would be applicable for any and all thermal treatment conditions? For that, the interpretation of survival curves may not be sufficient, since survival curves developed under any given set of conditions can give an indication of thermal behavior for only that one microbial strain, and the applicability of any single kinetics model is affected both by important within-species variations and by differences in thermal treatment conditions. It might be possible to develop such broadly applicable models by assessing accurate qualitative and quantitative knowledge of the main biochemical processes that occur in microbial cells during the process of cellular inactivation. Achievement of this could be brought about by the application of new methods of analysis, such as DSC, which provide quantitative data on the reactions that occur when microorganisms are submitted to heat. However, the applicability of currently available kinetics models, whatever their complexity, remains very restricted once the parameter values are determined.

As safety is the main criterion in disinfection processes, conservative models should be preferred. For example, Abraham *et al.* (1990) and then Le Jean *et al.* (1994) concluded that with the use of Shull's model, activation of dormant spores could occur by a slower reaction rate than subsequent inactivation of the activated spores. In addition to allowing a more satisfying fit of the model to the observed population dynamics, this conclusion regarding the model suggested that dormant spores may remain present in the population for as long a time period as the existence of activated spores is observed. This leads us to contest the use of the conventional model, not only for its failure in describing the true shape of the survival curves, but also because it does not account for the possible existence of a viable dormant spore fraction at the end of the thermal treatment. Also, it is preferable to elaborate models under conservative assumptions that limit the risk of overestimating the lethality. For example, the use of Shull's model with the assumption of a direct inactivation of dormant spores, which is not supported by qualitative experimental observations, may lead to an underestimation of the residual level of dormant viable spores. In fact, it must be noticed that even when those models are used that exhibit a good mathematical fit, there is no guarantee of maximum safety.

References

Ababouch, L., Dirka, A. & Busta, F. F. (1987). Tailing of survivor curves of clostridial spores heated in edible oil. *Journal of Applied Bacteriology*, **62**, 503–11.

Abraham, G., Debray, E., Candau, Y & Piar, G. (1990). Mathematical model of thermal destruction of *Bacillus stearothermophilus* spores. *Applied and Environmental Microbiolology*, **56**, 3073–80.

Alper, T., Gillies, N. E. & Elking, M. M. (1960). The sigmoid survival curve in radiobiology. *Nature*, **186**, 1062.

Anderson, W. A., Hedges, N. D., Jones, M. V. & Cole, M. B. (1991). Thermal inactivation of *Listeria monocytogenes* studied by differential scanning calorimetry. *Journal of General Microbiology*, **137**, 1419–24.

Barillère, J. M., Dubois, C. & Bidan, P. (1985a). Thermoresistance de souches de levures isolées de vin. 2 – Théories de la thermorésistance – Modèles mathématiques. *Science Aliments*, **5**, 607–17.

Barillère, J. M., Mimouni, A., Dubois, C. & Bidan, P. (1985*b*). Thermoresistance de souches de levures isolées de vin. 1 – Influence des conditions experimentales sur la forme des courbes de survie. *Science Aliments*, **5**, 365–78.

Belliveau, B. H., Beaman, T. C., Pankratz, H. S. & Gerhardt, P. (1992). Heat killing of bacterial spores analyzed by differential scanning calorimetry. *Journal of Bacteriology*, **174**, 4463–74.

Berry, R. M. Jr & Bradshaw, J. G. (1980). Heating characteristics of condensed cream in celery soup in a steritort: heat penetration and spore count reduction. *Journal of Food Science*, **45**, 869–74

Berry, R. M. Jr & Bradshaw, J. G. (1986). Comparison of sterilization values from heat penetration and spore count reduction in agitating retorts. *Journal of Food Science*, **51**, 477–93.

Bigelow, W. D. (1921). The logarithmic nature of thermal death time curves. *Journal of Infectious Diseases*, **29**, 528–36.

Bigelow, W. D., Bohart, G. S., Richardson, A. C. & Ball, C. O. (1920). *Heat Penetration in Processing Canned Foods*, Bulletin No. 16L. Washington, DC: National Canners Association.

Bigelow, W. D. & Esty, J. R. (1920). Thermal death point in relation to time of typical thermophilic organisms. *Journal of Infectious Diseases*, **27**, 602.

Briggs, A. (1960). The resistance of spores of the genus *Bacillus* to phenol, heat and radiation. *Journal of Applied Bacteriology*, **29**, 490–504.

Burton, H. A., Perkin, A. G., Davies, F. L. & Underwood, H. M. (1977). Thermal death kinetics of *B. stearothermophilus* spores at ultra high temperatures. III: Relationship between data from capillary tube experiments and from UHT sterilizers. *Food Technology*, **12**, 149–61.

Cerf, O. (1977). Tailing of survival curves of bacterial spores. *Journal of Applied Bacteriology*, **42**, 405–15.

Chick, H. (1910). The process of disinfection by chemical agencies and hot water. *Journal of Hygiene*, **10**, 237.

David, R. D. J. & Merson, R. L. (1990). Kinetic parameters for inactivation of *Bacillus stearothermophilus* at high temperatures. *Journal of Food Technology*, **55**, 488–92.

Davies, F. L., Underwood, H. M., Perkin, A. G. & Burton, H. (1977). Thermal death kinetics of *Bacillus stearothermophilus* spores at ultra high temperatures. I. Laboratory determination of temperature coefficients. *Journal of Food Technology*, **12**, 115–29.

Doutsias, G. (1974). Etude de la thermorésistance d'une souche de *Pichia etchellsi* isolée de saumure d'olives vertes. D.E.S. diploma, Université de Dijon, France.

Esty, J. R. & Meyer, K. F. (1922). The heat resistance of spores of *C. botulinum* and allied anaerobes. *Journal of Infectious Diseases*, **31**, 650.

Feehery, F. E., Munsey, D. T. & Rowley, D. B. (1987). Thermal inactivation and injury of *B. stearothermophilus* spores. *Applied and Environmental Microbiology*, **53**, 365, 370.

Fernandez, P. S., Ocio, M. J., Sanchez, T. & Martinez, A. (1994). Thermal resistance of *Bacillus stearothermophilus* spores heated in acidified mushroom extract. *Journal of Food Protection*, **57**, 37–41.

Guttman, L., Wilks, S. S. & Hunter, J. S. (1971). *Introductory Engineering Statistics.* New York: John Wiley and Sons.

Hayakawa, K. I., Schnell, P. G. & Dick, H. D. (1969). Estimating thermal death time characteristics of thermally vulnerable factors by programmed heating of sample solution or suspension. *Food Technology,* **23,** 1090–4.

Heppel, N. J. (1985). Comparison of residence time distribution of water and milk in an experimental UHT sterilizer. *Journal of Food Engineering,* **4,** 71.

Ingraham, T. R. & Marier, P. (1964). Activation energy calculation from a linearly-increasing-temperature experiment. *Canadian Journal of Chemical Engineering,* **44,** 161–3.

Ingraham, T. R. & Marier, P. (1965). Activation energy calculation from a linearly-increasing-temperature experiment. Part II *Canadian Journal of Chemical Engineering,* **45,** 354.

Jones, M. C. (1968). The temperature dependence of the lethal rate in sterilization calculations. *Journal of Food Technology,* **3,** 31–8.

Jonsson, U., Snygg, B. G., Härnulv, B. G. & Zacharisson, T. (1977). Testing two models for the temperature dependence of the heat inactivation rate of *Bacillus stearothermophilus* spores. *Journal of Food Science,* **5,** 1251–2, 1263.

Katzin, L. I., Sandholdzer, S. A. & Strong, M. E. (1943). Application of the decimal reduction time principle to a study of the resistance of coliform bacteria to pasteurization. *Journal of Bacteriology,* **45,** 265–72.

Labuza, T. P. (1980). Enthalpy/entropy compensation in food reactions. *Food Technology,* **34,** 67–77.

Lee, R. E. & Guilbert, C. A. (1918). On the application of the mass law to the process of disinfection being a contribution to the mechanistic theory as opposed to the vitalistic theory. *Journal of Physical Chemistry,* **22,** 348–72.

Le Jean, G., Abraham, G., Debray, E., Candau, Y. & Piar, G. (1994). Kinetics of thermal destruction of *Bacillus stearothermophilus* spores using a two reaction model. *Food Microbiology,* **11,** 229–41.

Lenz, M. K. & Lund, D. B. (1980). Experimental procedures for determining destruction on kinetics of food components. *Food Technology,* **34,** 51–5.

Mackey, B. M. & Derrick, C. M. (1987). The effect of prior heat shock on the thermoresistance of *Salmonella thompson* in foods. Letters in *Applied Microbiology,* **5,** 115–18.

Mackey, B. M., Miles, C. A., Parsons, S. E. & Seymour, D. A. (1991). Thermal denaturation of whole cells and cell components of *Escherichia coli* examined by differential calorimetry. *Journal of General Microbiology,* **137,** 2371–4.

Mackey, B. M., Miles, C. A., Parsons, S. E. & Seymour, D. A. (1993). Thermal denaturation and loss of viability in *Escherichia coli* and *Bacillus stearothermophilus. Letters in Applied Microbiology,* **16,** 56–8.

Maeda, Y., Nogushi, S. & Koga, S. (1978). Thermal analysis of the spores of *Bacillus cereus* with special reference to heat activation. *Canadian Journal of Microbiology,* **24,** 1331–4.

Maeda, Y., Teramoto, Y. & Koga, S. (1975). Calorimetric study on heat activation of *Bacillus cereus* spores. *Journal of General and Applied Microbiology,* **21,** 119–22.

Matsuda, N., Komaki, M. & Matsunawa, K. (1981). Thermal death characteristics of spores of *Clostridium botulinum* 62A and *Bacillus stearothermophilus* subjected to non-isothermal heat treatment. *Journal of the Food and Hygiene Society of Japan*, **22**, 125–34.

Michels, L., Spiliotis, V. & Etoa, F. X. (1985). Nouvelle méthode de dénombrement de bactéries sporulées: Application à la mesure de la thermorésistance. *Annales des Falsifications et de l'Expertise Chimique* **78**, 171–181.

Miles, C. A., Mackey, B. M., Parsons, S. E. & Seymour, D. A. (1986). Differential scanning calorimetry of bacteria. *Journal General Microbiology*, **132**, 939–52.

Moats, W. A. (1971). Kinetics of thermal death of bacteria. *Journal of Bacteriology* **105**, 165–171

Moats, W. A., Dabbah, R. & Edwards. V. M. (1971). Interpretation of non-logarithmic survival curves of heated bacteria. *Journal Food Science*, **36**, 523–6.

Ocio, M. J., Fernandez, P. S., Alvarruiz, A. & Martinez, A. (1994). Comparison of TDT and Arrhenius models for rate constant inactivation predictions of *Bacillus stearothermophilus* heated in mushroom-alginate substrate. *Letters in Applied Microbiology*, **19**, 114–17.

Perkin, A. G., Burton, H., Underwood, H. M. & Davies, F. L. (1977). Thermal death kinetics of *Bacillus stearothermophilus* spores at ultra high temperatures. II. Effect of heating period on experimental results. *Journal of Food Technology* **12**, 131–48.

Powers, E. L. (1962). Consideration of survival curves and target theory. *Physics in Medicine and Biology*, **7**, 3.

Press, W. H., Flannery, B. P., Teulkosky, S. A. & Wetterling, W. T. (1986). *Numerical Receipes*. Cambridge: Cambridge University Press.

Reichart, O. (1981). Experimental method for the determination of the thermal death parameters of microorganisms in a continuous system. *Acta Alimentaria*, **12**, 35–53.

Rodriguez, A. C., Smerage, G. H., Teixeira, A. A. & Busta, F. F. (1988). Kinetics effects of lethal temperatures on population dynamics of bacterial spores. *Transactions of the American Society of Agriculture Engineers*, **31**, 1594.

Rodriguez, A. C., Smerage, G. H., Teixeira, A. A., Lindsay, J. A. & Busta, F. F. (1992). Population model of bacterial spores for validation of dynamic thermal processes. *Journal of Food Process Engineering*, **15**, 1–30.

Sapru, V., Teixeira, A. A., Smerage, G. H. & Lindsay, J. A. (1992). Predicting thermophilic spore population dynamics for U. H. T. sterilization processes. *Journal of Food Science*, **57**, 1248–57.

Sapru, V., Teixeira, A. A., Smerage, G. H. & Lindsay, J. A. (1993). Comparison of predictive models for bacterial spore population resources to sterilization temperatures. *Journal of Food Science*, **58**, 223–8.

SAS Institute (1985). *SAS © Users guide: Statistics*. Cary, North Carolina: SAS Institute.

Scharer, J. M. (1965). Thermal death behavior of bacterial spores. PhD thesis, University of Pennsylvania.

Shull, J. J., Cargo, G. T. & Ernst, R. R. (1963). Kinetics of heat activation and of thermal death of bacterial spores. *Applied Microbiology*, **11**, 485–7.

Shull, J. J. & Ernst, R. R. (1962). Graphical procedure for comparing death of *Bacillus stearothermophilus* spores in saturated and superheated steam. *Applied Microbiology*, **10**, 452–7.

Spiliotis, V. (1983). Survie des spores de *Bacillus stearothermophilus*: influence de l'activité de l'eau et du pH du milieu de suspension pendant la stérilisation thermique et après conservation à temperature ambiante. Thèse de doctorat de troisième cycle, Université Paris-VII.

Stern. J. A. & Proctor, B. E. (1954). A micro-method and apparatus for the multiple determination of rates of destruction of bacteria and bacterial spores subjected to heat. *Food Technology*, **8**, 139.

Stumbo, C. R. (1948). A technique to study the resistance of bacterial spores to temperatures in the higher range. *Food Technology*, **2**, 228.

Swartzel, K. R. (1984). A continuous flow procedure for kinetics data generation. *Journal of Food Science*, **49**, 803.

Tobias, J., Heirreid, E. O. & Ordal, Z. J. (1953). A study of milk pasteurization at high temperatures. *Journal of Dairy Science*, **36**, 356–2.

Tobias, J. O., Kaufman, O. W. & Tracy, P. H. (1955). Pasteurization equivalents of high temperature short time heating with ice mix. *Journal of Dairy Science*, **38**, 959–68.

Tyler, S. A. & Dipert, M. H. (1962). On estimating the constants of the multi-hit curve using a medium speed digital computer. *Physics in Medicine and Biology*, **7**, 201.

Van Uden, N. & Vidal-Leira, M. M. (1976). Thermodynamic compensation in microbial thermal death – studies with yeasts. *Archives of Microbiology*, **108**, 293–8.

Wang, D. I.-C., Scharer, J. & Humphrey, A. E. (1964). Kinetics of death of bacterial spores at elevated temperatures. *Applied Microbiology*, **12**, 451–4.

Index

Note: Page references in bold type indicate the major chapter headings.

Acanthamoeba castellani 156–7, 158, 321
Acanthamoeba polyphaga 157, 158
Acanthamoeba rhysoides 321
acidulants 203–4, 204, 208
Acinetobacter 57, 166, 169
acquired immune deficiency syndrome (AIDS/HIV) virus 226
and enteric infections 81, 116
surface survival 224, 230, 233, 241
transmission of 10, 12, 15, 21
and UV radiation 216
actinometry 332, 335
biological 336, 337–8, 339
chemical 338–40, 344
adenoviruses/*Adenovirus*
characteristics of 78, 79, 80
detection 85, 88
surface survival 232, 241, 250
UV susceptibility 319, 320
Aeromonas 77
A. hydrophila 322
aerosols 4, 11, 17, 22, 25–6
disinfection methods **215–23**
HEPA filtration 220–1
UV inactivation 217–19
and food contamination 198
and sewage 18, 20
aflatoxins 197
aggregation, cellular 141, 144, 146, 153, 159, 163, 165–7, 173, 175
and UV inactivation 317, 326, 327, 357
air cleaning units 217–18
Alcare 249

alcohol 293
algae, growth in food 202
Alicaligenes faecalis 171
allergies 19
Alphavirus 15
American Type Culture Collection (ATCC) 171
ammonium disinfectants 143–4, 166, 171, 244, 248, 249
ampicillin resistance 273, 275–6
analysis of variance, use of 60, 61
analytic models 60–5
animals
as disease reservoirs 13–18, 23, 25, 259
epidemic model in 9–10
water contamination by 89–90, 93, 101, 102
anthrax 26
antibiotics 33
natural 203
resistance to 170, 272–6
antibodies 7–8, 107–8
in microbe detection 86, 87, 88, 89
antiseptics 164–5, 224, 245–6, 289–90
AOAC Sporicidal Test 294–6, 304
Aquaress 249
Arenavirus 26
Arrhenius relationship 377–9, 386–8, 389–90
arthropod vectors 4, 13, 14–16, 21
Aspergillus 12, 26, 220
Astrovirus 22, 100

398   *Index*

Bacillus 58
spores 201, 294–5
thermal inactivation 374, 380–4,
390–1
UV susceptibility 320, 322, 325,
338
B. anthracis 26
B. cereus 374
B. licheniformis 374
B. megaterium 168, 169, 175
B. pantothenticus 374
B. stearothermophilus 374, 380–1,
383–4, 387, 390–1
B. subtilis 168, 174, 374, 382–3
effect of UV 218, 274
bacteria
colonization of body 55–9
modeling strategies 60–7
resistance to water disinfectants
140–92
toxins 116, 197–8
transmission routes of
aerosols 25–7
control by surface disinfection 24,
258–84
direct physical contact 13–16
environmental non-water related
21–8
food 24–5, 26
infection probabilities 109
intermediate host 28–9
medical devices 21, 23
surfaces 24
water 16–22
UV susceptibility 317–18, 322
see also spores
bacteriocins 203
bacteriophages 91, 93, 172–3
f2 321, 329–30, 331, 359
MS-2 154–5, 159, 320–2, 325, 336–8
Qβ 321, 338
T7 154, 159
Bacteroides 57, 58, 59
Bacti-Stat 249
batch culture 65–6
Beer–Lambert law 323
benzalkonium 175
Bifidobacterium 58
biofilm 161–5, 169
Bioprep 249
biopreservation 205, 209
birds 13, 17, 102
blindfolded bowler analogy 107–13

Bordetella pertussis 26
Borrelia burgdorferi 15
botulism 12, 198
Branhamella catarrhalis 57
Brevibacterium 57
bromine 150, 166, 167, 174
Brucella 26, 195
Bunsen–Roscoe law 316

Calicivirus 22, 27, 100
foodborne 195, 197, 199
Camel 165
Campylobacter/campylosis 17, 19, 24
foodborne 26, 195, 197
infection probabilities 114
minimum infectious dose 265
on surfaces 261, 263, 267
waterborne 22, 76–7, 83, 100
C. jejuni 78, 156, 158
UV susceptibility 317–18
Candida 15
C. albicans 57
C. parapsilopsis 329–30, 331
canning processes 200–1
capsid 168, 169
capsule 141, 159, 161–5, 169, 170, 173
carbolic acid 293
cattle 10, 17, 28, 29, 226
cefoxitin 306
cell
action of disinfectants on 147–50
protection from 159–69, 170
photoreactivation 356–7
see also aggregation
cell–recycle culture 66
cetrimide 164, 174, 175, 249
chickenpox 16, 26, 27
chiggers 14
children, disease transmission 23,
79–81, 227–8, 238, 266–71
Chlamydia 27
C. psittaci 26
C. trachomatis 13–14, 24
chloramines 152, 157, 159, 166
bacterial resistance to 163–4, 167, 169
disinfectant action 140, 147, 148–9,
158
chlorhexidine 168, 169, 174, 175, 249
chlorine 32
and disinfection of medical devices
293–4
as food preservative 203–4
in water disinfection 33, 34, 127, 140

mechanism of action 146–8
  resistance to 142, 150–60, 162–9,
    171, 176–7
chlorine dioxide 140, 143–4, 149, 171,
    172
  resistance to 164, 166, 169, 173, 177
4-chloro–3,5-xylenol 249
chlorophenol 167–8, 169, 175
cholera *see Vibrio cholerae*
Citrobacter freundii 156
Clostridium 58, 59
  C. botulinum 12, 374
    food poisoning 198
  C. dificile 267, 272–3
  C. perfringens 12, 201
    as microbe indicator 92–3
  C. sporogenes 294–5, 374
  C. tetani 12
clothing 31–2, 215–16, 270, 273
cluster analysis 61
coagulants 32, 127, 140
Coccidioides 12, 26
colicins 168, 172–3
coliphages *see* bacteriophages
colonization, microbial **55–71**
Colorado tick fever 15
Coltivirus 15
common cold 26, 27
compartmental models 5–10, 30, 205,
    227–8
condoms 32
conjunctivitis 26, 79, 226
continuous culture 66–7
cooking in food processing 197, 198,
    200, 201, 207, 208
copepods *see* crustaceans
Coronavirus 22, 26, 100
corrosion 141, 150–1, 158
Corynebacterium 26, 57, 59
coxsackieviruses 83
  A 79, 80
  B 79, 80
  detection 85, 87
  infection probabilities 113, 114–15
  surface survival 232
  UV susceptibility 318–19, 320
crabs 112
crops/vegetables 202
  contaminated 16, 17, 20, 22, 24
    by chemicals 193–4
  diseases from 195–6, 199
crustaceans

contamination of 20, 22, 24, 102–3,
    195, 197, 201
  in water distribution systems 112,
    155, 156
cryptosporidiosis 17, 18, 19, 21, 81
Cryptosporidium
  cyst collection 86, 87
  cyst UV susceptibility 320
  and disinfectants 267
  infection probabilities 125–6
  in water 75, 81–4, 91–2, 106
  C. parvum 22, 26, 27, 99, 100
    infection probabilities 113, 114,
    116
  minimum infectious dose 109
cutaneous ecosystem *see* skin
Cyclops 155–6, 158
Cyclospora 113
Cytomegalovirus 15, 241
cytoplasm 147

dairy products 24, 195, 198, 200, 202
Daphnia 156, 158
day-care centers 266–71
decontamination 287, 289
Deltavirus 15
dengue 14, 15
detergents 32, 168, 172, 172–3
Dettol 165, 249
diabetes 80
diapers, and disease 267–9, 270, 271
diarrhea 267, 270
  bacterial 77, 272
  protozoan 81
  viral 19, 26, 79, 226
differential scanning calorimetry
    (DSC) 376, 392
dinoflagellates 194, 197
diphtheria 26
direct transmission 4, 5, 10–16
disease
  prevention 28–37
  portals of entry 3–4
  risk estimation 103, 106
  transmission modeling 5–10, 30, 34–7
  see also disinfection; transmission
disinfection 3, 32–3, 228
  of aerosols **215–23**
  bacterial resistance to **140–92**
    cell-mediated protection 159–69
    effect of growth conditions 142,
      163–4, 168–77
    inactivation kinetics 142–50

disinfectant evaluation 294–6
  mechanisms of action 150–68
  relative efficiencies of 143–4,
    246–50
  and food preparation 25
  of medical devices **285–310**
  Poisson distribution analysis 166–7
  of sewage 20
  of surfaces
    and bacterial control **258–84**
    and viral transmission 244–50
  thermal inactivation **369–96**
  viral resistance to 159, 167, 169
  of water 32–3, 35, 37, 127, 141–50
  bacterial resistance to **140–92**
  see also ultraviolet light
DNA, UV susceptibility 315, 316–17,
    319
dogs 21
drug resistance 272–6, 306
drying 201–3, 208, 229–30
ducted air disinfection 217, 220
dyes 168, 172–3
dysentery *see Entamoeba; Shigella*

ear 4
echoviruses/*Echovirus* 79–80, 83, 87
  infection probabilities 109, 113, 115
  resistance to disinfectants 167, 168,
    169
  surface survival 232, 241
  UV susceptibility 318, 320
eggs 195, 197, 201, 202
encephalitis 13, 14, 15, 22, 100
endemic disease 99, 100, 103, 127, 128
  modeling 3, 6–7
endoscopes 23, 287, 290, 304–7
Entamoeba 80
  E. histolytica 22, 100, 158
    and disinfectants 144, 146
enteritis 26, 195
Enterobacter aerogenes 163, 169
Enterobacter agglomerans 156, 163, 169
Enterobacter cloacae 153, 156, 162–3,
    169
Enterococcus faecalis,
    vancomycin-resistant (VRE) 273,
    275–6
Enterococcus faecium, vancomycin/
    ampicillin-resistant (VAREC) 273,
    275–6
enteroviruses/*Enterovirus* 11–12, 226
  characteristics of 78–80

concurrent infections 117
detection 85, 87–8
Enterovirus-70 226, 241
environment transmitted 26, 27
and pH 112
prediction of 91, 92–3
surface survival 230, 232, 241
and UV 318–20
water transmission 11, 22, 84, 100,
    124–5
see also coxsackieviruses;
    echoviruses; polioviruses
envelope, cellular 142, 147, 169–70,
    172–6
environment
  and disease transmission 3, 4, 11,
    16–28
  see also aerosols; food; medical
    devices; surfaces; water
enzymes 149, 160, 202, 204
epidemic disease 16, 99, 100
  modeling 3, 5–10
Escherichia 26
  E. coli 58
    and disinfectants 144, 160
    resistance to disinfectants 169, 171,
      172, 174, 177
      cellular alterations 167, 168, 169
      particle association 152
      vector association 156, 158
    foodborne 195
    surface contamination 264, 267
    thermal inactivation of 374
    UV susceptibility 218, 315,
      317–18, 329–31, 357, 359
    waterborne 76–7, 78, 83
    infection/illness/death
      probabilities 114
ethanol 249
ethylene oxide 288–91, 297–8, 300–1,
    307–8
Eubacterium 58, 59
eye 4, 19, 79

factor analysis 61
feces
  disinfection studies 152, 154, 155
  food contamination 14, 21, 199
  land contamination 20
  microbial content 78–9, 108
  surface contamination 268, 270
  viral discharge 228–9

water contamination 16, 17, 18,
   89–90, 102
Filovirus 26
filters
   HEPA 216, 218, 220–1
   for microbe collection 86, 87, 88, 93
   in water treatment 126, 127, 140
fimbriae 165, 166
fish 102, 195, 197, 199, 226
Flavivirus 15
Flavobacterium 142, 177
fleas 14
flies 14, 21, 28
flocculation 140
fluorescein isothiocyanate (FITC) 87
fomites *see* surfaces
food
   infectious disease from 4, 11,
      193–212, 227, 261–6
   contamination sources 5, 14, 16,
      24–5, 102, 196–208
      chemical 194–5
      nature of 26–7, 194–6
      risk modeling 205–8
   preserving quality of 200–5
Food and Drug Administration (FDA):
   regulation of disinfection 286,
   296–304
foot-and-mouth disease 25
formaldehyde 171, 292–3
Francisella tularensis/tularemia 21
   and biting flies 14, 15, 38
   and food 26, 38, 195
   and water 22, 38, 99, 100
freezing, and pathogens 201–2, 203,
   208
frozen foods 198, 202
fruit 24, 199, 202
fungi 12, 20, 24, 25, 57
   growth in food 196, 202, 203, 205
   resistance to disinfection 294
   toxins 197
fungicides 194
Fusobacterium 58, 59

Gallionella 151
gamma irradiation 32, 201, 308
Gardnerella vaginalis 59
gas gangrene 12
gases as preservatives 203–4, 208
gastroenteritis 75, 76, 78
   bacterial 12
   protozoan 320

viral
   aerosol transmitted 27
   foodborne 195, 201
   surfaceborne 11, 226, 228, 241
   waterborne 16, 18, 19, 21–2, 78–9,
      99, 100, 101, 125
gastrointestinal tract 4
   diseases of 14, 16, 24, 77, 78, 100–1
   see also diarrhea; gastroenteritis
   microbial ecosystem 58
generalized estimation equations
   (GEE) 64–5
genitourinary tract 4
Giardia/giardiasis 91
   cyst collection 86–7
   and disinfectants 267
   foodborne 24, 25, 196, 199
   infection probabilities 83–4, 109,
      125–6
   waterborne 17, 18, 21, 75, 80–1, 92,
      106
   G. lamblia 26–7
      cyst UV susceptibility 320, 321
      infection probabilities 109, 113
      minimum infectious dose 109
      in water 22, 82, 100
   G. muris 321
glutaraldehyde 171, 292, 305, 306, 308
gonorrhea 10, 15
granular activated carbon (GAC)
   153–4, 158, 164
growth phase, and disinfectant
   resistance 171, 175, 177
Guillain–Barré disease 78

hair follicles 56
hands
   and disease transmission 260, 262,
      264, 266
   viral 224, 229, 231, 236–8, 241–4
   washing of 32, 270, 271, 273
      formulations for 245–6, 247–50
heart, diseases of 79, 80, 226
heat
   in food preparation 200
   sterilization 288–91
   as water treatment 126–7
heavy metals 168, 172–3, 193
Helicobacter pylori 317–18
helminths 28, 155, 196, 202
Hemophilus 57
hemorrhagic fever 25, 26
Hep-2 cell-associated virus 154

HEPA filters 216, 218, 220–1
hepatitis 154
  transmission 15
    by food 27, 201
    by medical devices 21
    by water 19, 22, 100
  A 12
    and disinfectants 267
    foodborne 20, 195
    infection probabilities 115
    on surfaces 11, 232, 236, 239, 249
    UV susceptibility 319, 320
    waterborne 75, 78, 79, 80
  B 13, 226, 232–3, 243
  E 99, 115, 116, 124
Hepatovirus 22, 27, 100, 195, 197
herbicides 194
herpes simplex 13
herpesviruses 233, 236, 241
hexachlorophene 249
high efficiency particulate filters
  (HEPA) 216, 218, 220–1
Histoplasma 12
  H. capsulatum 26
histoplasmosis 21
horizontal transmission 10
host susceptibility 3–4, 5–9, 14, 226–7,
  259–60, 268
host–microbe ecosystems 55–9
  modeling of 59–67
Hyalella azteca 156
Hycolin 165
hydraulic indices 314, 349, 350–1
hydrogen peroxide 146, 170, 200
hydrogen sulphide 141

ice 22, 195–6
immunity 5–9, 122
immunization 28, 29–31
immunofluorescence assays 196
indicator organisms 90–1, 92–3
  and UV 314, 316–17, 321–2, 336–8
indirect transmission 4, 5, 10–12, 16–28
  see also aerosols; food; medical
    devices; surfaces; water
infant botulism 198
infant salmonellosis 263
infection control 31–3, 260, 269–71,
  286–8
  see also disinfection
infectious disease *see* disinfection;
  transmission
influenza/influenzaviruses 11, 26, 226

surface survival 233, 237
insect vectors 4, 13, 14, 155
insecticides 194
intermediate hosts 28, 29
intestinal tract, host–microbe
  ecosystem 55
iodine 33, 127, 247–8
  resistance to 165, 167, 171, 177
iron compounds 141, 151
irradiation 201, 208, 300, 308
isopropanol 249
Izal 165

keratoconjunctivitis 23
Klebsiella 164
  K. aerogenes 264
  K. oxytoca 156
  K. pneumoniae 58
    disinfectant resistance 151, 156,
      158, 163, 169, 173, 177
  K. terigena 318, 322, 325

Lactobacillus 57, 58, 59, 65, 205
Lambert's law 342
Legionella 26, 27, 157, 158
  L. gormanii 156
  L. pneumophila 12, 22, 102, 317–18
Lentivirus 15
leprosy 31
Leptospira interrogans 22, 26
leptospirosis 17, 19, 21
lipopolysaccharides 170, 175, 176
Listeria 26, 195
livestock 17, 28
  epidemic model 9–10
logistic regression 64
Lyme disease 14, 15
Lymphocryptovirus 15
Lyssavirus 15

malaria 99
malnutrition 116
Mastaadenovirus 22, 26
measles 16, 26, 116
meat 195–6, 198, 201, 202
medical devices
  and disease transmission 21–3, 26,
    27, 227, 246, 259
  disinfection/sterilization of **285–310**
  effects of device type 304–7
  evaluation 294–6
  FDA regulation 296–304
  methods 291–4

standards for 307–8
meningitis 22, 24, 26, 79, 100
menstrual cycle 58–9, 62
methicillin resistance 273–5
microbial colonization, modeling
    55–71
microbial fate 103
microbial risk assessment *see* risk
    assessment
Micrococcus 57, 59
M. luteus 218
microwaves 32, 201
minimum infectious dose 108–10, 123,
    283–9
mixed-effects models (MIXMOD) 63
models
    compartmental disease transmission
        5–10, 30, 205, 227–8
    disinfection/inactivation kinetics
        144–6, 316–35
    disinfection model for the home
        265
    first-order kinetics models 143
    mixed second-order models 324–7
    multi-target models 145, 324–6,
        327–8, 329–31, 334–5
    series-event models 145, 324–6,
        328–9, 329–31, 334–5
    thermal death 375–92
host–microbe ecosystems
    analytic 60–5
    simulation 60, 65–7
risk assessment 106–26
    foodborne 205–8
    waterborne dose–response models
        81–5
UV dose 351–6
UV intensity 314, 323–4, 340–8
waterborne epidemic model 35–7
molluscs
    and disease 20, 22, 24, 195, 197
    prevention 200–1
mononucleosis 15
Monte-Carlo simulation method 389
Moraxella 164, 169
Morbillivirus 26
Morrill dispersion index 350, 353
mortality, disease related 6–7
mosquitoes 14, 19
mouth: microbial ecosystem 57
multi-drug-resistant pathogens 272–6
multi-target models 145, 324–6, 327–8,
    329–31, 334–5

mumps 16, 226
muskrats 17
Mycobacterium boris 218
Mycobacterium chelonae absicessus
    306–7
Mycobacterium chelonei 171
Mycobacterium fortuitum 171
Mycobacterium tuberculosis
    animal sources 25, 26
    disinfection methods for 217, 220,
        221
    quarantine 31
    var. *bovis* 304
Mycoplasma hominis 59
mycoplasmas 57

Naegleria 22, 157, 321
    N. gruberi 167
nappies *see* diapers
natural community models 240
Neisseria 57, 59
    N. gonorrhoeae 15, 175
nematodes 141, 150, 155, 158
neonates 56
Newcastle disease 25
nonparametric tests 61
Norwalk virus 79, 80, 85, 101, 195
nosocomial infections 228, 272–8,
    285–90
Notavirus 85
nucleic acid hybridization probes 196
nucleic acids
    denaturation 148–50
    UV susceptibility 315, 319

Omp proteins 172
Onchocerca volvulus 28, 29
operating theaters 215
oral ecosystem 57
oronasal tract 55
Orthohepadnavirus 15
Orthopoxvirus 26
oxygen 91, 146–7, 172, 582
ozone
    as disinfectant 140, 149–50
    resistance to 152, 167, 169
    as food preservative 203–4

pandemics 16, 110
parainfluenzaviruses 26, 234, 236
paralysis 79
Paramyxovirus 26
paratyphoid 22, 100

particle association 141, 151–5, 158, 159, 163, 358
pasteurization 200
pathogen monitoring, risk assessment **75–98**
PCR (polymerase chain reaction) 87, 88–9, 196, 241
penicillin 168, 174
Peptostreptococcus 57, 58, 59
peroxydisulfate-*t*-butanol 339
pertussis 16, 26
pH
  and aggregation 165
  and disinfection 292
  and disinfection resistance 143
  and microbial growth in food 202, 203
phage sensitivity testing 160
phage-typing tests 174
pharyngoconjunctival fever 18, 19, 22
phenolic disinfectants 244, 248, 293, 306
2-phenoxyethanol 168, 175
photoreactivation 356–7
physical contact 4, 13–16
pigs 10, 28, 29
plague 13, 14, 15, 25, 26
plankton 156
pleurodynia 79
Pneumocystis carinii 26
pneumonia 27
poliomyelitis 19, 27, 30, 226
polioviruses 11
  infection probabilities 115
  minimum infectious dose 108–9, 123
  and surfaces 230, 234, 249
  UV susceptibility 318–19, 320
  and water disinfectants 144
    resistance to 154, 155, 159, 160, 169, 176
  waterborne 12, 79, 80
polychlorinated biphenyls 194
polymer production, microbial 141, 159
polymerase chain reaction (PCR) 87, 88–9, 196, 241
populations, modeling disease in 5–10, 33–7
porins 170
potassium ferrioxalate 339, 344
poultry 24, 195, 197, 263
pregnancy 99, 116, 259–60
preservatives 200–5
  listed 204
Prevotella 65

primary transmission 5
Propionibacterium acnes 57, 59
protein denaturation 147–9
protozoa
  colonization of body 57, 58
  and disinfectants 143
    resistance to 150, 156–7, 158
  foodborne 24–5, 27, 196, 202, 208
  infection probabilities 113, 114, 116, 119, 125–6
  and medical devices 23, 24, 25–6
  minimum infectious dose 109, 110
  and pH 112
  risk estimation 118–21
  UV susceptibility 320–1
  waterborne 4, 17–22
    cysts/oocyst levels 85–7, 89, 104–6
    indicators of 90, 91, 93
pseudoinfection 23
Pseudomonas 23
P. aeruginosa 306
  resistance to disinfectants 165, 166, 167, 169, 175, 177
  in swimming pools 164, 171
P. alicaligenes 171, 177
P. cepacia 175
P. flurescens 218
P. multivorans 174
P. paucimobilis 164, 169
P. pickettii 164, 169
P. viscosa 385
pseudorabies 10
psittacosis 26

quarantine 31

rabies 10, 13, 14, 15
racoons 10, 13, 14
random-effects models 63
regression
  anlysis 62–5, 90–1
  trees 61–2
reinfection rates 117–18
Reiter's syndrome 78
reoviruses 88
  UV susceptibility 319, 320
reservoirs (of disease) 3–4, 13, 16–18, 101, 259–60, 268
residence time distribution (RTD) 338, 348–56
respiratory diseases 19, 26, 27, 79, 267
respiratory syncytial virus 234, 236, 237
retorting 200–1

rheumatic diseases 226
rhinoviruses/*Rhinovirus* 26, 27, 226, 228
  surface contamination 234–5, 236,
    237, 241, 247–8
Rickettsia 14, 15
rinderpest 10
risk assessment
  of foodborne disease 205–8
  from contaminated water **75–98,
    99–139**
    methods 75–89
    hazard identification/exposure
      76–89
    modeling 81–2, 83, 85, 106–26
    probabilities of illness/death
      113–16
    pathogen monitoring programs
      89–94
  of transmission by surfaces 261–78
RNA, UV susceptibility 319
Rocky mountain spotted fever 15
rodenticides 194
rodents 13, 17, 18
rotaviruses/*Rotavirus* 226
  concurrent infections 117
  and disinfectants 248, 267
  surface contamination 11, 26, 227–8,
    235, 236–8, 241
  UV susceptibility 318–19, 320, 325
  waterborne 18, 22, 78–9, 100, 101
    detection of 85
    infection probabilities 83, 84, 113,
      115
    minimum infectious dose 108–9
RTD 338, 348–56
rubella 15, 226
Rubivirus 15, 26
Ruminococcus 58

Saccharomyces cerevisiae 322, 338, 374
Salmonella/salmonellosis 14
  foodborne 24, 26, 195–6, 197, 201
  minimum infectious dose 265
  surface transmission 261–2, 263, 267
  waterborne 76–7, 78, 155, 158
    infection probabilities 83, 84
  S. anatum 374, 385
  S. paratyphi 22, 100
  S. senthenberg 374
  S. thompson 374
  S. typhi(typhoid) 5, 16, 19, 21–2, 76,
    100
    infection probabilities 113, 114

  levels in water 105
  minimum infectious dose 109
  protection from disinfectants 153
  severity of infection 77, 78
  typhoid endemics 3, 33–4
  UV susceptibility 318, 322
sanitation 242–50, 267
  in food production/preparation
    198–9, 200
  see also hands, washing of
Sarcinia lutea 218
Savlon 165, 174–5, 249
Schistosoma 29
Scrub Stat IV 249
scrub typhus 14
secondary transmission 5
sedimentation 140
sepsis 23
septicemia 12
Septisol 249
series-event models 145, 324–6, 328–9,
  329–31, 334–5
Serratia marcescens 218
serum-associated bactericidal agents
  175–6
sewage, treatment 18, 20, 88, 89–90,
  102
sexually transmitted diseases 10, 13, 32,
  247
shellfish *see* crustaceans; molluscs
Shigella/shigellosis
  foodborne 21, 26, 195
  surface transmission 267
  vector associated 158
  waterborne 22, 75, 76–7, 78, 100
    probability of infection 83
  S. dysenteriae 317–18
  S. flexneri 114
  S. sonnei 153, 156
simian immunodeficiency virus 14
simulation models 60, 65–7, 240,
  246
skin 4, 11, 13, 19, 25
  decontamination 32
  see also hands, washing of
  host–microbe ecosystem 55, 56–7
smallpox 11, 16, 24, 26, 30
snails 28, 29
soap 32, 245
sparganosis 27, 29, 195
Sphaerotilus 151
spirochetes 58
Spirometra 27, 29, 195

spores 201
  sporicidal tests 294–6, 304
  thermal inactivation 371, 373, 374,
    380–4, 390–1
Staphylococcus 57, 65, 171
  food poisoning 198
  S. aureus 59
    disinfectant resistance 165, 169
    foodborne 201
    methicillin resistant (MRSA)
      273–5
    penicillin resistant 168, 174
    surface contamination 264
    UV inactivation 315
  S. epidermidis 58, 59
steam sterilization 290, 291, 298, 300,
    307
sterilization 228, 285–310
  assurance levels (SAL) 297–9, 307
  see also disinfection
stratum corneum 56
streptococcal pharyngitis 25
Streptococcus 57, 58, 59
  food contamination 199
  S. faecalis 168, 174, 374
  S. salivarius 171
surfaces 4, 11, 24, 26, 27
  bacterial control by disinfection of
    258–84
    risk assessment 261–78
  food contamination from 198
  viral transmission 24, 26, **224–57**
    interruption of 244–50
    and modeling 226–40
    surface survival/transfer 229–38,
      240–4
  see also medical devices
surfactant 166
swimming pools 18, 150, 160, 164, 171,
    174
swine 10, 29
syphilis 13, 15

T-test, use of 60, 62
Taenia 27, 28, 29, 195–6
tapeworms 28, 29, 195–6
temperature
  and aggregation 165
  and disinfection 292
  and disinfection resistance 143, 171,
    172, 173, 175
  and foodborne disease 205
  as microbe indicator 90–1

and microbe survival 91–2, 103
and microbial growth in food 202,
    203
and virus survival 229
tetanus 12
Tetrahymena pyriformis 156, 157, 158
thermal abuse 25, 203
thermal inactivation 200–1, **369–96**
  kinetics experiments 370–5
  modeling 375–90
    applications 385–90
    exponential death 376–9
    non-exponential death 379–84
ticks 14
toxic substance theory 146–7
toxins, bacterial 116, 197–8
toys 11
  and bacterial transmission 268,
    269–70, 271
  and viral transmission 238, 242–3
transmission **3–54**
  endemic 7
  epidemic 6, 8, 9, 30
  modeling 5–10
  physical barriers to 34, 35, 37
  prevention of 3, 28–37
    effectiveness 33–7
  routes 10–28
    by environmental: water 16–21,
      22; non water 21–8
    by food 24–5, 26–7
    by intermediate host 28–9
    by physical contact 13–16
    medical devices 21–3, 26
    by intermediate host 28–9
    via surfaces 24, 26, **224–57**, 259–61
  see also aerosols; food; medical
    devices; surfaces; water
Treponema pallidum 15
Trichinella 27, 195–6
Trichomonas vaginalis, UV
  susceptibility 321
Triclosan 249
tuberculation 141, 150–1, 158
tuberculosis *see Mycobacterium
  tuberculosis*
tularemia *see Francisella tularensis*
turbidity 103
  and disinfection 141, 151–3, 154
  as microbe indicator 92
  and UV inactivation 317, 358
typhoid *see Salmonella typhi*
typhus 14, 15

ultraviolet light (UV) disinfection 32, 91, 200, 229
  adverse effects 216, 218–19
  of aerosols (UVGI) 216–19
  of water **313–68**
    batch inactivation 322–31
    continuous flow 314, 331–56
      actinometry 332, 335–40
      flow dynamics 332–4
      hydraulic indices 314, 349, 350–1
      intensity modeling 323–48
      kinetics 334–5
      residence time (RTD) 338, 348–56
    dose 314, 316–17, 351–6
    inactivation kinetics/models 314, 316–35
    indicators 316–17, 321–2
    intensity 314, 337
      modeling 314, 323–4, 340–8
      and UV dose 351–6
    interferences 314–15, 317, 356–9
    lamps/sources 340–8, 358–9
    mathematical nomenclature 359–60
    microbial susceptibility 315–22
Ureoplasma urealyticum 59
urine 16, 17
urogenital tract, host–microbe ecosystem 55

vaccination 7, 9, 30–1, 80
vaccinia virus 235
vagina
  microbial ecosystem 58–9
  modeling 62–5
vancomycin resistance 273, 275–6
Varicellovirus 26
vectorborne disease 4, 14–16, 155–8
vehicleborne disease 4
Veillonella 58, 59
ventilation systems 215, 220–1
vertical transmission 10
Vibrio
  foodborne 22, 195, 197, 199
  V. cholerae/cholera
    acid resistance 112
    endemic 102–3
    foodborne 21, 26
    and shellfish 20, 102–3, 156, 201
    infection probabilities 113, 114, 116
  minimum infectious dose 109, 110
  pandemics 16, 110
  UV susceptibility 317–18, 322
  in water 19, 21, 22, 76, 99–100, 102–3
    levels 105
  V. vulnificus 12
viruses
  detection of 85, 87–8
  discharge 228–9
  foodborne 202, 208
  infection probabilities 109
  resistance to disinfectants 154–5, 159, 160, 167–76
  transmission routes of
    aerosols 25–7
    direct physical contact 13–16
    food 24–5, 27
    medical devices 21, 23
    surfaces 24, **224–57**
    water 16–22
    see also aerosols; food; medical devices; surfaces; water
  UV susceptibility 318–20

washing 32
wastewater *see* water
water
  contamination indicators 90–3
  as disease transmission route 4–5, 11–12, 16–22
  distribution systems 150–1, 161–2, 163, 169, 177–8
    contamination 126
    vectors in 155–8
  in food production 193, 195, 196, 197, 198
  infectious diseases from 22, 84, 99–106
    hospitalization rates 75–6, 77, 78, 81–2
    risk assessment **75–98, 99–139**
      probability of illness/death 113–16
      ingestion rate 107, 118, 123–4
  microbial monitoring 89–94
  treatment 32–7, 83, 92–3, 100, 126–8
    disinfectants and resistance **140–92**
    pathogen monitoring 89–94
    risk reduction 119–21, 125
    UV disinfection **313–68**
Wilcoxon rank sum test 62

wounds 4, 11, 12, 13, 19–20
  and food contamination 25,
    199
yeasts 57, 58, 196–7
yellow fever 15

Yersinia 77, 158
  Y. enterocolitica 78, 153, 156, 173,
    177, 317–18
  Y. pestis 15, 26

zoonoses 14–15